Edited by
János Fischer and David P. Rotella

Successful Drug Discovery

Related Titles

Li, J., Corey, E.J. (eds.)

Drug Discovery

Practices, Processes, and Perspectives

2013

Print ISBN: 978-0-470-94235-2, also available in digital formats

Huang, X., Aslanian, R.G. (eds.)

Case Studies in Modern Drug Discovery and Development

2012

Print ISBN: 978-0-470-60181-5, also available in digital formats

Fischer, J., Ganellin, C.R., Rotella, D.P. (eds.)

Analogue-based Drug Discovery III

2012

Print ISBN: 978-3-527-33073-7, also available in digital formats

Ravina, E.

The Evolution of Drug Discovery

From Traditional Medicines to Modern Drugs

2011

Print ISBN: 978-3-527-32669-3

Fischer, J., Ganellin, C.R. (eds.)

Analogue-based Drug Discovery II

2010

Print ISBN: 978-3-527-32549-8, also available in digital formats

Edited by János Fischer and David P. Rotella

Successful Drug Discovery

Verlag GmbH & Co. KGaA

The Editors

János Fischer
Gedeon Richter Plc.
Department of Medicinal Chemistry
Gyömröi ut 30
1103 Budapest
Hungary

David P. Rotella
Montclair State University
Department of Chemistry & Biochemistry
1, Normal Ave, Montclair
New Jersey 07043
United States

All books published by **Wiley-VCH** are carefully produced. Nevertheless, authors, editors, and publisher do not warrant the information contained in these books, including this book, to be free of errors. Readers are advised to keep in mind that statements, data, illustrations, procedural details or other items may inadvertently be inaccurate.

Library of Congress Card No.: applied for

British Library Cataloguing-in-Publication Data
A catalogue record for this book is available from the British Library.

Bibliographic information published by the Deutsche Nationalbibliothek
The Deutsche Nationalbibliothek lists this publication in the Deutsche Nationalbibliografie; detailed bibliographic data are available on the Internet at http://dnb.d-nb.de.

© 2015 Wiley-VCH Verlag GmbH & Co. KGaA, Boschstr. 12, 69469 Weinheim, Germany

All rights reserved (including those of translation into other languages). No part of this book may be reproduced in any form – by photoprinting, microfilm, or any other means – nor transmitted or translated into a machine language without written permission from the publishers. Registered names, trademarks, etc. used in this book, even when not specifically marked as such, are not to be considered unprotected by law.

Print ISBN: 978-3-527-33685-2
ePDF ISBN: 978-3-527-67844-0
ePub ISBN: 978-3-527-67845-7
Mobi ISBN: 978-3-527-67846-4
oBook ISBN: 978-3-527-67843-3

Typesetting Laserwords Private Limited, Chennai, India
Printing and Binding Markono Print Media Pte Ltd., Singapore

Printed on acid-free paper

Advisory Board Members

Klaus P. Bøgesø
(Lundbeck, Denmark)

Kazumi Kondo
(Otsuka, Japan)

John A. Lowe III
(JL3Pharma LLC)

Barry V.L. Potter
(University of Bath, UK)

Contents

Preface *XIII*
List of Contributors *XVII*

Part I   General Aspects *1*

1   **Serendipitous Target-Based Drug Discoveries** *3*
        János Fischer and David P. Rotella
1.1   Introduction *3*
1.2   Recent Examples of Target-Based Drug Discovery *4*
1.3   Serendipitous Target-Based Drug Discoveries *7*
1.4   Drospirenone (Contraceptive with Anti-aldosterone Activity) *7*
1.4.1   Summary of Drospirenone Discovery *8*
1.5   Escitalopram (Selective Serotonin Reuptake Inhibitor Antidepressant) *9*
1.5.1   Summary of the Escitalopram Discovery *11*
1.6   Ezetimibe (Inhibitor of Cholesterol Absorption) *11*
1.6.1   Summary of Ezetimibe Discovery *13*
1.7   Lamotrigine (Discovery of a Standalone Drug for the Treatment of Epilepsy) *13*
1.7.1   Summary of Lamotrigine Discovery *15*
1.8   Omeprazole (Proton Pump Inhibitor Acid-Suppressive Agent) *15*
1.8.1   Summary of Omeprazole Discovery *16*
1.9   Outlook *17*
        Acknowledgments *17*
        List of Abbreviations *17*
        References *18*

2   **Drug Discoveries and Molecular Mechanism of Action** *19*
        David C. Swinney
2.1   Introduction *19*
2.2   Mechanistic Paradox *19*
2.3   Molecular Mechanism of Action *20*

2.3.1	The Primary Driver of an Optimal MMOA is the Potential for Mechanism-Based Toxicity	22
2.3.2	Details of MMOA are not Captured by IC_{50} and K_I	22
2.3.3	Metrics, Biochemical Efficiency	24
2.4	How MMOAs were Discovered	25
2.4.1	MMOAs of Medicines Approved by the USFDA Between 1999 and 2008	27
2.5	Case Study: Artemisinin	30
2.6	Summary	31
	List of Abbreviations	32
	References	32

Part II   Drug Class 35

3   Insulin Analogs – Improving the Therapy of Diabetes 37
John M. Beals
3.1	Introduction	37
3.2	Pharmacology and Insulin Analogs	38
3.3	Chemical Description	39
3.4	Rapid-Acting Insulin Analogs (Prandial or Bolus Insulin)	41
3.5	Long-Acting Insulin Analog Formulations (Basal Insulin)	48
3.6	Conclusions and Future Considerations	54
	List of Abbreviations	55
	References	55

Part III   Case Histories 61

4   The Discovery of Stendra™ (Avanafil) for the Treatment of Erectile Dysfunction 63
Koichiro Yamada, Toshiaki Sakamoto, Kenji Omori, and Kohei Kikkawa
4.1	Introduction	63
4.2	Discovery of Avanafil	65
4.2.1	Differentiation Strategies to Develop a New Drug	65
4.2.2	Discovery of Isoquinoline Derivatives from Isoquinolinone Lead	65
4.2.3	Scaffold-Hopping Approaches from the Isoquinoline Leads	67
4.2.3.1	Monocyclic Type A Series: Tetrasubstituted Pyrimidine Derivatives	68
4.2.3.2	Monocyclic Type B Series: Trisubstituted Pyrimidine Derivatives	68
4.2.3.3	SAR of Substitution at the 2-Position of the Pyrimidine Ring (**15**)	70
4.2.3.4	SAR of Substitution at the 4-Position of the Pyrimidine Ring (**16**)	70
4.2.3.5	SAR of Substitution at the 5-Position of the Pyrimidine Ring (**19, 20**)	73
4.2.3.6	Core Structure Modifications of the Pyrimidine Nucleus of Avanafil	73

4.3	Pharmacological Features of Avanafil	*75*
4.3.1	PDE Inhibitory Profiles	*75*
4.3.2	*In Vivo* Pharmacology	*76*
4.3.2.1	Potentiation of Penile Tumescence in Dogs	*76*
4.3.2.2	Influence on Retinal Function in Dogs	*79*
4.3.2.3	Influence on Hemodynamics in Dogs	*79*
4.3.2.4	Influence on Nitroglycerin (NTG)-Induced Hypotension in Dogs	*80*
4.4	Clinical Studies of Avanafil	*81*
4.5	Conclusion	*83*
	List of Abbreviations	*83*
	References	*83*

5	**Dapagliflozin, A Selective SGLT2 Inhibitor for Treatment of Diabetes** *87*	
	William N. Washburn	
5.1	Introduction	*87*
5.2	Role of SGLT2 Transporters in Renal Function	*88*
5.3	O-Glucoside SGLT2 Inhibitors	*89*
5.3.1	Hydroxybenzamide O-Glucosides	*90*
5.3.2	Benzylpyrazolone O-Glucosides	*94*
5.3.3	*o*-Benzylphenol O-Glucosides	*95*
5.4	*m*-Diarylmethane C-Glucosides	*97*
5.4.1	Synthetic Route	*98*
5.4.2	Early SAR of C-Glucoside Based SGLT2 Inhibitors	*99*
5.4.3	Identification of Dapagliflozin	*102*
5.5	Profiling Studies with Dapagliflozin	*105*
5.6	Clinical Studies with Dapagliflozin	*108*
5.7	Summary	*108*
	List of Abbreviations	*109*
	References	*110*

6	**Elvitegravir, A New HIV-1 Integrase Inhibitor for Antiretroviral Therapy** *113*	
	Hisashi Shinkai	
6.1	Introduction	*113*
6.2	Discovery of Elvitegravir	*114*
6.2.1	HIV-1 Integrase and Diketo Acid Inhibitors	*114*
6.2.2	Monoketo Acid Integrase Inhibitors and Elvitegravir	*116*
6.3	Conclusion	*121*
	List of Abbreviations	*123*
	References	*123*

7	**Discovery of Linagliptin for the Treatment of Type 2 Diabetes Mellitus** *129*
	Matthias Eckhardt, Thomas Klein, Herbert Nar, and Sandra Thiemann
7.1	Introduction *129*
7.2	Discovery of Linagliptin – High Throughput Screening Hit Optimization *130*
7.3	Rationalization of DPP-4 Inhibition Potency by Crystal Structure Analysis and Studies of Binding Kinetics *139*
7.4	Basic Physicochemical, Pharmacological, and Kinetic Characteristics *141*
7.5	Preclinical Studies *143*
7.5.1	Glucose Regulation by Linagliptin *143*
7.5.2	Effects of Linagliptin on the Kidney *145*
7.6	Clinical Studies *146*
7.6.1	Clinical Pharmacokinetics *146*
7.6.2	Clinical Pharmacodynamics *148*
7.6.2.1	Inhibition of DPP-4 *148*
7.6.2.2	Effects on Glucagon-Like Peptide-1 and Hyperglycemia *148*
7.6.3	Clinical Use in Special Patient Populations *148*
7.6.3.1	Patients with Renal Impairment *148*
7.6.3.2	Patients with Hepatic Impairment *150*
7.6.4	Cardiovascular Safety *150*
7.7	Conclusion *151*
	List of Abbreviations *151*
	References *152*
8	**The Discovery of Alimta (Pemetrexed)** *157*
	Edward C. Taylor
	List of Abbreviations *175*
	References *176*
9	**Perampanel: A Novel, Noncompetitive AMPA Receptor Antagonist for the Treatment of Epilepsy** *181*
	Shigeki Hibi
9.1	Introduction *181*
9.1.1	Competitive Receptor Antagonists *182*
9.1.2	Noncompetitive Receptor Antagonists *182*
9.2	Seeds Identification by High Throughput Screening (HTS) Assays *183*
9.3	Structure and Activity Relationship (SAR) Study Starting from the Unique Structure of Seed Compounds *184*
9.3.1	Introduction of Conjugated Aromaticity *184*
9.3.2	Discovery of 1,3,5-Triaryl-1H-pyridin-2-one Template *184*
9.3.3	Optimization of 1,3,5-Triaryl-1H-pyridin-2-one Derivatives *185*

9.4	Pharmacological Properties of Perampanel; Selection for Clinical Development *187*	
9.4.1	The Pharmacological Evaluation of Perampanel *187*	
9.4.2	The Pharmacokinetic Evaluation of Perampanel *189*	
9.5	Clinical Development of Perampanel *189*	
9.5.1	Phase I *189*	
9.5.2	Phase II and Phase III *190*	
9.6	Conclusion *190*	
	List of Abbreviations *190*	
	References *191*	
10	**Discovery and Development of Telaprevir (Incivek^{TM}) – A Protease Inhibitor to Treat Hepatitis C Infection** *195*	
	Bhisetti G. Rao, Mark A. Murcko, Mark J. Tebbe, and Ann D. Kwong	
10.1	Introduction *195*	
10.1.1	Crystal Structure of NS3/4A Protease *196*	
10.1.2	Assays *196*	
10.2	Discussion *197*	
10.2.1	Substrate-Based Inhibitor Design *197*	
10.2.2	Structure-Based Inhibitor Optimization *200*	
10.2.3	Pre-Clinical Development *206*	
10.2.4	Clinical Development *206*	
10.3	Summary *207*	
	List of Abbreviations *208*	
	References *209*	
11	**Antibody–Drug Conjugates: Design and Development of Trastuzumab Emtansine (T-DM1)** *213*	
	Sandhya Girish, Gail D. Lewis Phillips, Fredric S. Jacobson, Jagath R. Junutula, and Ellie Guardino	
11.1	Introduction *213*	
11.2	Molecular Design of T-DM1 *214*	
11.3	Strategies for Bioanalysis *216*	
11.4	Strategies for Chemistry and Manufacturing Control *218*	
11.5	Nonclinical Development *219*	
11.6	Clinical Pharmacology *220*	
11.7	Clinical Trials and Approval *222*	
11.8	Summary *224*	
	List of Abbreviations *225*	
	References *226*	

Index *231*

Preface

The International Union of Pure and Applied Chemistry (IUPAC) has supported this new book project; the book has key inventors describe their new drug discoveries and aims to study the general aspects of successful drug discoveries and the optimization in a drug class.

The book *Successful Drug Discovery* is a continuation of the three volumes of *Analogue-based Drug Discovery* where analogs of existing drugs were the focus. The new book has a broader scope because it includes both pioneer drugs and their analogs and spans all the important types of small-molecule-, peptide-, and protein-based drugs.

The editors thank the advisory board members: Klaus P. Bøgesø (Lundbeck, Denmark), Kazumi Kondo (Otsuka, Japan), John A. Lowe III (JL3Pharma LLC, USA), and Barry VL Potter (University of Bath, UK). Special thanks are due to the following reviewers who helped both the authors and the editors: Klaus Bøgesø, Derek Buckle, Matthias Eckhardt, Arun Ganesan, William Greenlee, Katalin Hornok, Roy Jefferis, Béla Kiss, Patrizio Mattei, Eckhard Ottow, John Proudfoot, Joerg Senn-Bilfinger, István Tarnawa.

The first volume of *Successful Drug Discovery* consists of three parts.

Part I: (General Aspects)

Serendipity is an important part of drug discovery that can be present at each step. The first chapter – written by the editors of this book – reveals special serendipitous cases of some important drug discoveries where the key lead molecule and the drug have been discovered on different targets.

David C. Swinney analyzed the molecular mechanism of action in drug discovery. Phenotypic and target-based drug discovery approaches are discussed from the viewpoint of pioneer drugs and analogs.

Part II: (Drug Class)

John M. Beals gives an overview of insulin analogs, which optimize PK/PD profiles but also provide sufficient stability for the treatment of type 1 and type 2 diabetes mellitus. Rapid-acting and long-acting analogs are summarized.

Part III: (Case Histories)

1) *Avanafil*
   Kiichiro Yamada, Tosiaki Sakamoto, Kenji Omori, and Kohei Kikkawa described the discovery of the new and highly selective PDE5 inhibitor with rapid onset of action. The researchers succeeded in a remarkably short period to create a structurally new drug.

2) *Dapagliflozin*
   William N. Washburn reports on the discovery of dapagliflozin which was the first SGLT2 inhibitor approved for the treatment of type 2 of diabetes. A C-glucoside side product that was characterized during early SAR studies serendipitiously became the lead structure that ultimately produced the drug.

3) *Elvitegravir*
   Hisashi Shinkai described the discovery of elvitegravir, a new HIV-1 integrase inhibitor by using a 4-quinolone-3-carboxylic acid scaffold. The once-daily single tablet of a combination of elvitegravir with a CYP3A4 inhibitor and a nucleoside reverse transcriptase inhibitor afforded potent and durable antiretroviral efficacy.

4) *Linagliptin*
   Matthias Eckhardt, Thomas Klein, Herbert Nar, and Sandra Thiemann have written about the discovery of linagliptin a highly potent and long-acting DPP-4 inhibitor. Linagliptin has a non-linear pharmacokinetic and nonrenal elimination profile, unique within the class of approved DPP-4 inhibitors.

5) *Pemetrexed*
   Edward C. Taylor gives an overview on the discovery of pemetrexed representing an excellent long term collaboration between academia and industry. The individual interest of the author in pterine chemistry provided a starting point for the discovery of pemetrexed, which is used for the treatment of pleural mesothelioma and non-small cell lung cancer.

6) *Perampanel*
   Shigeki Hibi described the discovery of perampanel, a novel, noncompetitive AMPA receptor antagonist for the treatment of epilepsy. Starting from HTS, optimization of a promising triaryl-$1H$-pyridin-2-one scaffold afforded perampanel, which is the first AMPA antagonist approved as an antiepileptic drug.

7) *Telaprevir*
   B. Govinda Rao, Mark A. Murcko, Mark J. Tebbe, and Ann D. Kwong have written the chapter on the discovery of telaprevir, a protease inhibitor to treat hepatitis C infection. Telaprevir, a NS3/NS4A serine protease inhibitor, was successfully used between 2011 and 2013 in triple combination with ribavirin and interferon.

8) *Trastuzumab emtansine*
   Sandhya Girish, Gail D. Lewis Phillips, Fredric S. Jacobson, Jagath R. Junutula, and Ellie Guardino described the discovery and development

of trastuzumab emtansine. The antibody drug conjugate represents a novel approach in treating patients with HER-2-positive cancer. The drug conjugate enables selective delivery of DM-1, a potent cytotoxic agent, into HER-2-positive target cells.

Last but not least the editors and authors thank the coworkers of Wiley-VCH, especially Dr. Frank Weinreich for the excellent cooperation.

June 2014
Budapest, Hungary
Montclair, NJ, USA

János Fischer
David P. Rotella

List of Contributors

John M. Beals
Lilly Research Laboratories
Lilly Corporate Center
Indianapolis, IN 46285
USA

Matthias Eckhardt
Boehringer Ingelheim Pharma
Division Research Germany
88397 Biberach an der Riss
Germany

János Fischer
Gedeon Richter Plc.
Department of Medicinal
Chemistry
Gyömröi ut 30
1103 Budapest
Hungary

Sandhya Girish
Genentech
1 DNA Way
South San Francisco, CA 94080
USA

Ellie Guardino
Genentech
1 DNA Way
South San Francisco, CA 94080
USA

Shigeki Hibi
Eisai Product Creation Systems
Strategic Operations
Koishikawa 4-6-10
Bunkyo-ku
Tokyo 112-8088
Japan

Fredric S. Jacobson
Genentech
1 DNA Way
South San Francisco, CA 94080
USA

Jagath R. Junutula
Genentech
1 DNA Way
South San Francisco, CA 94080
USA

and

Cellerant Therapeutics
1561 Industrial Rd
San Carlos, CA 94070
USA

Kohei Kikkawa
Mitsubishi-Tanabe
Pharmacology Research
Laboratories II
2-2-50 Kawagishi
Toda 335-8505
Japan

Thomas Klein
Boehringer Ingelheim Pharma
Division Research Germany
88397 Biberach an der Riss
Germany

Ann D. Kwong
InnovaTID
125 CambridgePark Drive
Suite 301
Cambridge, MA 02140
USA

Gail D. Lewis Phillips
Genentech
1 DNA Way
South San Francisco, CA 94080
USA

Mark Murcko
Department of Biochemical
Engineering
MIT & Disruptive Biomedical,
LLC
520 Marshall Street
Holliston, MA 01746
USA

Herbert Nar
Boehringer Ingelheim Pharma
Division Research Germany
88397 Biberach an der Riss
Germany

Kenji Omori
Nagoya University
Graduate School of
Pharmaceutical Sciences
Furo-cho
Chikusa-ku
Nagoya 464-8601
Japan

Bhisetti G. Rao
Drug Design Expert LLC
19 Minute Man Ln
Lexington, MA 02421
USA

David P. Rotella
Montclair Sate University
Department of Chemistry &
Biochemistry
Montclair, NJ 07043
USA

Toshiaki Sakamoto
Mitsubishi-Tanabe
Medicinal Chemistry Research
Laboratories II
2-2-50 Kawagishi
Toda 335-8505
Japan

Hisashi Shinkai
JT Inc.
Central Pharmaceutical Research
Institute
1-1 Murasaki-cho
Takatsuki
Osaka 569-1125
Japan

David C. Swinney
Institute for Rare and Neglected
Diseases Drug Research
897 Independence Avenue
Suite 2C
Mountain View, CA 94043
USA

Edward C. Taylor
Princeton University
Department of Chemistry
Princeton, NJ 08544
USA

and

288 Western Way
Princeton, NJ 08540
USA

Mark J. Tebbe
Tebbe Consulting, LLC
282 Renfrew Street
Arlington, MA 02476
USA

Sandra Thiemann
Boehringer Ingelheim Pharma
Corporate Division Medicine
55216 Ingelheim am Rhein
Germany

William N. Washburn
(retired from
Bristol-Myers-Squibb)
120 Pleasant Valley Rd
Titusville
NJ 08560-1909
USA

Koichiro Yamada
Mitsubishi-Tanabe
Medicinal Chemistry Research
Laboratories II
2-2-50 Kawagishi
Toda 335-8505
Japan

Part I
General Aspects

1
Serendipitous Target-Based Drug Discoveries
János Fischer and David P. Rotella

1.1
Introduction

Breakthrough drug discoveries – based on a molecular biological target – can significantly improve therapy for disease. For example, captopril (discovered in 1976) is a pioneer angiotensin-converting enzyme (ACE)-inhibitor used for treatment of essential hypertension. Subsequent compounds acting on the same target (e.g., enalapril, lisinopril, and perindopril) are used for the same purpose. An alternative and complementary treatment for hypertension involves use of angiotension II receptor antagonists. Losartan (discovered in 1986) was the first compound in this class and was followed by several additional molecules (e.g., valsartan, telmisartan, and irbesartan). Treatment of hypertension by these mechanisms provided physicians with additional options to consider as a part of combination therapy or when other possibilities such as diuretics and/or β-blockers are unsatisfactory. For the treatment of obstructive airway diseases several short and long-acting β-2-adrenoreceptor agonists (e.g., salbutamol, formoterol, and salmeterol) that act directly on lung tissue to improve airway function have been discovered. Antimuscarinics selective for M1 and M3 receptors such as tiotropium bromide (discovered in 1989) were found to be effective for treatment of chronic obstructive pulmonary disease (COPD). These distinct mechanisms of action can be used in combination to treat COPD and other complex airway disorders. Imatinib (discovered in 1992) is a BCR-Abl tyrosine kinase inhibitor for the treatment of chronic myeloid leukemia (CML) that demonstrated the concept of targeted chemotherapy substantially improving survival in this difficult to treat disease. In the field of metabolic diseases, sitagliptin (discovered in 2001) was the first dipeptidyl peptidase-IV (DPP-IV) inhibitor for the treatment of type 2 diabetes. The recognition that inhibition of this enzyme could prolong the serum half life of glucagon-like peptide-1 (GLP-1), a peptide hormone that helps tightly regulate blood sugar without substantial risk of hypoglycemia, provided physicians, and patients with another effective mechanistic option for treatment of type 2 diabetes.

Successful Drug Discovery, First Edition. Edited by János Fischer and David P. Rotella.
© 2015 Wiley-VCH Verlag GmbH & Co. KGaA. Published 2015 by Wiley-VCH Verlag GmbH & Co. KGaA.

1.2
Recent Examples of Target-Based Drug Discovery

In contemporary drug discovery, a key feature to help maximize the chance for success (i.e., marketing approval) is a clear understanding of the molecular target and mechanism of action for a drug candidate. Contributions to this volume describe the rationale for target selection, association(s) with the disease, and in some cases clinical biomarkers that provide critical information on target engagement as well as the correlation between dose and effect pharmacokinetic/pharmacodynamic (PK/PD) effects. This approach allows the discovery team to test a hypothesis for a first-in-class drug candidate. For a follow-on program where previous experience provides the necessary background, it may prove necessary to investigate aspects such as selectivity or adverse events that were not appreciated or were incompletely understood with the initial molecule.

For example, the discovery of dapagliflozin, the first sodium-glucose transporter type 2 (SGLT2) inhibitor approved for use in the treatment of type 2 diabetes, had a clear mechanism and target. By preventing reabsorption of glucose in the kidney, lower blood sugar could result. Approximately 90% of glucose reabsorption occurs in the kidney, and by promoting glucose excretion, lowered blood sugar should result. Lowering blood glucose is an indicator of target engagement that is also a clinically relevant biomarker for efficacy. This mechanism of action has a low risk of hypoglycemia because it is independent of insulin secretion, providing clinicians with a useful option to use in combination with metformin or insulin. Clinical trials revealed that SGLT2 inhibition could also lead to weight loss and improved plasma lipid profiles. Each of these unanticipated (based on the mechanism of action) effects are welcome benefits in type 2 diabetic patients because of other comorbidities including obesity, atherosclerosis, and hyperlipidemia. Combination studies of SGLT2 inhibitors with other diabetes treatments are underway, including use with DPP4 inhibitors. Given the complexity of type 2 diabetes, addition of another validated mechanism to treat the disease, especially one that can be used effectively in combination with others, is a true advance in the treatment of a serious and widespread disease.

The introduction of trastuzumab emtansine for treatment of metastatic breast cancer represents a particularly interesting example of a combination of a small molecule cytotoxic agent DM1 and the antibody trastuzumab that recognizes the human epidermal growth factor receptor 2 (HER2) receptor specifically expressed in breast cancer cells. In spite of the utility of the antibody alone for treatment of the disease, not all HER2 positive tumors respond, and some patients become refractive. While trastuzumab exerts its effect via more than one mechanism, the appeal of targeted delivery of a potent cytotoxic agent has potential advantages for therapeutic efficacy. Selective delivery only to cancer cells continues to be one of the limitations associated with use of cytotoxic agents in oncology that is most difficult to overcome. DM1 is a microtubule binding agent that is 3–10 times more potent in vitro compared to the well studied derivative maytansine, and up to 500 times more potent compared to the widely used taxanes. High potency for the

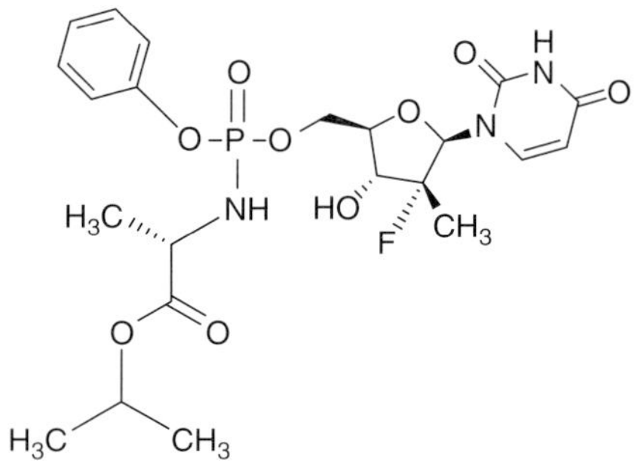

Figure 1.1 Sofosbuvir (Sovaldi).

small molecule target was recognized to be an important goal because of dose limiting toxicities. This antibody–drug conjugate employs two distinct and complementary mechanisms to achieve these goals, providing patients and clinicians facing metastatic breast cancer with a potentially useful option.

The approval of sofosbuvir (Sovaldi®) (Figure 1.1) in 2013, a viral ribonucleic acid (RNA) polymerase inhibitor used in combination with ribavarin represents the first all oral therapy for treatment of hepatitis C viral (HCV) infection. This chronic disease was initially treated with a combination of injectable interferon and ribavarin and was associated with significant adverse events, and it was also inconvenient for patients because one of the drugs had to be injected and the extended therapeutic course had to be an extended one (∼52 weeks) to achieve maximal efficacy. Against this therapeutic backdrop for a disease that affects over 150 million people worldwide, a number of alternative approaches with specific molecular targets associated with the virus are being investigated. Foremost among these are two proteases, NS3A/NS4A, along with viral RNA polymerase. Polymerase approaches for viral diseases are being very actively investigated; however, they frequently suffer from low potency because the molecule must be converted to a triphosphate derivative in cells. The first step in this process is slow and rate limiting, which results in reduced efficacy. Sofosbuvir is a monophosphate prodrug of a modified nucleoside that can bypass the slow initial phosphorylation step. The molecule is rapidly converted to a triphosphate derivative, which is a potent inhibitor of the viral enzyme. Clinical studies with sofosbuvir revealed that a sustained viral response could be achieved rapidly (in as short as 24 weeks) in combination with ribavarin. This substantially reduces the time course of therapy, is associated with fewer adverse events, and can be more easily administered because it is an all-oral regimen.

Treatment of rheumatoid arthritis is dominated by a number of biologics that require periodic injections. These therapies, while beneficial and effective, are recognized to be less convenient compared to oral administration of a small molecule. A number of approaches have been investigated to identify suitable molecular targets and small molecules for this purpose. In 2012, the approval of tofacitinib (Figure 1.2) represented the first small molecule since methotrexate to be approved for treatment of this crippling progressive disorder. Tofacitinib inhibits Janus kinase 3 (JAK3), an intracellular tyrosine kinase that plays a role

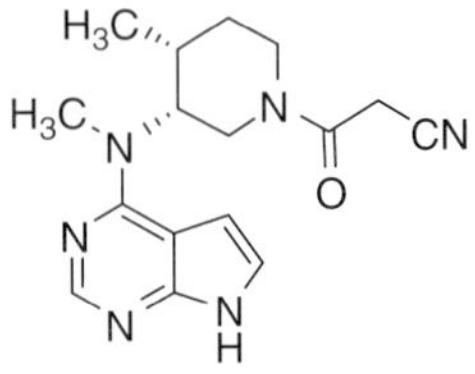

Figure 1.2 Tofacitinib (Xeljanz).

in signal transduction associated with a number of proinflammatory cytokines. JAK3 is localized in lymphocytes; analysis of plasma in rheumatoid arthritis patients showed that high levels of the kinase were present in the synovial fluid. This link between pathology and the site of action (i.e., the joint) provided reasonable assurance that inhibition of the enzyme represented a mechanistically reasonable approach for treatment of this disease. Kinase selectivity was a key objective to address because inhibition of JAK1 and/or JAK2 was undesirable because of potential toxicity. Clinical studies revealed that the molecule was effective as monotherapy and had an acceptable safety profile. More recently, Phase 2 and 3 combination studies with methotrexate showed efficacy in patients who did not respond to other anti-TNF (tumor necrosis factor) therapies such as adalimumab.

In early 2014, a novel small molecule was approved for treatment of psoriatic arthritis. Apremilast (Figure 1.3) is an inhibitor of phosphodiesterase 4 (PDE4), a well known enzyme that has been studied for a number of years. It was well known that inhibition of this cyclic nucleotide hydrolase elevated local concentration of cyclic AMP, an important second messenger that reduced circulating levels of inflammatory cytokines. These proteins are well known to be elevated in a number of types of arthritis. A number of PDE4 inhibitors were evaluated clinically and failed for both safety and efficacy reasons. Apremilast demonstrated potential disease modifying effects in patients with psoriatic arthritis, and is continuing to undergo clinical evaluation for osteoarthritis and rheumatoid arthritis.

These examples illustrate the strong connection between a target, mechanistic hypothesis, and measurable pharmacologic effects. While this approach to drug discovery represents a proven approach that can be successful, it does not guarantee success from a preclinical/discovery perspective or in the clinic. For example, inhibition of cyclooxygenase 2 (COX2), while efficacious for treatment of inflammatory disorders, also had undesirable cardiovascular effects. As outlined in the following paragraphs, serendipity and unexpected results play a

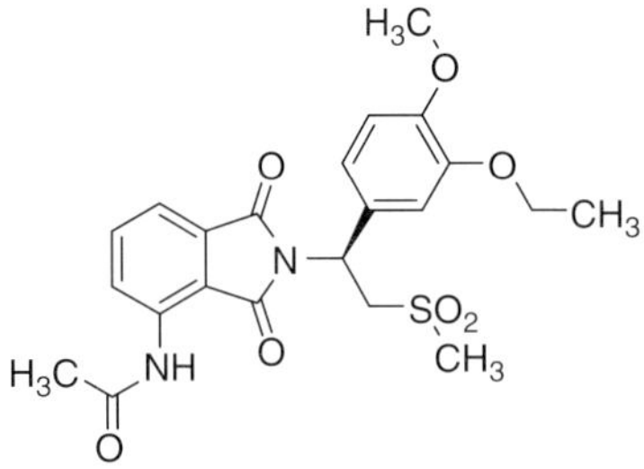

Figure 1.3 Apremilast (Otezla).

role in the discovery of novel therapeutic agents. Given the complexity of the undertaking, it is perhaps not surprising, however, that as a scientist one must be prepared to recognize and take advantage of these opportunities, using them to advance a project into and through the clinic to benefit patients.

1.3
Serendipitous Target-Based Drug Discoveries

As described in the introduction, target-based discoveries have an important role to play in new drugs. In some special cases, however, the initial drug target only helps to find a lead molecule, but the final product has a different mechanism of action, that is, a different target. These serendipitous target-based drug discoveries are discussed in this chapter using the following drug discoveries as examples in alphabetic order: drospirenone, escitalopram, ezetimibe, lamotrigine, and omeprazole.

1.4
Drospirenone (Contraceptive with Anti-aldosterone Activity)

Aldosterone (Figure 1.4) is the most potent natural mineralocorticoid hormone. It has an important role in the regulation of fluid and electrolyte balance and blood pressure. Spironolactone (Figure 1.5) is the first successful aldosterone antagonist that has been used since 1959 as a diuretic agent for the treatment of edema, liver cirrhosis, and as an antihypertensive drug.

Spironolactone causes hormone-related side effects such as menstrual irregularity, gynecomastia, and impotence due to its low receptor selectivity. Between the mid-1970s and the mid-1980s, a new search started for novel aldosterone antagonists with a primary goal to identify structures free from the unwanted effects seen with spironolactone.

Wiechert and coworkers synthesized more than 600 steroid derivatives [1] and investigated their anti-aldosterone activity. Compounds with fused cyclopropane rings displayed remarkable activity, and spirorenone (Figure 1.6) was selected as a clinical candidate.

The chemical name of spirorenone is: (6β,7β); 15β,16β-dimethylen-3-oxo-17α-pregna-1,4-dien-21,17-carbolactone. Spirorenone differs from spironolactone in

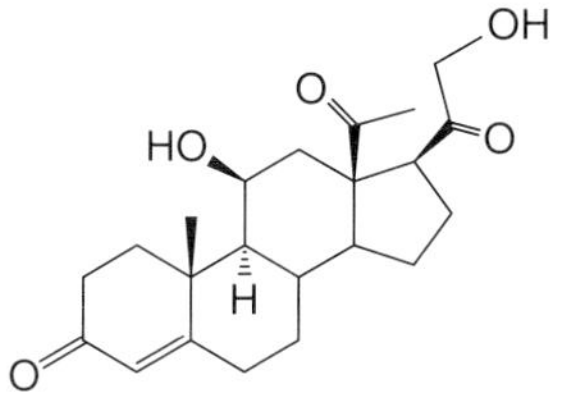

Figure 1.4 Aldosterone.

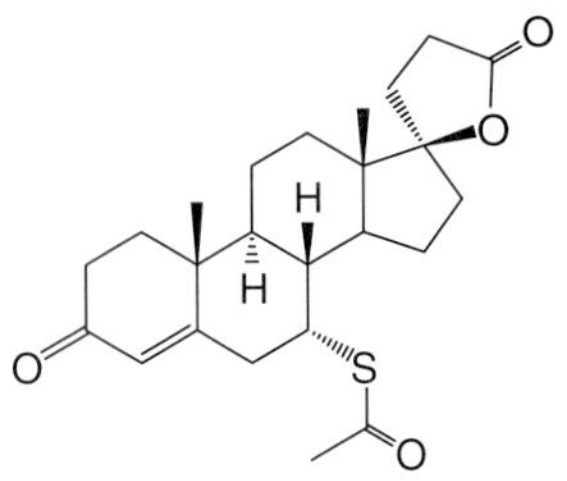

Figure 1.5 Spironolactone.

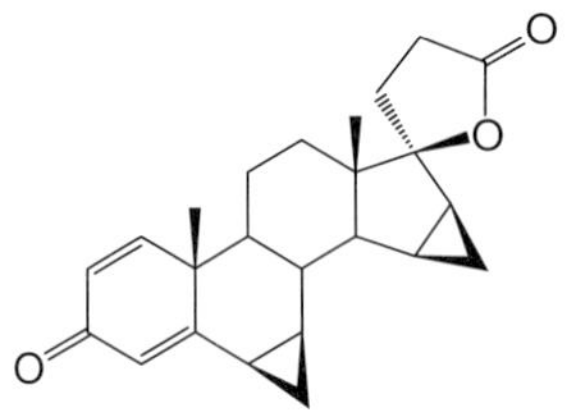

Figure 1.6 Spirorenone.

that an additional double bond at C-1 in ring A and two cyclopropane rings fused to rings B and D are present while the 7-acetylthio-substituent is missing.

Spirorenone proved to have five times higher anti-aldosterone activity compared to spironolactone, and its hormonal side effects were much lower. It was serendipitously found that low doses of spirorenone resulted in reduction of the testosterone level in men. This unexpected effect derived from the metabolic transformation of spirorenone to 1,2-dihydrospirorenone (drospirenone) (Figure 1.7), which was found to be an orally active progestin with anti-androgen properties. Unlike other species, spirorenone is metabolized to drospirenone in humans and in monkeys by the action of the enzyme Δ^1-hydrase. Drospirenone is used as a successful contraceptive.

1.4.1
Summary of Drospirenone Discovery

The initial step in the target-based drug discovery of drospirenone afforded the rather active lead molecule, spirorenone, as a potent aldosterone antagonist. The clinical studies of spirorenone led to the discovery of drospirenone, which was also identified as a metabolite of spirorenone. Drospirenone is an orally active

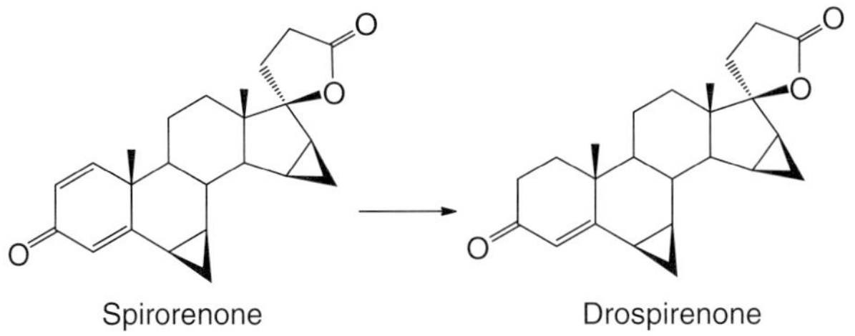

Figure 1.7 Formation of drospirenone from spirorenone.

progestin with notable anti-androgen and anti-mineralocorticoid properties, and has been developed as a contraceptive.

1.5 Escitalopram (Selective Serotonin Reuptake Inhibitor Antidepressant)

Researchers at Lundbeck [2] wanted to prepare a trifluoromethyl-substituted derivative of the tricyclic antidepressant melitracen. Instead of the expected analog a novel phthalane derivative was formed (Figure 1.8).

Melitracen acted as a norepinephrine (NE) reuptake inhibitor (unselective, due to action on various postsynaptic receptor sites), and serendipitously, it was found that the structurally different phthalane derivative also had this mechanism of action. Analog design revealed that the unsubstituted N-des-methyl analog, talopram (Figure 1.9), was – in contrast to melitracene – a highly selective and very potent NE uptake inhibitor.

The clinical trial of talopram was halted in Phase II because of its psychomotor side effects. On the recommendation of Professor Arvid Carlsson, a new research

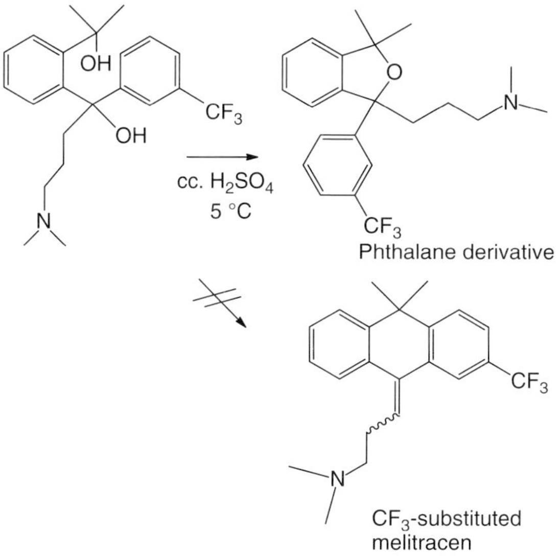

Figure 1.8 Unexpected formation of a phthalane derivative.

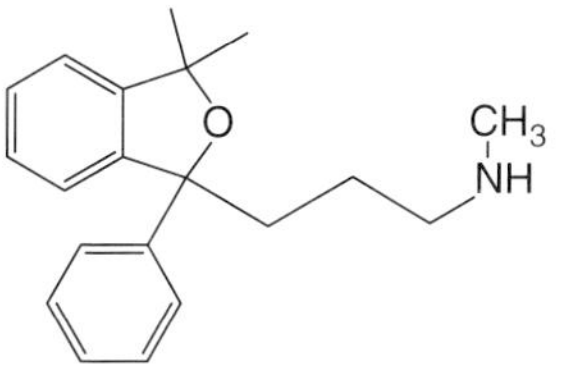

Figure 1.9 Talopram.

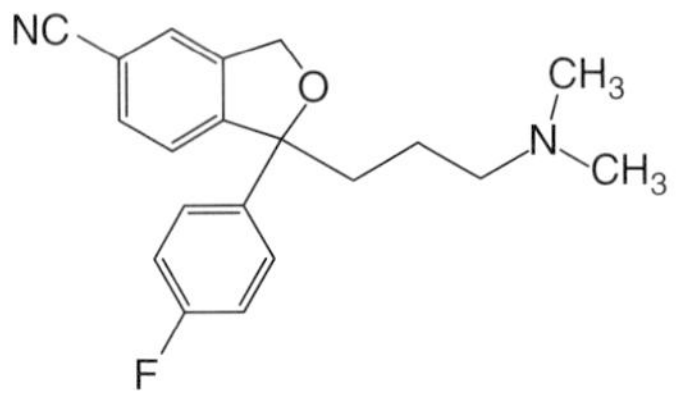

Figure 1.10 Citalopram.

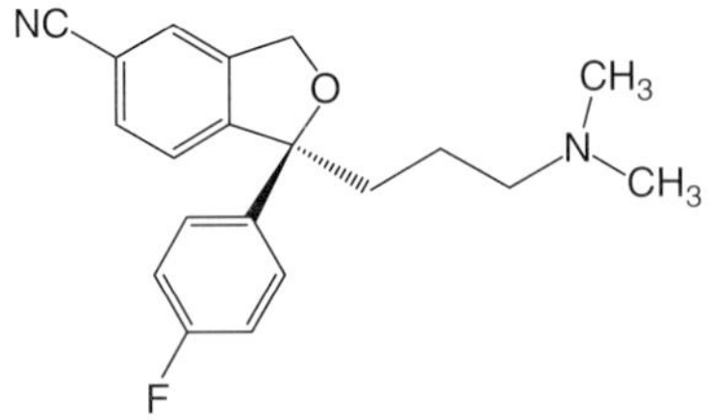

Figure 1.11 Escitalopram.

program started at Lundbeck in 1971. The research was aimed at finding selective 5-hydroxytryptamine (5-HT) uptake inhibitors, and it was assumed that these drugs would not have the above activating profile.

It seems paradoxical that talopram, a highly selective NE uptake inhibitor was used as a lead for the design of selective serotonin reuptake inhibitors (SSRIs), but some analogs of talopram from the Lundbeck compound collection had dual 5-HT/NE inhibition. This research afforded citalopram (Figure 1.10) which has a 5-cyano group on the isobenzofuran ring and is unsubstituted in the 3-position, has a 4-fluoro substituent at the phenyl group, and a dimethylamino group in its side chain. Citalopram became a successful antidepressant as a selective serotonin reuptake inhibitor.

Pure enantiomers of citalopram became available in 1988. The biological activity of racemic citalopram resides in the S-citalopram, which has a primary (inhibitory) (S1) site at the serotonin transporter (SERT) protein and a secondary allosteric binding site (S2) that modulates the binding properties at the primary site. R-citalopram is not a neutral component in racemic citalopram, and it counteracts the effects of escitalopram (Figure 1.11). The precise mechanism is not clear, but R-citalopram counteracts the association of escitalopram with the S1 site, perhaps, through its interaction with the SERT allosteric S2 site.

Retrospectively, it is interesting to analyze the structure–activity relationships of the direct analogs of citalopram. A compound deriving from the citalopram synthesis as an impurity (Figure 1.12) had one of the highest affinities at the S1 site, whereas it was inactive at the allosteric site (S2). It was a fortunate step to select citalopram for development and not a derivative (such as this by-product) which has no allosteric activity.

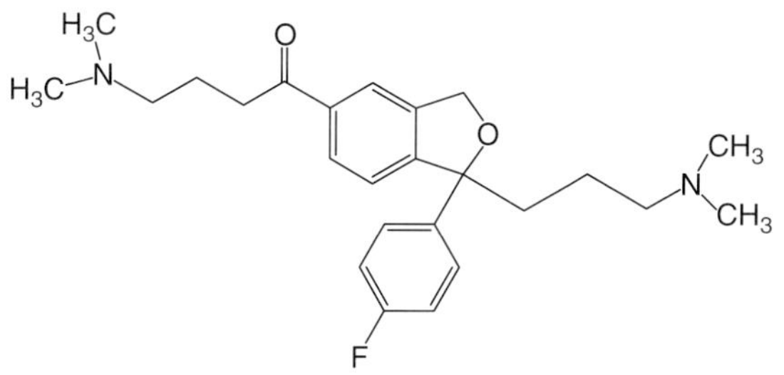

Figure 1.12 Synthesis byproduct of citalopram.

1.5.1
Summary of the Escitalopram Discovery

It was serendipitous that citalopram, a selective serotonin inhibitor was discovered from the selective NE uptake inhibitor talopram. It was fruitful that citalopram was selected for the development, because S-citalopram is also active as an allosteric modulator at SERT, whereas more potent compounds later proved to be inactive at this receptor site. When the selection was made there was no knowledge of the allosteric site of SERT.

1.6
Ezetimibe (Inhibitor of Cholesterol Absorption)

Cholesterol in the serum has two major sources: it is either endogenously synthesized or it has a dietary origin. The statins are excellent cholesterol-lowering drugs and they decrease cholesterol levels by inhibiting cholesterol biosynthesis. Ezetimibe is the only successful drug which inhibits the absorption of dietary cholesterol. Two recently published books describe the discovery of ezetimibe. It is described as a standalone drug [3] and as an example of small-molecule drug discovery [4]. In this chapter we discuss the role of the targets in its discovery.

Research at Schering-Plough (now Merck) focused on inhibitors of the enzyme acyl-Coenzyme A cholesterol acyltransferase (ACAT). This protein is located in the endoplasmic reticulum and forms cholesteryl esters from cholesterol, that is, the enzyme has two substrates: acyl-CoA and cholesterol and the enzyme-catalyzed reaction affords CoA and cholesteryl ester. The esterification of cholesterol mediated by ACAT is important for the absorption of cholesterol. Inhibition of this enzyme blocks the absorption of intestinal cholesterol.

The ACAT inhibitors SA 58035 (Sandoz) and CI 976 (Parke-Davis) were used as lead molecules (Figure 1.13).

The two known ACAT inhibitors are rather different and have only one common feature, open-chain amides. Researchers at Schering–Plough synthesized 2-azetidinone derivatives (compounds A and B) as the simplest ring-closure derivatives (Figure 1.14).

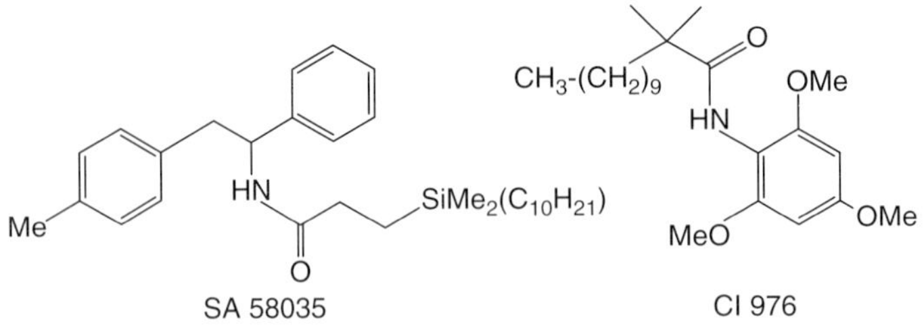

Figure 1.13 ACAT inhibitor lead molecules.

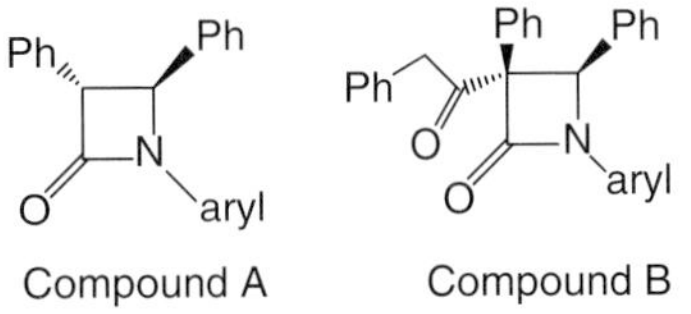

Figure 1.14 Structures of compounds A and B.

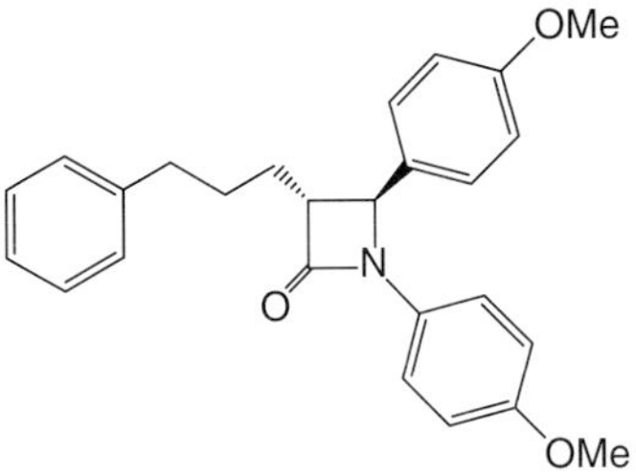

Figure 1.15 SCH 48461.

Compound A was inactive but compound B showed activity in hamsters. Several derivatives were synthesized and investigated *in vitro* (ACAT assay: inhibition of the esterification of cholesterol by oleic acid) and *in vivo* (inhibition of plasma cholesterol and accumulation of cholesteryl ester in 7 day cholesterol-fed hamsters).

The *in vitro* and *in vivo* activities showed a poor correlation; therefore, only the *in vivo* results were followed and SCH 48461 (Figure 1.15) was shown to reduce the serum cholesterol in clinical trials.

SCH 48461 was found to be rapidly and completely metabolized in animals and the metabolite mixture had a higher activity of cholesterol absorption than did the clinical candidate. The most probable metabolites were synthesized and tested *in vivo* to identify the most potent metabolite of SCH 48461 (Figure 1.16). Further optimization of this compound by introducing two fluorine atoms to block metabolism afforded ezetimibe. (Figure 1.17) which was 50 times more potent than SCH 48461 in the cholesterol-fed hamster assay. Following successful clinical trials ezetimibe was introduced in the market in 2002. It was only in 2004 that Schering-Plough researchers discovered its mechanism of action. The molecular target of ezetimibe is the NPC1L1 (Niemann-Pick C1-like1) protein, which is a critical mediator of cholesterol absorption.

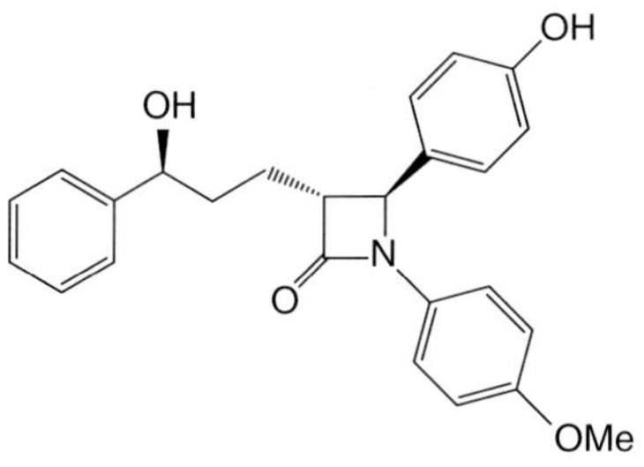

Figure 1.16 Potent metabolite of SCH 48461.

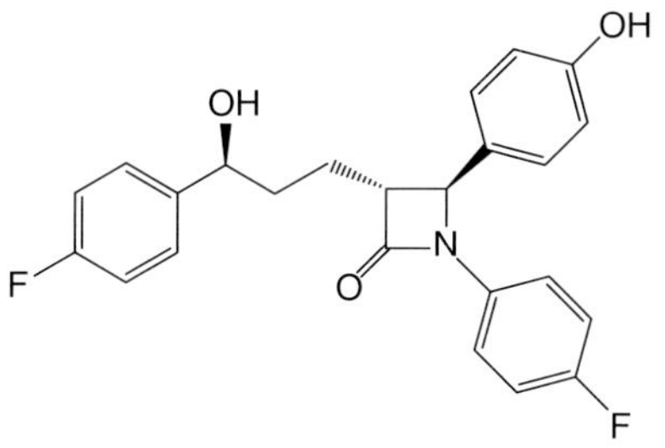

Figure 1.17 Ezetimibe.

1.6.1
Summary of Ezetimibe Discovery

A research program into cholesterol absorption inhibitors started with inhibitors of ACAT enzyme. Poor correlation between *in vitro* and *in vivo* assays led to the abandonment of the *in vitro* studies, and the initial azetidinone derivatives were optimized in animal tests to give a clinical candidate. It was further optimized following analysis of metabolites and direct analogs to afford ezetimibe whose mechanism of action was elucidated years after its introduction.

1.7
Lamotrigine (Discovery of a Standalone Drug for the Treatment of Epilepsy)

The discovery of lamotrigine as a standalone drug was recently described in a book chapter [3].

The discovery process was partly serendipitous, based on the observation that folic acid produces epileptogenic foci and that antiepileptic drugs (Figure 1.18) such as phenytoin, phenobarbital, and carbamazepine have antifolate properties. Thus, a correlation between antiepileptic and antifolate properties was assumed.

In 1973, Burroughs Wellcome (now GSK) researchers examined the anticonvulsant activity of their investigational antifolate drugs [5]. Pyrimethamine was developed in 1950 for both the treatment and prevention of malaria. It was found to have some anticonvulsant and antifolate activity, and it was increased in the analog BW 99U (Figure 1.19).

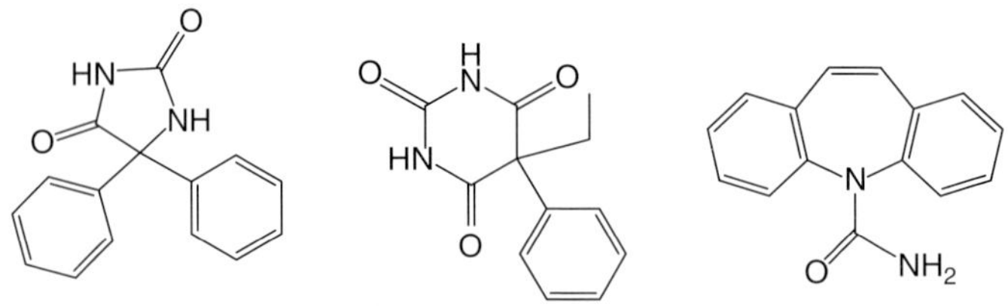

Phenytoin Phenobarbital Carbamazepine

Figure 1.18 Antiepileptic drugs: phenytoin, phenobarbital, and carbamazepine.

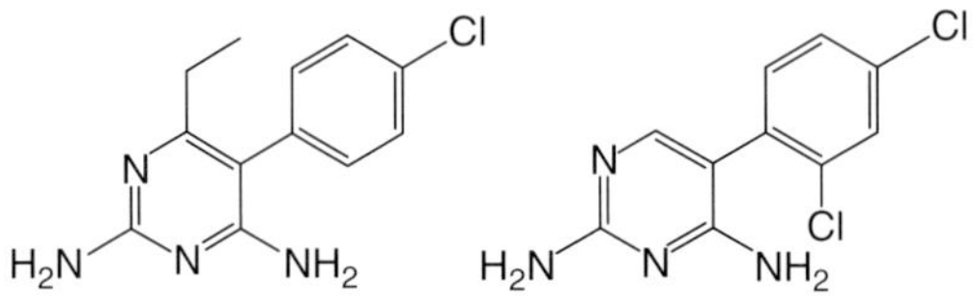

Pyrimethamine BW 99U

Figure 1.19 Pyrimethamine and BW 99U.

A series of structurally analogous phenyltriazines were synthesized in order to enhance anticonvulsant activity. One of these, BW 288U, had good anticonvulsant activity, although it was a mediocre antifolate. Further optimization afforded the most active analog, lamotrigine, which had a very weak antifolate activity. Lamotrigine (Figure 1.20) was at least as potent as phenytoin and long-acting. It was introduced in 1990 for the treatment of partial and secondary generalized seizures.

Lamotrigine is the first medication besides topiramate used in the treatment of seizures associated with the Lennox–Gastaut syndrome, a severe form of epilepsy. It was noticed that patients reported a higher level of happiness and mastery, or perceived internal locus of control, independent of seizure control. It received approval in 2003 for the treatment of bipolar I disorder, the first drug to do so since lithium.

In vitro studies demonstrated that lamotrigine primarily acts as a blocker of voltage sensitive sodium ion channels.

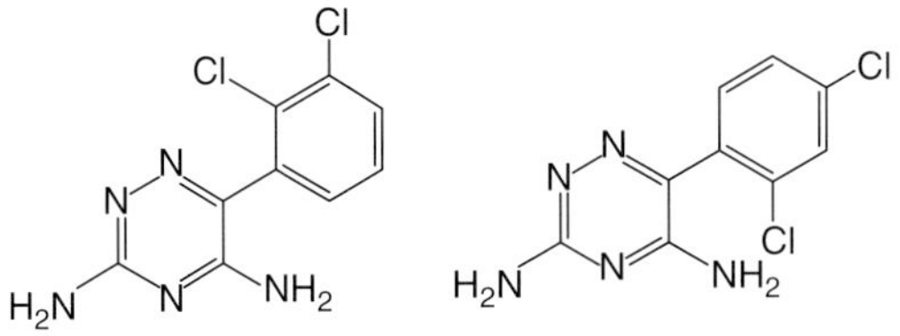

Lamotrigine BW 288U

Figure 1.20 Lamotrigine and BW 288U.

1.7.1
Summary of Lamotrigine Discovery

An erroneous working hypothesis, namely the hypothetical correlation between antiepileptic drugs and antifolate acidity serendipitously helped to select pyrimethamine, an antimalaria drug, as a lead molecule to search for new anticonvulsive agents among its analogs. This research work afforded the successful antiepileptic lamotrigine, which has a different mechanism of action.

1.8
Omeprazole (Proton Pump Inhibitor Acid-Suppressive Agent)

The discovery of proton-pump inhibitors has been described in a chapter of a book in 2006 [6].

The treatment of gastroesophageal reflux disease (GERD) and peptic ulcers had a breakthrough during the late 1970s with cimetidine and other analogous H_2 receptor antagonists. The second breakthrough was the introduction of proton-pump inhibitors at the end of 1980 and in the 1990s. Proton-pump inhibitors have a higher efficacy and a longer duration of action.

The success story of proton pump inhibitors goes back to antigastrin research which started in the 1960s. The peptide hormone, gastrin, is produced in the lower stomach and it stimulates the production of hydrochloric acid in the parietal cells. After the structure elucidation of gastrin (1964), ICI researchers synthesized about a thousand substances similar to the gastrin molecule but none could be developed. Servier and Searle also synthesized the gastrin analogs CMN 131 and SC-15396 (antigastrin) (Figure 1.21).

The molecules of Servier and Searle had weak antisecretory effects in animals, but because of toxicity they were not developed. Researchers at Hässle used CMN 131 as the lead molecule, and derivatives, which had no thioamide moiety, were designed in order to avoid its toxic effects. A new animal model, the conscious gastric fistula dog, was a very important tool in the successful research to give timoprazole (Figure 1.22).

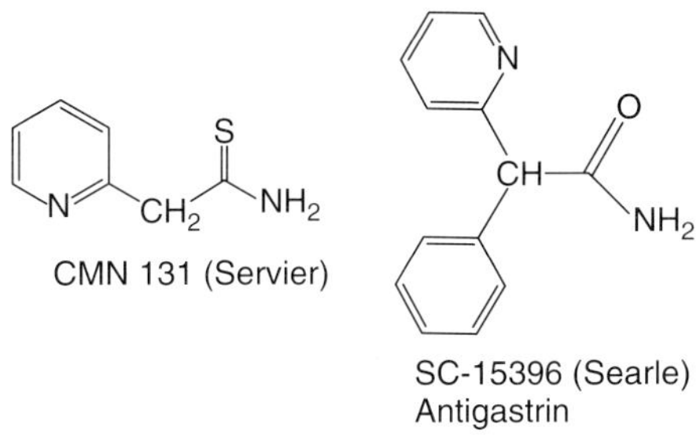

Figure 1.21 Gastrin receptor blockers.

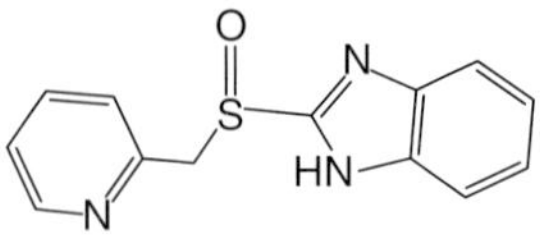

Figure 1.22 Structure of timoprazole.

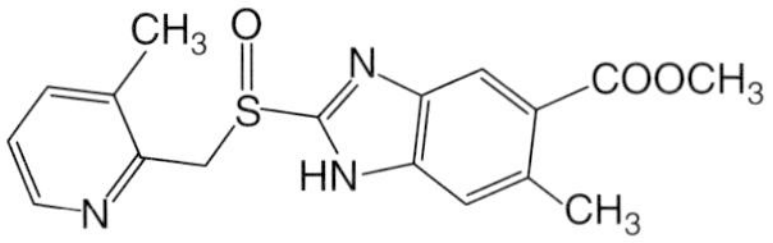

Figure 1.23 Structure of picoprazole.

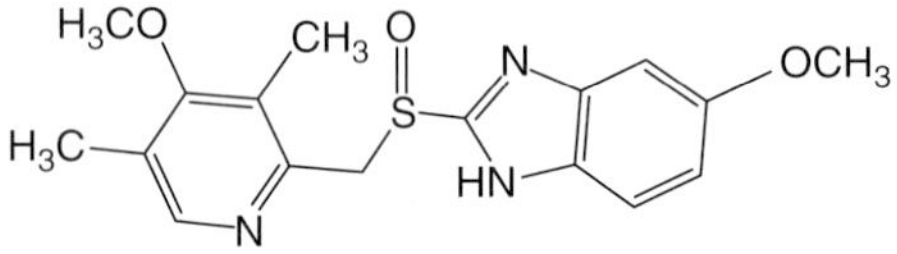

Figure 1.24 Structure of omeprazole.

Timoprazole caused enlargement of the thyroid gland and research continued to study substituted derivatives of timoprazole. Picoprazole (Figure 1.23) was the most potent antisecretory compound in this series without thyroid effects and it was a clinical candidate in 1976.

Picoprazole, however, caused necrotizing vasculitis in beagle dogs, but this side effect was observed only in dogs treated with a particular antiparasitic drug.

At the end of 1970s the proton pump (an H^+, K^+-ATP-ase) was discovered which regulates acid secretion in parietal cells and in 1983 substituted benzimidazoles, such as timoprazole were found to inhibit the proton pump.

Optimization continued using *in vitro* techniques and a great number of new substituted benzimidazoles were studied. The product of this research was omeprazole (Figure 1.24) synthesized in 1979. Following a long development process omeprazole was introduced in the market in 1988.

1.8.1
Summary of Omeprazole Discovery

Antigastrin projects started in 1967 and the discovery process lasted 4 years to give a lead compound of thioamide structure (CMN 131). The lead compound, however, was too toxic and heterocyclic compounds, where no thioamide moiety was present, were designed. The new molecules were tested in conscious gastric fistula dogs to give picoprazole, and its further optimization with the help of *in vitro* studies of proton pump inhibition afforded the successful pioneer drug, omeprazole.

1.9
Outlook

In this short survey we have discussed five very important drug discoveries where the starting targets helped the discovery process, but the drug product had a different mechanism of action.

The following general remarks can be made for target-based drug discovery research:

1) In target-based drug discovery, it is very useful to start phenotypic screening as early as possible, and if there are insufficient correlations between the two approaches, then phenotypic screening should be preferred.
2) The above very successful drug discoveries demonstrate that even an inappropriate target can be very useful to generate an important lead molecule.
3) It can happen that the mechanism of action of a drug will be discovered in a late phase of the drug research, or only after the introduction of the drug.

Acknowledgments

We thank Klaus P. Bøgesø, Arun Ganesan, William Greenlee, Eckhard Ottow, and Jörg Senn-Bilfinger for carefully reading and reviewing the manuscript.

List of Abbreviations

ACAT	acyl-coenzyme a Cholesterol acyltransferase
ACE	angiotensin-converting enzyme
COPD	chronic obstructive pulmonary disease
COX	cyclooxygenase
DM1	mertansine, a derivative of maytansine, cytotoxic agent
DPP	dipeptidyl peptidase
GLP-1	glucagon-like peptide-1
HCV	hepatitis C virus
HER2	human epidermal growth factor receptor 2
5-HT	5-hydroxytryptamine (serotonin)
JAK	Janus kinase
NE	norepinephrine
NPC1L1	Niemann-Pick C1-like1
PDE	phosphodiesterase
RNA	ribonucleic acid
SERT	serotonin transporter
SGLT-2	sodium-glucose transporter type 2
SSRI	serotonin reuptake inhibitor
TNF	tumor necrosis factor

References

1. Wiechert, R. (2006) in *Analogue-Based Drug Discovery* (eds J. Fischer and C.R. Ganellin), Wiley-VCH Verlag GmbH, Weinheim, pp. 395–400.
2. Bøgesø, K.P. and Sánchez, C. (2013) in *Analogue-Based Drug Discovery*, Vol. III (eds J. Fischer, C.R. Ganellin, and D. Rotella), Wiley-VCH Verlag GmbH, Weinheim, pp. 269–293.
3. Fischer, J., Ganellin, C.R., Ganesan, A., and Proudfoot, J. (2010) in *Analogue-Based Drug Discovery*, Chapter 2, Vol. II (eds J. Fischer and C.R. Ganellin), Wiley-VCH Verlag GmbH, Weinheim, pp. 29–59.
4. Ganellin, C.R. (2013) in *Introduction to Biological and Small Molecule Drug Research and Development*, Chapter 15 (eds C.R. Ganellin, R. Jefferis, and S. Roberts), Elsevier, pp. 399–416.
5. Sneader, W. (1996) *Drug Prototypes and Their Exploitation*, John Wiley & Sons, Ltd, Chichester, p. 601.
6. Lindberg, P. and Carlsson, E. (2006) in *Analogue-Based Drug Discovery*, Chapter 2 (eds J. Fischer and C.R. Ganellin), Wiley-VCH Verlag GmbH, Weinheim, pp. 81–113.

János Fischer is a Senior Research Scientist at Richter Plc. (Budapest, Hungary). He received his education in Hungary with MSc and PhD degrees in organic chemistry at Eotvos University of Budapest with Professor A. Kucsman. Between 1976 and 1978, he was a Humboldt fellow at the University of Bonn with Professor W. Steglich. He has worked at Richter Plc. since 1981 where he participated in the research and development of leading cardiovascular drugs in Hungary. His main current interest is analog-based drug discovery. He is the author of some 100 patents and scientific publications. In 2012, he was reelected Titular Member of the Chemistry and Human Health Division of IUPAC. He received an honorary professorship at the Technical University of Budapest.

David P. Rotella is the Margaret and Herman Sokol Professor of Medicinal Chemistry at Montclair State University. He earned a BS Pharm degree at the University of Pittsburgh (1981) and a PhD (1985) at The Ohio State University with Donald. T. Witiak. After postdoctoral studies in organic chemistry at Penn State University with Ken S. Feldman, he was an assistant professor at the University of Mississippi. David worked at Cephalon, Bristol-Myers, Lexicon, and Wyeth where he was involved in drug discovery projects related to neurodegeneration, schizophrenia, cardiovascular, and metabolic disease.

2
Drug Discoveries and Molecular Mechanism of Action
David C. Swinney

2.1
Introduction

What is the role of mechanism in drug discovery? What molecular mechanisms of action (MMOAs) are used by therapeutically useful medicines? This chapter addresses these questions by beginning with a synopsis of the MMOA [1] and then commenting on the analysis of the MMOAs of medicines approved by the US FDA between 1999 and 2008 [2].

2.2
Mechanistic Paradox

How a drug works is defined as its mechanism of action. What mechanism of action means, its value, and impact can be very different depending on one's perspective and intention.

1) *Practitioner, medical doctor*: A medical doctor is interested in a mechanistic understanding that differentiates medicines for specific indications. For example, the simple fact that many antihypertensives, antidiabetic, or antibacterials use different mechanisms enables a practitioner to experimentally determine which is the most effective and best tolerated for individual patients. At this level the minimum information necessary is that medicines use different mechanisms of action for a similar indication.
2) *Clinical development*: Mechanistic understanding helps with dose setting, provides biomarkers for prediction of response, and can help with patient selection. Mechanistic PK/PDs (pharmacokinetic/pharmacodynamics) models are critical for accurate translation of dose to effect [3].
3) *Discovery research:* Mechanistic understanding is the central feature of discovery that contributes to testable hypothesis, empirical assays, and molecular descriptors to identify and optimize drug candidates.

Successful Drug Discovery, First Edition. Edited by János Fischer and David P. Rotella.
© 2015 Wiley-VCH Verlag GmbH & Co. KGaA. Published 2015 by Wiley-VCH Verlag GmbH & Co. KGaA.

4) *Funders:* Funders of drug discovery, including private ventures, government, or senior management in a pharma company, require metrics to measure progress, risk, and chance of success. A mechanistic hypothesis provides a basis for the investment and decision making.

Who *does not* care about mechanism? Mechanism is not important to:

- *Patients:* Patients are rarely concerned about the mechanism, and the majority lack the education to understand even the most basic details. Their concern is does it work or not and what are the side effects. Will it make them feel better?
- *Regulatory agencies:* A medicine is required to be effective at a safe dose. There is no requirement for mechanistic understanding.

Mechanistic understanding is very important to those who do drug discovery research, providing knowledge of disease pathobiology and biomarkers and strategies to discover new medicines for the disease. Patients, regulators, and even funders do not require knowledge of the details. An analogy is fixing a broken automobile; you take it to a mechanic and trust that the mechanic will get it running properly. The details are not important, for example, how they adjusted the carburetor, only that the automobile is working effectively.

Mechanistic information provides a basis for communication of the value of a therapeutic approach. For example, scurvy was described by Hippocrates (about 460 to 380 BC), and herbal cures for scurvy including oranges and lemons have been known in many native cultures, but lack of communication of mechanistic knowledge prevented acceptance for a long time. Scurvy is a horrible disease due to ascorbic acid deficiency. Scurvy was at one time common among sailors, pirates, and others aboard ships as they were at sea longer than perishable fruits and vegetables could be stored, subsisting instead only on cured and salted meats and dried grains, and by soldiers similarly deprived of these foods for extended periods. While the curative effects of citrus fruit for scurvy were known in 1437, it took centuries for this to be widely recognized as a treatment for scurvy [4].

Modern drug discovery uses scientific knowledge to help identify therapeutically useful medicines in a timely method. Mechanistic understanding helps at many levels, but the paradox is that the value and type of mechanistic understanding depends on the context and the desired outcome; one definition of mechanism, such as a drug target, is neither adequate nor accurate for all. In the following text an aspect of mechanistic understanding, MMOA, useful to discovery research will be summarized.

2.3
Molecular Mechanism of Action

Knowing the parts of an efficient machine – a watch, an automobile, or a computer – is not enough to describe how it works. The parts must collaborate in precise ways to provide accurate time, reliable transportation, or processed information.

2.3 Molecular Mechanism of Action | 21

No mechanism based toxicity
Minimal affect on dose response curves

Mechanism-based toxicity
Modify shape of dose response curves

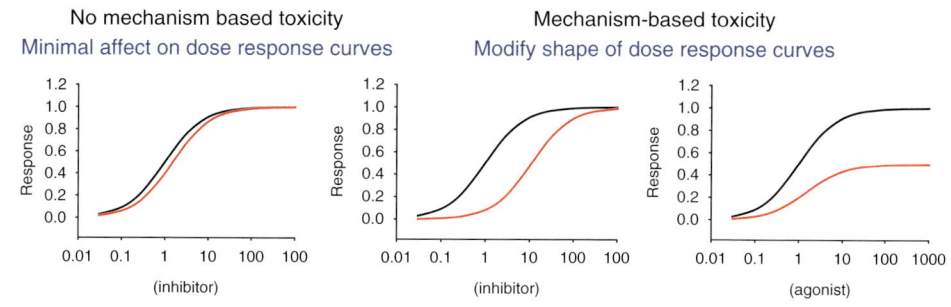

Full agonist irreversible inhibitors insurmountable antagonists noncompetitive inhibitors slow dissociation

Rapidly reversible inhibitors uncompetitive partial agonists functionally selective receptor modulators use dependent channel blockers

Figure 2.1 A primary driver of the impact of binding mechanism on the therapeutic index is the potential for mechanism-based toxicity. The curves show the relationship of concentration to binding versus function. When there is no mechanism-based toxicity, the binding should be efficiently coupled to function and the concentration–response curves will optimally be overlapping (a). When there is potential for mechanism-based toxicity the functional curves may be shifted to the higher concentrations to limit mechanism-based toxicity (b). This is what would be expected with rapidly reversible competitive inhibitors at equilibrium. A decrease in maximal response as seen with partial agonists is another mechanism to minimize mechanism-based toxicity (c). (From Ref. [2] with permission.)

Biology is infinitely more complex. The phrase "molecular mechanism of action" describes the way that biological parts collaborate to provide an effective and safe medicine [5].

The MMOA of a medicine connects specific molecular interactions to the response. Pharmacological action begins with an interaction between two molecules. This point is well recognized as the basis for pharmacological action. Ehrlich noted in 1913 that a substance will not work unless it is bound *"corpora non agunt nisi fixata"* [6]. However, binding alone is not always sufficient for a substance to communicate the desired message to physiology. For example, two similarly structured molecules can bind to an enzyme with similar affinity; however, only one will bind in a manner suitable for the catalytic reaction. Two similarly structured molecules can bind to a receptor with similar affinity; however, one will initiate the response (agonist) whereas the other will block the response (antagonist).

MMOA will influence the dose–response relationships through the interactions that translate binding to the pharmacological response. For example, a rapidly reversible competitive inhibitor must compete with the endogenous substrate/ligand for occupancy of the binding site. The endogenous competitive effectors will shift a dose–response curve to higher doses (Figure 2.1, left curves) [7]. Higher concentrations of drug will be required for equivalent occupancy and to achieve the desired pharmacological response.

2.3.1
The Primary Driver of an Optimal MMOA is the Potential for Mechanism-Based Toxicity

An underappreciated feature of an optimal MMOA is that it provides better safety margins and an improved therapeutic index. Much of current drug discovery is focused on specific targets, affinity, and occupancy of the compound at the target and thermodynamic selectivity at similar binding sites. Off-target safety risk is empirically evaluated. Interestingly, physiological MMOAs use spatial and temporal regulation to ensure efficacy and safety. For example, a key feature of T-cell receptor (TCR) signaling in response to MHC presented antigens is the binding kinetics of the TCR-MHC-antigen complex [8, 9]. Mammalian DNA polymerases use kinetic proofreading to increase selectivity and fidelity [10]. In pharmacology the strategy is to minimize the safety issues while maintaining sufficient efficacy; partial agonists, such as varenicline [11], the fast off-rates of memantine from the NMDA (*N*-methyl-D-aspartate) receptors [12], and the atypical anti-psychotics, such as clozapine, from the D_2 receptor are thought to contribute to a tolerable response [13].

The most significant factor for selection of a MMOA is the potential for mechanism-based toxicity (on-target toxicity) [14, 15]. A drug with no potential for mechanism-based toxicity will maximize its therapeutic index via mechanisms in which drug binding is efficiently coupled to physiology. In general, drugs used against nonhuman targets, as is the case for most anti-infective agents, will not have mechanism-based toxicity and these mechanisms are superior in this context. These mechanisms are generally contraindicated for targets in which there is mechanism-based toxicity. This is a principal driver for the desire for slow dissociation rates, long residence times, and irreversible inhibition. Kinetic mechanisms that are irreversible, insurmountable, noncompetitive, full agonist, or slow dissociating will be safer because lower drug concentrations are required for efficacy. Lower drug concentrations minimize off-target toxicity and result in an increased therapeutic index.

For most human targets there is potential for mechanism-based toxicity. Therefore, the challenge is to identify MMOAs that minimize the potential for mechanism-based toxicity while retaining the desired response. Competition with endogenous effector, surmountable with rapid kinetics, uncompetitive inhibition, partial agonism, functional selectivity, and allosteric partial antagonism are mechanisms in which the physiological environment can help to shape the dose–response curves to minimize mechanism-based toxicity while retaining sufficient drug efficacy.

2.3.2
Details of MMOA are not Captured by IC_{50} and K_I

A challenge for drug discovery is that many of the molecular features important to an optimal MMOA are not captured by IC_{50} and K_I values in target-based screening assays. This is exemplified in a recent report on the molecular features that

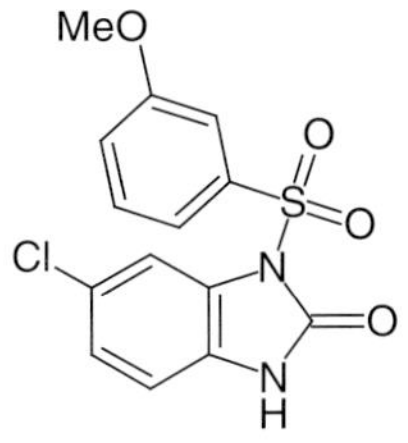

Figure 2.2 Benzimidazol-2-one derivatives that bind to HIV-1 reverse transcriptase.

differentiate binding between functionally different ligands for the β_2-adrenergic receptor (β_2AR) [16]. Wacker and colleagues found no discernible structural difference in binding between the inverse agonist and the antagonist. β_2AR bound to pharmacologically distinct ligands (antagonists and inverse agonists) have virtually identical backbone conformations in the crystal structures suggesting that the conformational changes capable of modifying signaling properties are very small, beyond the resolution of the obtained data. Alternatively, the major effect of inverse agonists, antagonists, and extrapolated to agonists on β_2AR is not in modifying a specific conformation with large conformational changes, but on minor structural changes and a significantly larger contribution from receptor dynamics.

Another example of MMOA not captured by K_I demonstrates the contribution of the association rate for the binding of benzimidazol-2-one derivatives (Figure 2.2) to HIV-1 reverse transcriptase (HIV RT) [17]. The lysine 103 to asparagine mutant (K103N) has a minimal influence on the bound conformation of NNRTIs, while it significantly affects the kinetics of the inhibitor-binding process. A detailed enzymatic analysis elucidating the molecular mechanism of interaction between benzimidazol-2-one derivatives and the K103N mutant of HIV RT showed that the loss of potency of these molecules toward the K103N RT was specifically due to a reduction of their association rate to the enzyme. Unexpectedly, these compounds showed a strongly reduced dissociation rate from the K103N mutant, as compared to the wild type enzyme, suggesting that, once occupied by the drug, the mutated binding site could achieve a more stable interaction with these molecules. Available structural and biochemical data, strongly support that the mutated N103 residue slows the k_{on} rate for inhibitor binding through the formation of a hydrogen bond with the Y188 side chain.

The important nuances of MMOA are not captured by IC_{50} and K_I in binding assays. This highlights the gap between the molecular and physiological approaches to drug discovery. MMOA provides an opportunity to bridge the gap by mapping molecular changes to physiological outcomes. Molecular descriptors, such as binding kinetics, combined with metrics, such as biochemical efficiency (BE), and discovery strategies based on increased understanding of the MMOA will help to bridge this gap.

2.3.3
Metrics, Biochemical Efficiency

BE is defined as how effectively the binding of an inhibitor to the target provides the desired pharmacological response [14, 15]. Quantitatively, BE is the ratio of the K_I obtained in a binding or enzyme assay compared to the IC_{50} in a physiologically relevant functional assay (BE = K_I/IC_{50}). A ratio of one indicates that binding is efficiently coupled to the physiological response. Approximately 80% of marketed drugs compete with endogenous effectors [14, 15]. The binding of a majority of approved medicines is efficiently coupled to the physiological response. The efficient coupling of many successful competitive drugs is due to additional mechanistic features, including slow off-rates and irreversibility, which limit competition with high concentrations of endogenous effectors. Binding kinetics provide a mechanism through which these medicines increase their therapeutic index by reducing the concentration of drug required for a pharmacological response. BE will directly influence the therapeutic index via the impact on the physiological drug concentrations required for pharmacodynamic response.

The physiological consequences of BE are exemplified in a publication by Yun and coworkers on the epidermal growth factor (EGF) receptor [18]. The diminished ATP affinity of the oncogenic EGF receptor mutants opens a "therapeutic window (index)," which renders them more easily inhibited relative to the wild-type EGF receptor and other kinases. The authors describe how resistance mutations in the ATP binding site restore the affinity of the kinase for ATP to wild-type levels. The mutations enable ATP to compete more effectively with the inhibitor. The work also provides an explanation for why EGF receptor covalent irreversible inhibitors are insensitive to the mutations. Due to a lack of competition with ATP, the dose–response curves for the irreversible inhibitors are not shifted to higher concentrations (Figure 2.1).

The challenge to overcome poor BE can make a specific target intractable by a specific mechanistic approach. Mourey *et al.* [19] described the pharmacologic properties of a benzothiophene MK2 inhibitor, PF-3644022, with good activity but poor BE (Figure 2.3). PF-3644022 is a potent freely reversible ATP-competitive compound that inhibits MK2 activity (K_I = 3 nM) with good selectivity when profiled against 200 human kinases. They noted that of the MK2 inhibitor chemotypes reported, few have submicromolar potency at inhibiting TNFα production in cells, perhaps because of poor physiochemical properties, poor cell permeability, poor BE, or inadequate enzyme potency. Of several MK2 chemotypes investigated, only the benzothiophenes have cellular IC_{50} values less than 500 nM. PF-3644022 is a highly permeable and potent MK2 inhibitor (K_I 3 nM), yet it exhibits poor BE with at least 30-fold weaker activity at inhibiting TNFα production in cells. Given that PF-3644022 is an ATP-competitive inhibitor, the shift in cellular potency may be caused by competition with high cellular concentrations of ATP (approximately 5 mM). The binding constant of MgATP for nonactive MK2 is 30 μM. The authors believe that the best K_I values achievable with MK2 are low nanomolar, because

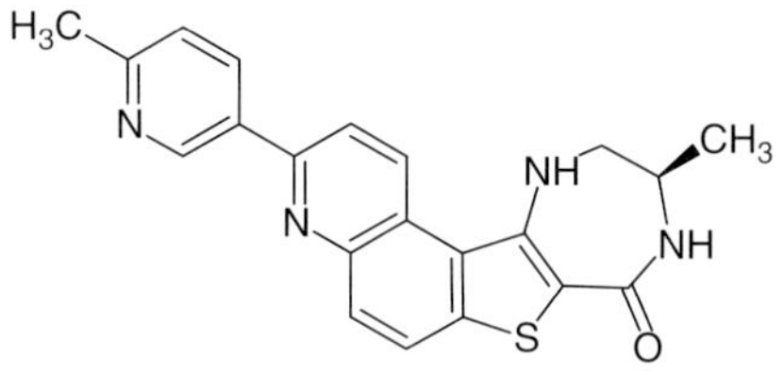

Figure 2.3 Benzothiophene MK2 inhibitor, PF-3644022.

they were unable to achieve further potency even after gaining additional interactions in the ATP pocket. They also developed several irreversible MK2 inhibitors as tool compounds that did in fact exhibit BEs near one, but had insufficient selectivity to explore as drug leads. Although the MK2 knockout mouse validated MK2 as a very attractive target for TNFα inhibition, the very low BE suggests a low probability of success developing MK2 inhibitors as drugs.

2.4 How MMOAs were Discovered

Drug discovery and development in the last quarter century has focused on the promise of molecular medicine to identify medicines to treat unmet medical needs by targeting specific gene products. For example, mutations or defects at specific molecular locations in human DNA were found to be responsible for some cancers, raising the hope of developing successful therapies tailored to individual patients. The gene to medicine approach has had success, as noted by imatinib (Gleevec) [20] and gefitinib (Iressa) [21], and encouraged researchers to follow the same path in the majority of drug discoveries.

The basis for pharmacological action and current practice in drug discovery is that a substance will not work unless it is bound, "corpora non agunt nisi fixata" as noted by Erhlich in 1913 [6]. As a consequence of this understanding, drug discovery and development in the last few decades has followed a simple paradigm; start with a potential drug target, identify a molecule that will bind with good affinity, optimize metabolism, and biopharmaceutical properties to achieve levels of the drug that maintain target occupancy through the dosing interval and then de-risk for safety.

This target-based approach is based on some very simple assumptions: (i) the target protein provides a therapeutic mechanism, (ii) reversible equilibrium binding is sufficient for efficacy, (iii) selectivity is achievable based on equilibrium binding at similar targets, and (iv) blood levels sufficient to cover the target will lead to clinical efficacy. Potential safety liabilities cannot be de-risked without empirical models, demonstration of clinical efficacy, and safety requirement studies.

Prior to the target-based drug discovery new medicines were discovered by evaluating different chemicals against phenotypes – an organism's observable characteristics – in authentic biological systems, such as animals or cells. There were many factors that influenced the shift from a phenotypic approach to a

2 Drug Discoveries and Molecular Mechanism of Action

target-based approach including the idea that a rational, measurable gene to clinic to registration progression would increase research and development (R&D) success and productivity. Rational, informed target-based drug discovery allows molecular tools of genetics, chemistry, and informatics to drive discovery and also provides criteria and boundaries for choosing patient populations, setting doses, and quantitative measurement of efficacy and toxicity. Unfortunately, this has not yet transformed the industry.

To investigate whether some strategies have been more successful than others in the discovery of new drugs, the discovery strategies and the MMOA for new molecular entities (NMEs) and new biologics that were approved by the US Food and Drug Administration between 1999 and 2008 were analyzed [2]. Out of the 259 agents that were approved, 75 were first-in-class drugs with new MMOAs, and out of these, 50 (67%) were small molecules, and 25 (33%) were biologics. The results also show that the contribution of phenotypic screening to the discovery of first-in-class small-molecule drugs exceeded that of target-based approaches – with 28 and 17 of these drugs coming from the two approaches, respectively – in an era in which the major focus was on target-based approaches (Figure 2.4).

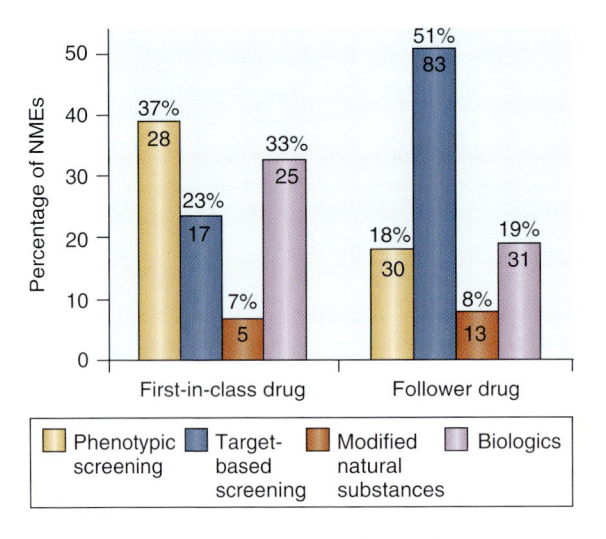

Figure 2.4 The distribution of new drugs discovered between 1999 and 2008, by the discovery strategy. The graph illustrates the number of new molecular entities (NMEs) in each category. Phenotypic screening was the most successful approach for first-in-class drugs, whereas target-based screening was the most successful for follower drugs during the period of this analysis. The total number of medicines that were discovered via phenotypic assays was similar for first-in-class and follower drugs – 28 and 30, respectively – whereas the total number of medicines that were discovered via target-based screening was nearly five times higher for follower drugs versus first-in-class drugs (83 to 17, respectively). (From Ref. [2] with permission).

Examples of first-in-class medicines discovered by phenotypic screening included those discovered using animal models such as ezetimibe (Zetia) for reducing levels of blood cholesterol, those discovered with cellular assays such as vorinostat (Zolinza), the first HDAC (histone deacetylase) inhibitor, that was reported to come from the observation that DMSO had an unexpected effect on cancer cells, and linezolid (Zyvox), an oxazolidinone antibiotic identified in antibacterial assays. Target-based successes included a number of tyrosine kinase inhibitors for cancer – gefitinib (Iressa) (EGFR), imatinib (Gleevec) (BCR-ABL), sorafenib (Nexavar) (Raf), and sunitinib (Sutent) (VEGFR/PDGFR) and antivirals – maraviroc (Selzentry) (CCR5), raltegravir (Isentress) (HIV integrase), and zanamivir (Relinza) (influenza neuraminidase) [2].

The *Nature Reviews* article proposed that lower productivity partly reflects target-based discovery's lack of consideration of the molecular complexities of how the drugs work [2]. These observations led to the proposal that the progression of drug discovery from an unmet medical need to best in class medicines is facilitated by the use of phenotypic assays to identify first-in-class medicines and their respective MMOAs. Progression correlates with an iterative increase in knowledge to specifically address a phenotype related to the unmet medical need. Early in the progression, the knowledge is achieved by empirical analysis. Ideally as more knowledge is gained to link a specific mechanism of modulation (MMOA) to the desired phenotype, drug discovery can focus efforts toward addressing specific hypotheses (Figure 2.5) [5].

The merits of phenotypic and target based discovery has been the subject of a number of recent articles [5, 22–24] and is not discussed further in this chapter.

2.4.1
MMOAs of Medicines Approved by the USFDA Between 1999 and 2008

The above analysis of discovery methods suggest that the success of phenotypic assays is, at least in part, due to the ability to discover new MMOAs that are *a priori* difficult to predict. What are the MMOAs of approved medicines? How do they work at the molecular level? The features of the MMOAs include the targets, when known, and how the binding couples to provide the functional response. The information on these features, while not being complete, is informative. Figure 2.6, from Swinney and Anthony review [2] shows the activities and targets for the first-in-class medicines approved between 1999 and 2008.

The majority of MMOAs of small-molecule first-in-class NMEs involved modulation of receptors or inhibiting the activity of enzymes. The pharmacological responses were often achieved by binding to the target protein to elicit a positive or negative response. Many different biochemical mechanisms mediated the drug response at the target. These included reversible, irreversible, and slow binding kinetics; competitive and uncompetitive interactions between physiological substrates/ligands and drugs; and inhibition, activation, agonism, partial agonism, allosteric activation, induced degradation, and others.

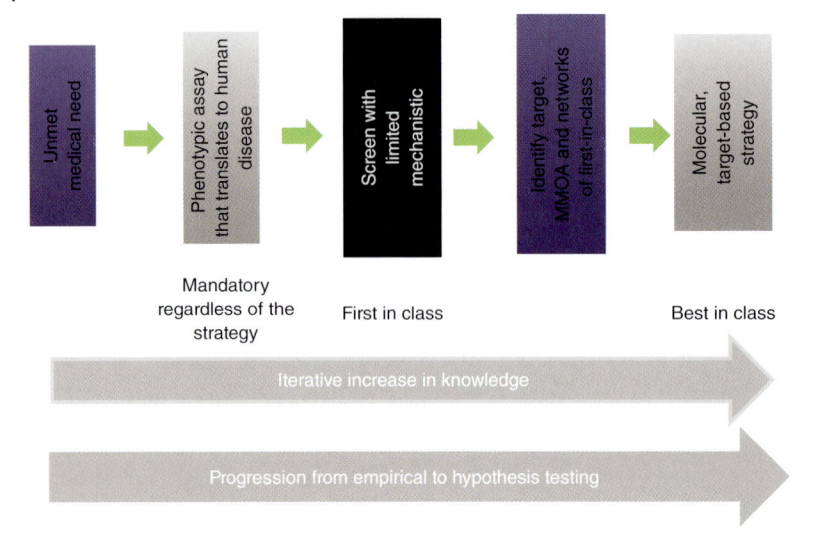

2 Drug Discoveries and Molecular Mechanism of Action

Figure 2.5 Progression of drug discovery from unmet medical need to best in class medicines. This figure is an oversimplified schematic highlighting the contribution of empirical approaches to first-in-class medicines, hypothesis driven approaches for best in class medicines, and the role of mechanistic understanding. Progression correlates with an iterative increase in knowledge to specifically address a phenotype related to the unmet medical need. Early in the progression the knowledge is achieved by empirical analysis. Ideally as more knowledge is gained to link a specific mechanism of modulation to the desired phenotype, drug discovery can focus efforts to address specific hypotheses. The relative timing of employing empirical versus hypothesis driven approaches is influenced by the validation of mechanistic understanding. (From Ref. [5] with permission.)

Importantly, simple equilibrium binding at the target was rarely sufficient for the translation of drug binding to the target to a therapeutically useful safe and effective response, a subtle aspect of how drugs work that is largely underappreciated. These results are consistent with the previous conclusion that "two components are important to the MMOA. The first component is the initial mass action-dependent interaction. The second component requires a coupled biochemical event to create a transition away from mass-action equilibrium" [14, 15].

The majority of the target classes for the small molecule first-in-class medicines, where the target was known, were enzymes and receptors, 22 and 10, respectively. Five NMEs targeted channels or receptors and the mechanism and targets are unknown for nine NMEs. It is tempting to speculate that the success of these target classes (enzymes, receptors, and channels) is due to the fact that these are proteins that carry out a function and that function is captured in the discovery assays, regardless of their being phenotypic or target-based. It is not simple mass action binding. Another contributing factor may be the role of these functions in the pathobiology. Further analysis is needed to support this speculation.

2.4 How MMOAs were Discovered | 29

Affect enzyme activity

Kinase inhibitors	Protease inhibitors	Other enzyme inhibitors	Microbial enzyme inhibitors	Enzyme cofactor
Gefitinib	Aliskiren	Azacitidine	Caspofungin	Sapropterin
Imatinib	Sitagliptin	Cilostazol	Linezolid	
Sorafenib	Bortezomib[b]	Fondaparinux	Raltegravir	
Sunitinib		Miglustat	Retapamulin	
Sirolimus[a]		Nitazoxanide	Zanamivir	
		Nitisinone		
		Orlistat		
		Vorinostat		

(a)

Affect receptor activity

Activate response	Inhibit response	Modulate response
Cinacalcet	Aprepitant	Fulvestrant[c]	Aripiprazole
Eltrombopag	Bosentan	Mifepristone	
Ramelteon	Conivaptan	Target nuclear receptors	
	Maraviroc		

(b)

Affect ion channel activity

Acamprosate
Memantine
Varenicline
Ziconotide

(c)

Affect transporter activity

Ezetimibe

(d)

Others

Aminolevulinic acid
Daptomycin
Nelarabine
Verteporfin

(e)

Unknown/unclear target/mechanism

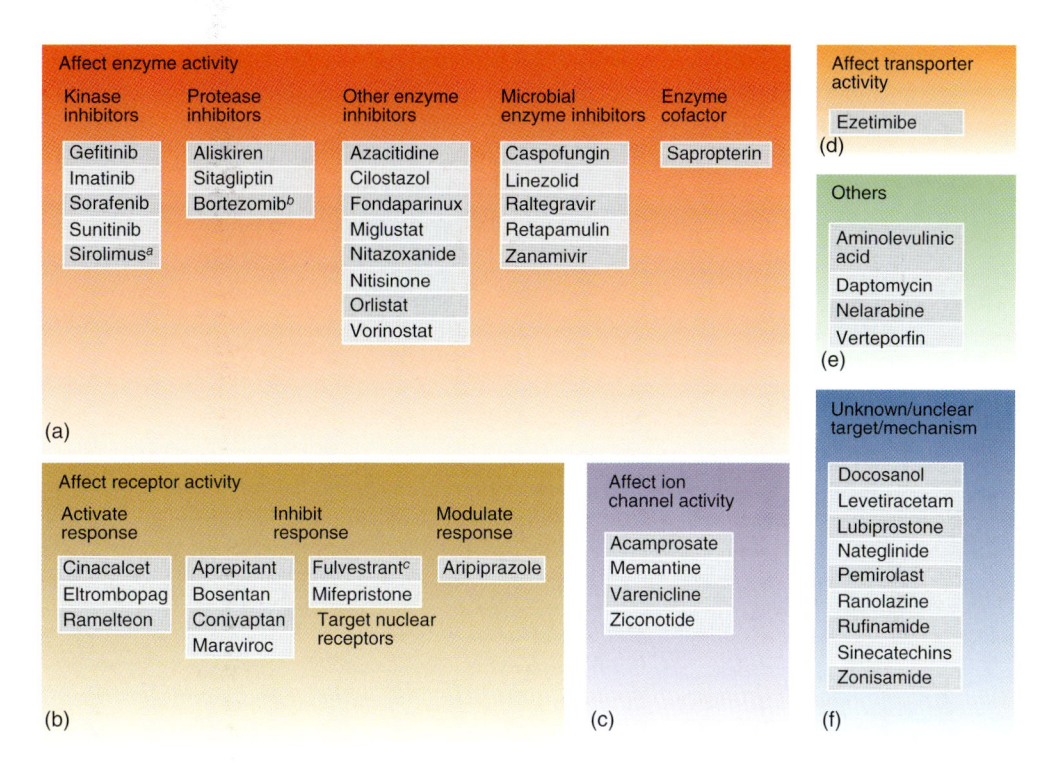

Docosanol
Levetiracetam
Lubiprostone
Nateglinide
Pemirolast
Ranolazine
Rufinamide
Sinecatechins
Zonisamide

(f)

Nature reviews | Drug discovery

Figure 2.6 Activities of first-in-class small-molecule new molecular entities. Nearly half (22 out of 50) of the first-in-class small-molecule drugs that were approved between 1999 and 2008 affected enzyme activity (a). The molecular mechanisms of action (MMOAs) of these drugs included reversible, irreversible, competitive, and noncompetitive inhibition, blocking activation, and stabilizing the substrate. The next largest group of targets (10 drugs) were receptors (b), most of which were G protein-coupled receptors. Their MMOAs included agonism, partial agonism, antagonism, and allosteric modulation. Two drugs – fulvestrant and mifepristone – targeted nuclear receptors. Four of the drugs targeted ion channel activity (c), their MMOAs included uncompetitive antagonism and partial agonism. One drug, ezetimibe, targeted the activity of a transporter (d). The remaining drugs had other activities (e), or unclear targets or MMOAs (f). Of the NMEs with other activities, two had a unique MMOA: verteporfin, a porphyrin that catalyzes the generation of reactive oxygen species and is used for photodynamic therapy; and daptomycin, which has an MMOA that involves disruption of bacterial membranes. [a]Sirolimus binds to the protein FKBP12 and the sirolimus–FKBP12 complex inhibits the kinase activity of mammalian target of rapamycin, whereas the other four kinase inhibitors target receptor tyrosine kinases. [b]Bortezomib inhibits the 26S proteasome – a multiprotein complex – by inhibiting the chymotryptic-like activity of the proteasome. [c]Fulvestrant acts by promoting receptor degradation. (From Ref. [2] with permission.)

The analysis and discussion provide insights into the diversity of MMOAs that provide the desired functional response. It is my contention that the MMOAs also are important to provide a safe response and thereby an optimal therapeutic index. The direct impact of MMOA on safety is demonstrated for a number of the NMEs. The strategies for discovery of the partial agonists, aripiprazole and varenicline, were designed to find molecules and MMOAs that provided an improved therapeutic index by minimizing mechanism-based toxicity. Specific examples that highlight the ability of rapid competitive kinetics to minimize mechanism toxicity are described for the NMDA receptor antagonist memantine by Lipton [12], for the D_2 dopamine receptor antagonist by Kapur and Seeman [13], and for the rapidly reversible NSAIDS such as ibuprofen [25]. With all of these approaches, long residency times resulted in undesirable side effects. The side effects were reduced with the rapidly reversible inhibitors. It is postulated that the rapid reversibility allows the drug to compete effectively with high levels of endogenous effectors produced by activation of the system. Lipton has highlighted this as a drug discovery strategy, using the NMDA receptor as an example, based on the principle that drugs and/or their targets should be activated by the pathological state that they are intended to inhibit [13].

2.5
Case Study: Artemisinin

Artemisinin is the most effective current therapy for the treatment of malaria (Figure 2.7). Artemisinin is selectively activated to a reactive species with selective cytotoxicity. Recently, Mercer and coworkers demonstrated that artemisinin endoperoxide bioactivation to carbon-centered radicals results in cytotoxicity only when it is mediated by heme or a heme-containing protein [26]. They hypothesized that artemisinin radicals act locally within the mitochondria to modify heme or heme-containing proteins, which results in the generation of reactive oxygen species (ROS) via electron transport chain (ETC) dysfunction and the induction of cell death via apoptosis. They presented a unified chemical rationale for the sensitivity of cancer cells, which have upregulated iron metabolism and heme synthesis to maintain continuous growth and proliferation, and primitive erythroblasts, defined as the embryotoxic target of the artemisinin, which have active mitochondria and heme synthesis [26].

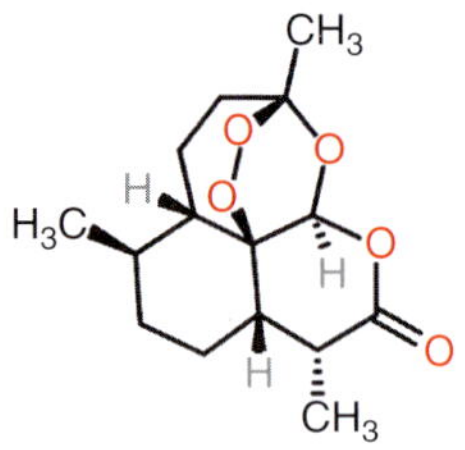

Figure 2.7 Structure of artemisinin.

The discovery of artemisinin is attributed to You-You Tu, at the Institute of Chinese Materia Medica, China Academy of Traditional Chinese Medicine in the early 1970s. An *Artemisia* extract showed a promising degree of inhibition against parasite growth, consistent with activity that had been reported for this species in "A Handbook of Prescriptions for Emergencies" by Ge Hong (Jin Dynasty, 284–346 AD). It is reported that the extracts were evaluated in mouse malaria *plasmodium berghei*. [27]. It is difficult to envision how based on our current knowledge and technologies we could *a priori* identify this MMOA in a target-based approach.

The value and details of mechanistic understanding clearly is context dependent. Current drug discovery and development is target centric, focused on a protein that binds a medicine as the definition of a medicine and the metrics and driver for drug discovery success. However, this is a very limited definition of mechanism that while being useful to help communicate mechanistic information across the multidiscipline aspects of drug discovery and development can limit progress of drug discovery. A bottle-neck for phenotypic approaches to drug discovery in many cases is identification of a target, but is it necessary to have a target? Will other indicators of mechanism suffice to help develop the drug, measure its success, and differentiate it from other drugs in a class? The exact target and mechanism for artemisinin is still not known, but it is clearly a very valuable medicine that is effectively used for the treatment of malaria. There is enough mechanistic information available to differentiate it from other anti-malarials. While mechanistic understanding is important for drug discovery, the type and value of the mechanistic understanding depends on the context and use. Successful drug discovery does not always need to be defined by a target. This highlights the mechanistic paradox that mechanism is critical to drug discovery success but its value is context specific. The mindset of current drug discovery and development must change its central feature from the target to a mechanistic understanding that is valuable and appropriate for each specific.

2.6 Summary

Mechanism of action is a central feature of drug discovery. Mechanism takes on different meanings depending on the stage of drug discovery, development, registration, and use. The MMOA of a medicine connects specific molecular interactions to the response. This feature of drug action is important for communicating a safe, therapeutically useful response and includes the drug target as well as the kinetic and conformational aspects that contribute to the communication of the desired response. MMOAs of medicines are difficult to predict *a priori*. This feature of drug discovery is proposed to contribute to the success of phenotypic drug discovery for the first-in-class medicines and target-based drug discovery for followers. Clearly a drug target should not be the exclusive definition of mechanism. The mindset of current drug discovery and development must change its central feature from the target to a mechanistic understanding that is valuable and appropriate for each specific project.

List of Abbreviations

BE biochemical efficiency
HDAC histone deacetylase
NMDA *n*-methyl-D-aspartate
MMOA molecular mechanism of action
PK/PD pharmacokinetic/pharmacodynamics
R&D Research and Development

References

1. Swinney, D.C. (2011) Molecular Mechanism of Action (MMOA) in drug discovery. *Annu. Rep. Med. Chem.*, **46**, 301–317.
2. Swinney, D.C. and Anthony, J. (2011) How were new medicines discovered? *Nat. Rev. Drug Disc.*, **10**, 507–519.
3. Ploeger, B.A., van der Graaf, P.H., and Danhof, M. (2009) Incorporating receptor theory in mechanism-based pharmacokinetic-pharmacodynamic (PK-PD) modeling. *Drug Metab. Pharmacokinet.*, **24**, 3–15.
4. Wikipedia *http://en.wikipedia.org/wiki/Scurvy* (accessed 31 January 2014).
5. Swinney, D.C. (2013) Phenotypic vs target-based drug discovery for first-in-class medicines. *Clin. Pharmacol. Ther.*, **93**, 299–301.
6. Ehrlich, P. (1913) Chemotherapeutics: scientific principles, methods, and results. *Lancet*, **182**, 445–451.
7. Swinney, D.C. (2008) Application of binding kinetics to drug discovery: translation to binding mechanisms to clinically differentiated medicines. *Pharm. Med.*, **22**, 23–34.
8. Lanzavecchia, A., Lezzi, G., and Viola, A. (1999) From TCR engagement to T cell activation: a kinetic view of T cell behavior. *Cell*, **96**, 1–4.
9. Marshall, C.J. (1995) Specificity of receptor tyrosine kinase signaling: transient versus sustained extracellular signal-regulated kinase activation. *Cell*, **80**, 179–185.
10. Hopfield, J.J. (1980) The energy relay: a proofreading scheme based on dynamic cooperativity and lacking all characteristic symptoms of kinetic proofreading in DNA replication and protein synthesis. *Proc. Natl. Acad. Sci. U.S.A.*, **77**, 5248–5252.
11. Coe, J.E. *et al.* (2005) Varenicline: an α4β2 nicotinic receptor partial agonist for smoking cessation. *J. Med. Chem.*, **48**, 3474–3477.
12. Lipton, S.A. (2006) Paradigm shift in neuroprotection by NMDA receptor blockade: memantine and beyond. *Nat. Rev. Drug Discov.*, **5**, 160–170.
13. Kapur, S. and Seeman, P. (2001) Does fast dissociation from the dopamine D2 receptor explain the action of atypical antipsychotics? A new hypothesis. *Am. J. Psychiatry*, **158**, 360–9.
14. Swinney, D.C. (2004) Biochemical mechanisms of drug action: what does it take for success. *Nat. Rev. Drug Disc.*, **3**, 801–808.
15. Swinney, D.C. (2006) Biochemical mechanisms of New Molecular Entities (NMEs) approved by United States FDA during 2001–2004: mechanisms leading to optimal efficacy and safety. *Curr. Top. Med. Chem.*, **6**, 461–478.
16. Wacker, D., Fenalti, G., Brown, M.A., Katritch, V., Abagyan, R., Cherezov, V.R., and Stevens, C. (2010) Conserved binding mode of human beta2 adrenergic receptor inverse agonists and antagonist

References

16. revealed by X-ray crystallography. *J. Am. Chem. Soc.*, **132**, 11443–11445.
17. Samuele, A., Crespan, E., Viterllaro, S., Monforte, A.-M., Logoteta, P., Chimirri, A., and Maga, G. (2010) Slow binding-tight binding interaction between benzimidazol-2-one inhibitors and HIV-1 reverse transcriptase containing the lysine 103 to asparagine mutation. *Antiviral Res.*, **86**, 268–275.
18. Yun, C.-.H., Mengwasser, K.E., Toms, A.V., Woo, M.S., Greulich, H., Wong, K.-K., Meyerson, M., and Eck, M.J. (2008) The T790M mutation in EGFR kinase causes drug resistance by increasing the affinity for ATP. *Proc. Natl. Acad. Sci. U.S.A.*, **105**, 2070–2075.
19. Mourey, R.J., Burnette, B.L., Brustkern, S.J., Daniels, J.S., Hirsch, J.L., Hood, W.F., Meyers, M.J., Mnich, S.J., Pierce, B.S., Saabye, M.J., Schindler, J.F., South, S.A., Webb, E.G., Zhang, J., and Anderson, D.R. (2010) A benzothiophene inhibitor of mitogen-activated protein kinase-activated protein kinase 2 inhibits tumor necrosis factor alpha production and has oral anti-inflammatory efficacy in acute and chronic models of inflammation. *J. Pharmacol. Exp. Ther.*, **333**, 797–807.
20. Capdeville, R., Buchdunger, E., Zimmermann, J., and Matter, A. (2002) Glivec (ST571, imatinib), a rationally developed, targeted anticancer drug. *Nat. Rev. Drug Disc.*, **1**, 493–502.
21. Barker, A.J. *et al.* (2001) Studies leading to the identification of ZD1830 (Iressa): an orally active, selective epidermal growth factor receptor tyrosine kinase inhibitor targeted to the treatment of cancer. *Bioorg. Med. Chem. Lett.*, **11**, 1911–1914.
22. Swinney, D.C. (2013) The contribution of mechanistic understanding to phenotypic screening for first-in-class medicines. *J. Biomol. Screen.*, **18**, 1186–1192.
23. Lee, J.A. and Berg, E.L. (2013) Neo-classic drug discovery: the case for lead generation using phenotypic and functional approaches. *J. Biomol. Screen.*, **18**, 1143–1155.
24. Plenge, R.M., Scolnick, E.M., and Altshuler, D. (2013) Validating therapeutic targets through human genetics. *Nat. Rev. Drug Disc.*, **12**, 581–594.
25. Swinney, D.C. (2006) Can binding kinetics translate to a clinically differentiated drug? From theory to practice. *Lett. Drug Des. Dis.*, **3**, 569–574.
26. Mercer, A.E., Copple, I.M., Maggs, J.L., O'Neill, P.M., and Park, B.K. (2011) The role of heme and the mitochondrion in the chemical and molecular mechanisms of mammalian cell death induced by the artemisinin antimalarials. *J. Biol. Chem.*, **286**, 987–996.
27. Liao, F. (2009) Discovery of artemisinin (qinghaosu). *Molecules*, **14**, 5362–5366.

David C. Swinney has a PhD in medicinal chemistry from the University of Washington, Seattle, and expertise in drug discovery, enzymology, and binding kinetics. He is currently founder and CEO of the 501c3, non-profit Institute for Rare and Neglected Diseases Drug Discovery, aka iRND3, located in Mountain View, CA. The mission of iRND3 is to discover new medicines for rare and neglected diseases. He has over 25 years of applied preclinical drug discovery experience (Roche, Syntex. iRND3) working to identify efficient drug discovery strategies that discover quality clinical candidates. He also believes in learning from past success to inform future success.

Part II
Drug Class

3
Insulin Analogs – Improving the Therapy of Diabetes
John M. Beals

3.1
Introduction

The discovery of insulin in 1921 by Banting and Best [1] introduced a life-saving therapy for patients with type 1 diabetes mellitus (T1DM) through the chronic exogenous administration of a parenteral therapeutic that enabled control of glucose and, indirectly, ketones. Soon afterwards, the manufacturing process to extract insulin from bovine and porcine pancreata was developed by the pharmaceutical company, Eli Lilly and Co. (Indianapolis, IN). Early insulin preparations improved the quality of life for individuals afflicted with T1DM and prolonged their life span; however, the early insulin replacement therapy had limitations. The formulation necessitated the patient to execute and endure multiple daily injections (MDIs), that is, >3. This immediately led to attempts to retard absorption from the subcutaneous (SC) depot to relieve patients of the arduous dosing regimen.

Prior to the compilation of our current knowledge of insulin structure and the advent of recombinant deoxyribonucleic acid (DNA) techniques and biomolecule production, insulin manufacturing innovation drove enhancements in purity while the formulation arena leveraged the unique biophysical chemistry of insulin. The early drug products (DPs) were crude, acidic formulations of relatively impure insulin, which yielded a short-acting insulin profile necessitating MDI therapy. These acid formulations were required to solubilize copurified proteins and protect insulin from degradation by residual pancreatic enzymes, but these acidic formulations compromised insulin stability and potency through deamidation of asparagine residues [2]. Over the next 60 years, the purity of the insulin extracts vastly improved, first by removal of the contaminating proteins and second by increased purity of insulin itself. This allowed the development of more stable, neutral insulin formulations for rapid-acting insulin administration as well as insulin formulations with protracted time action. Fast-acting formulation development focused on improving chemical stability by moving from acidic to neutral formulations [3]. The addition of excess zinc and/or protamine

Successful Drug Discovery, First Edition. Edited by János Fischer and David P. Rotella.
© 2015 Wiley-VCH Verlag GmbH & Co. KGaA. Published 2015 by Wiley-VCH Verlag GmbH & Co. KGaA.

to the formulations created protracted time action profile [4, 5]. These newer DPs reduced challenges with compliance to an MDI regimen [2, 6, 7].

This cadre of new DPs with fast-acting, intermediate-acting, and very long-acting time actions provided clinicians a means to tailor MDI regimens to minimize glucose excursions with the fewest number of injections. Beginning in the 1980s and later in the 1990s, clinical trials designed to assess the impact of intensive insulin therapy, for example, UK Prospective Diabetes Study (UKPDS) and Diabetes Control and Complications Trial (DCCT), unambiguously demonstrated that intensive therapy lowered microvascular complications (i.e., diabetic retinopathy, diabetic nephropathy, and diabetic neuropathy). The tight control achieved with intensive insulin therapy in these trials utilized both animal-sourced and human insulin DPs, including both fast-acting and protracted insulin formulations, with the former used for post-prandial control and the latter used for basal control. Although the value of tight control, from a microvascular complications perspective, was demonstrated; the trial highlighted notable limitations with the existing exogenous SC insulin therapy, that is, increased hypoglycemia and weight gain [8]. The challenges with recapitulating appropriate physiological post-prandial and basal insulin profiles with the insulin formulations of this era have been extensively reviewed [9, 10]. The advent of recombinant DNA technology, coupled with a broader understanding of the biophysical properties of insulin, enabled the development of insulin analogs with altered physiochemical properties, relative to human insulin, and improved pharmacokinetic (PK) profiles upon their SC administration.

Four major advances laid the foundation for the recombinant expression and engineering of insulin, to create DPs whose SC injections more accurately recapitulated the time action profiles of endogenous insulin secretion. (i) In the early 1950s, the primary structure of bovine insulin was determined, providing detailed information of the composition of insulin [11–14]. Since then, numerous insulin sequences have been determined and the insights gained from the evolutionary record are driving protein engineering of insulin today [15, 16]. (ii) In the mid-1960s, the physiology of insulin synthesis and secretion was determined, providing insight into the importance of proinsulin [17, 18]. (iii) In the late-1960s, the determination of the crystal structure of the zinc porcine insulin hexamer laid the foundation for protein engineering to affect both insulin self-association as well as establish insights for future insulin formulation [19, 20]. (iv) In the late-1970s and early-1980s, the cloning and expression of the human insulin gene allowed insulin production as an active pharmaceutical ingredient (API) on an industrial scale sufficient to treat the growing population of patients with diabetes [21].

3.2
Pharmacology and Insulin Analogs

Endogenous insulin secretion in a nondiabetic person is parsed into two distinct and different profiles (i) insulin response to a meal, referred to as the prandial

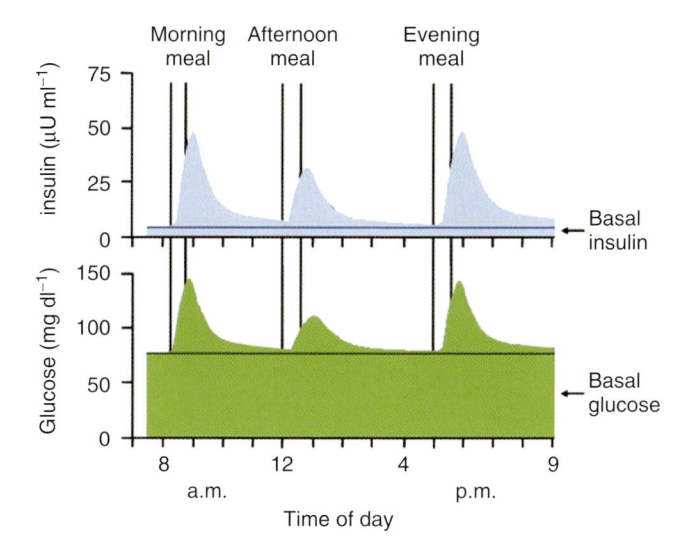

Figure 3.1 A schematic representation of glucose and the corresponding insulin secretion profiles, during a 24-h period, in nondiabetic individuals. (Adapted from Ref. [22].)

insulin profile and (ii) background insulin secretion that is continually secreted between meals and during the nighttime hours, referred to as basal insulin profile (Figure 3.1). Specifically, endogenous prandial insulin secretion yields peak insulin levels in the range of 60–80 µU mL^{-1} whereas basal insulin secretion ranges typically range from 5 to 15 µU mL^{-1} [23].[1] Thus, exogenously administered insulin needs to reproduce two vastly different time action profiles. Consequently, considerable effort has been expended on attempts to develop insulin formulations and analogs that meet the PK and pharmacodynamic (PD) requirements of each condition, while trying to eliminate hypoglycemia and weight gain that has been observed in intensive therapy. This was impossible to achieve with the traditional formulation preparation of animal and human insulin, because attempts at aggressive control of one condition, adversely affects the other.

3.3
Chemical Description

To understand the protein engineering advances introduced to modulate insulin PK profiles for SC administration, it is essential to understand the physiochemical properties that the evolutionary process selected, for this pancreatic-secreted hormone. Human insulin, a 51-amino acid protein, is composed of two polypeptide chains, a 21 amino acid A-chain and a 30 amino acid B-chain, stabilized by two interchain disulfide bonds, Cys^{A7}-Cys^{B7} and Cys^{A20}-Cys^{B19}, and one

---

1) One International Unit (IU) of human insulin, by definition, is the activity contained in 0.03846 mg of the international standard for human insulin.

3 Insulin Analogs – Improving the Therapy of Diabetes

G–I–V–E–Q$_5$–C–C–T–S–I$_{10}$–C–S–L–Y–Q$_{15}$–L–E–N–C$_{20}$–N

F–V–N–Q–H$_5$–L–C–G–S–H$_{10}$–L–V–E–A–L$_{15}$–Y–L–V–C–G$_{20}$–E–R–G–F–F$_{25}$–Y–T–P–K–T$_{30}$

Species	A^8	A^{10}	A^{21}	B^3	B^{28}	B^{29}	B^{30}	B^{31}	B^{32}
Human	Thr	Ile	Asn	Asn	Pro	Lys	Thr		
Pork	Thr	Ile	Asn	Asn	Pro	Lys	Ala		
Beef	Ala	Val	Asn	Asn	Pro	Lys	Ala		
Insulin lispro	Thr	Ile	Asn	Asn	Lys	Pro	Thr		
Insulin aspart	Thr	Ile	Asn	Asn	Asp	Lys	Thr		
Insulin glulisine	Thr	Ile	Asn	Lys	Pro	Glu	Thr		
Insulin glargine	Thr	Ile	Gly	Asn	Pro	Lys	Thr	Arg	Arg
Insulin detemir	Thr	Ile	Asn	Asn	Pro	Lys^a			
Insulin degludec	Thr	Ile	Asn	Asn	Pro	Lys^b			

^{a}Lys^{B29}(N-tetradecanoyl)
^{b}Lys^{B29}-[N^6-[N-(15-carboxy-1-oxopentadecyl)-L-γ-glutamyl]-L-lysine]

(a)

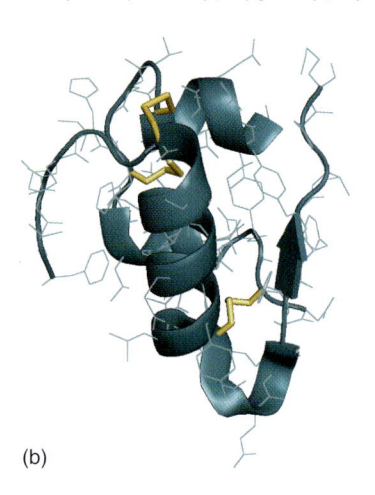

(b)

Figure 3.2 (a) The primary sequence of human insulin is composed of two polypeptide chains, a 21 amino acid A-chain and a 30 amino acid B-chain. An intrachain disulfide bond stabilizes the A-chain and the A- and B-chains are linked through two interchain disulfide bonds. Interestingly, the amino acid sequence of insulin for >100 vertebrate species has been determined with ~35% of the sequence highly, if not fully, conserved during evolution (Gly^{A1}, Ile^{A2}, Val^{A3}, Cys^{A6}, Cys^{A7}, Cys^{A11}, Tyr^{A19}, Cys^{A20}, Leu^{B6}, Cys^{B7}, Gly^{B8}, Leu^{B11}, Val^{B12}, Glu^{B13}, Cys^{B19}, Gly^{B23}, and Phe^{B24}, Phe^{B25}) [15, 16, 24]. A preponderance of these residues are implicated directly in receptor binding (Ile^{A2}, Val^{A3}, Tyr^{A19}, Gly^{B23}, and Phe^{B24}) or indirectly through conformational stabilization (Leu^{B6}, Gly^{B8}, Leu^{B11}, Glu^{B13}, and Phe^{B25}). The strong conservations of Leu^{A13} and Leu^{B17} support involvement in receptor binding. The table inset highlights sequence differences between important species in insulin therapy and subsequent analogs. (b) The secondary and tertiary structure of human insulin monomer depicted with the side chains (shown in lines), secondary structure (shown in ribbons and arrows), and the disulfides (shown in yellow) [25]. Coils denote the α-helical domains and the flat arrow denotes the β-sheet domain that stabilizes the dimer interface created by the self-assembly of two insulin monomers. The monomeric conformation depicted is that of the T-state wherein the B1–B8 are in extended conformation.

intrachain disulfide bond between residues Cys^{A6}-Cys^{A11} (Figure 3.2) [25]. Insulin is produced in the β-cells of the pancreas as a single-chain protein with both the A and B chains connected by a connecting peptide (C-peptide). After folding into the correct conformation, with appropriately paired disulfide bonds, the C-peptide is removed by enzymes, leaving the final two-chain insulin molecule.

Insulin evolved to self-associate into non-covalent dimers through favorable hydrophobic interactions at the C-terminus of the B-chain [26] that in turn stabilized an inter-molecular β-sheet. Physiologically, insulin is stored as a zinc-containing hexamer in the β-cells of the pancreas, wherein, three noncovalent insulin dimers assemble to form an insulin hexamer stabilized by two zinc ions. The zinc ions are coordinated through an octahedral lattice created, in part, by three His^{B10} residues [25, 27] (Figure 3.3). Numerous roles have been proposed for the evolutionarily conserved, self-association tendencies of insulin, including; (i) chemical and thermal stabilization of the molecule during the intracellular vacuole storage [29, 30], (ii) protection of the monomeric insulin from fibrillation *in vivo* [31], (iii) substitution for chaperone-assisted stabilization and folding during intracellular expression [32], and/or (iv) essential for secretory trafficking [33].

The ability of insulin to self-associate into discrete hexamers in the presence of zinc and phenolic preservatives, for example, phenol or *m*-cresol, has and still is exploited in the preparation of commercial insulin formulations. These heterologous ligands impart conformational changes to the hexamer that enhance the chemical and physical stability. The binding of the phenolic ligands stabilizes a conformational change that occurs at the N-terminus of the B-chain of monomers assembled in the hexamer, shifting a conformational equilibrium in residues Phe^{B1} to Gly^{B8} from an extended structure (T-state) to an α-helical structure (R-state) (Figure 3.4). The phenolic ligand binding occurs in a cavity between monomers of adjacent dimers and is stabilized by hydrogen bonds with the carbonyl oxygen of Cys^{A6} and the amide proton of Cys^{A11} as well as numerous van der Waals contacts [35]. This conformational change is referred to as the T ↔ R transition and is exploited to increase the chemical, physical, and thermal stability of insulin in commercial preparations [36, 37].

Although zinc and phenolic preservatives may enhance chemical, physical, and thermal stability of commercial formulations, they can alter the SC absorption of insulin, that is, delaying prandial (fast-acting) insulin and prolonging basal (slow-acting) insulin. Therefore, thoughtful consideration is required to design analogs and formulations that not only provide appropriate PK/PD profiles but also provide sufficient stability for therapeutic use.

3.4
Rapid-Acting Insulin Analogs (Prandial or Bolus Insulin)

As noted in the introduction, initial insulin formulations necessitated acidic conditions to mitigate proteolysis, but these formulations were subject to severe

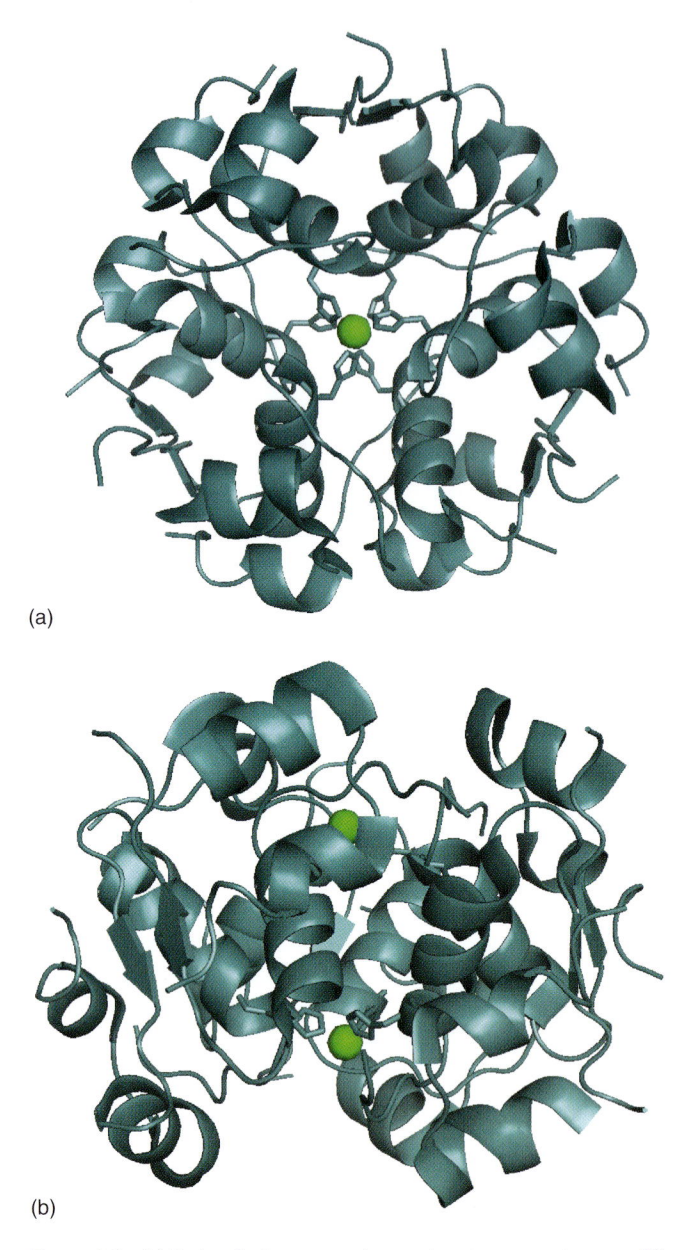

Figure 3.3 (a) T_6 insulin hexamer, observed down the axis of symmetry created with six insulin monomers (teal) assembled through the trimerization of three dimers by two zinc ions (green). Zinc-induced hexamerization yields a hexamer that behaves like the dimer of trimers. Three His[B10] residues, one from each insulin monomer composing the zinc-centric trimer, coordinate the zinc ion. (b) T_6 insulin hexamer, observed perpendicular to the axis of symmetry; note the exposure of the zinc ions in the T_6 hexamer [28].

3.4 Rapid-Acting Insulin Analogs (Prandial or Bolus Insulin)

(a)

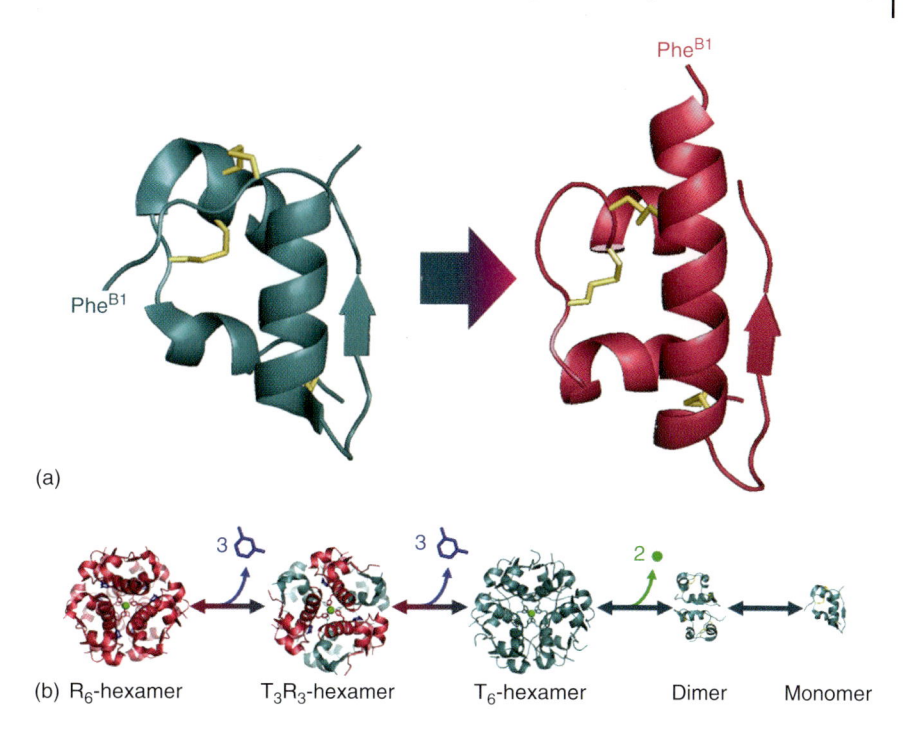

(b) R$_6$-hexamer   T$_3$R$_3$-hexamer   T$_6$-hexamer   Dimer   Monomer

Figure 3.4 (a) The T (teal) and R (rouge) states of insulin are a result of a ligand-dependent conformational change that occurs in Phe^{B1}-Gly^{B8} region of the insulin monomer. In the T-state, the Phe^{B1}-Gly^{B8} region is in an extended conformation and in the R-state, this region is in an α-helical conformation. (b) The self-association of human insulin enabled in the presence of heterologous ligand-induced, zinc (green), and phenolic preservative (blue). Phenolic preservative binds to six pockets on the insulin hexamer stabilizing an R$_6$ conformation that increases hexamer stability by burying the zinc ions and reduces chemical instability, e.g., deamidation, by stabilizing Phe^{B1}-Gly$_{B8}$ region in helical conformation [28, 34]. The binding of the preservative is complex and essential for creation of a pharmaceutical product with suitable chemical and physical stability.

chemical instability due to the inherent primary structure of insulin, specifically the acid-induced deamidation of Asn^{A21}, which is associated with significant potency loss during prolonged storage. Minimization of the chemical instability of these prandial formulations was achieved through the development of neutral, zinc-stabilized solutions wherein insulin was chemically stabilized by the addition of zinc (2–4 Zn ions per insulin) in the presence of phenolic preservatives, primarily phenol, and/or *m*-cresol. Under these formulation conditions, these excipients induced the R$_6$ hexameric conformation that decreased the accessibility of residues involved in deamidation, Asn^{B3} and Asn^{A21}, and decreased high molecular weight polymer formation [2, 6]. The resulting glucodynamic profile of neutral, regular formulations, when administered subcutaneously, showed peak insulin activity between 2 and 3 h with a maximum duration of 5–8 h.

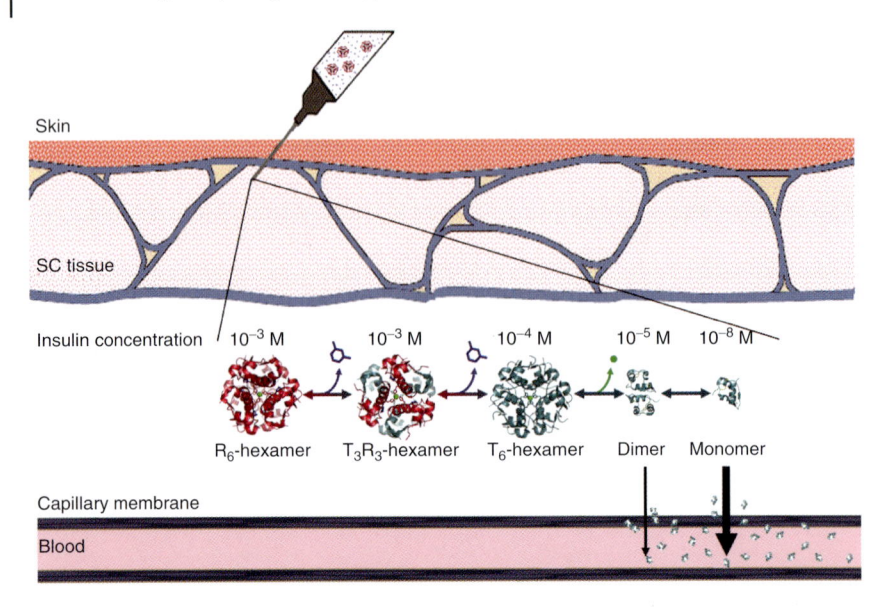

Figure 3.5 The time-dependent, diffusion-enabled dissociation of formulated human insulin hexamers in the SC space [38]. The surrounding SC tissue absorbs the hydrophobic phenolic preservative (blue) creating the T_6-hexamer with accessible zinc ions. Diffusion from the site creates a concentration gradient that destabilizes the T_6-hexamer allowing zinc (green) to diffuse from the hexamer and the creation of dimers and monomers that can be readily absorbed. Based on the work of Porter and colleagues, regular formulations of insulin are predominately absorbed across the capillary membrane (~80%) with the remainder absorbed by the lymphatic system (20%) [39]. Moreover, the extent of lymphatic absorption is a function of molecular weight, with small molecular weight species being more amenable to capillary absorption [40]. Based on the rate and molecular weight-based dependence on absorption, it is inferred that monomers and dimers are the primary species capable of absorption across the capillary membrane.

These neutral, stable formulations improved the glucodynamic profiles over the acidic formulations [3]; however, the glucodynamic profiles from these formulations were still sufficiently protracted that the physiological response to postprandial glucose excursions was challenging because, it required risky premeal administration, ~30–60 min prior to eating, in an attempt to recapitulate insulin levels needed to facilitate glucose absorption. The protracted glucodynamic profile of insulin is attributed to the time-dependent dissolution of the insulin hexamers into dimeric and/or monomeric insulin species, which are capable of capillary-mediated absorption from the SC depot. This dissociation of the insulin hexamer is achieved by the absorption of the excess preservative by the surrounding SC tissue and time-dependent diffusion of the insulin hexamer from the site of injection. This effective reduction in the concentration of insulin induces the release of zinc, enabling the breakdown into dimers and monomers (Figure 3.5) [38]. This sequence of hexameric dissociation is supported by the capillary/lymphatic transport studies that demonstrated a relationship between molecular weight and dose recovery in the popliteal lymph following SC injection [41, 42]. Results suggest that the slower lymphatic transport system accounts for only ~20% of the absorption of Humulin R (a neutral pH, soluble insulin formulation) from the interstitium with the remaining balance of insulin is predominately absorbed through capillary diffusion [39].

The recognition that the self-association properties, which nature evolved in insulin for endogenous production and release, may not be optimal for the exogenous SC administration of insulin led to the design of monomeric insulin analogs, that is, insulin with significantly diminished dimerization potential, that reduce the time-dependent diffusion required for hexamer dissociation into monomers. It was hypothesized that a more rapid absorption profile would enable a more convenient premeal timing for dose administration by the patient, that is, just prior to eating, while recapitulating an endogenous-like glucodynamic response that mimics postprandial glucose excursions. To enable formulations with improved time action profiles, researchers focused on the creation of monomeric insulin analogs for prandial therapy by weakening the dimer self-association properties of insulin [43–45].

The first commercially available rapid-acting analog was $Lys^{B28}Pro^{B29}$-human insulin, referred to as insulin lispro, which is the API of the formulation Humalog® (Eli Lilly and Co., Indianapolis, IN) (Figure 3.6). Insulin lispro exploits a sequence inversion at positions B28 and B29 to produce an insulin analog with reduced self-association behavior compared to human insulin [43–45]. This reduced dimer self-association property translates into a notably faster time-action profile, with a peak activity of ~1 h [46]. However, the development of a functionally useful commercial insulin therapy requires the development of a stable formulation with 18–36 month shelf-life stability under refrigerated conditions (2–8 °C) and 7–28 day stability under ambient in-use conditions (up to 30 °C). The creation of monomeric insulin lispro by its nature disrupted the self-association properties used to enable the necessary shelf-life and in-use stability needed in commercial insulin formulations. Surprisingly, it was discovered that the monomeric analog insulin lispro can be stabilized in a phenolic preservative/zinc-dependent hexameric complex in the formulation; thus, providing the requisite chemical and physical stability needed in a commercial insulin preparation, without retarding its rapid time-action [47]. Insights from the crystal structure of the insulin lispro hexamer [26] suggest that the loss of critical stabilizing hydrophobic interactions at the dimer interface coupled with a hexamerization process dependent upon *both* zinc and phenolic preservatives are contributory to this paradoxical behavior [28, 47].

The hexameric complex of insulin lispro is only stable under formulation conditions, that is, in the presence of zinc and millimolar concentrations of preservative. These formulation conditions impart conformational changes that yield the necessary solution stability. Yet, the weak hexamerization property in the absence of preservatives allows the exploitation of the rapid absorption of the phenolic preservative at the injection site that destabilizes the hexamer and facilitates the

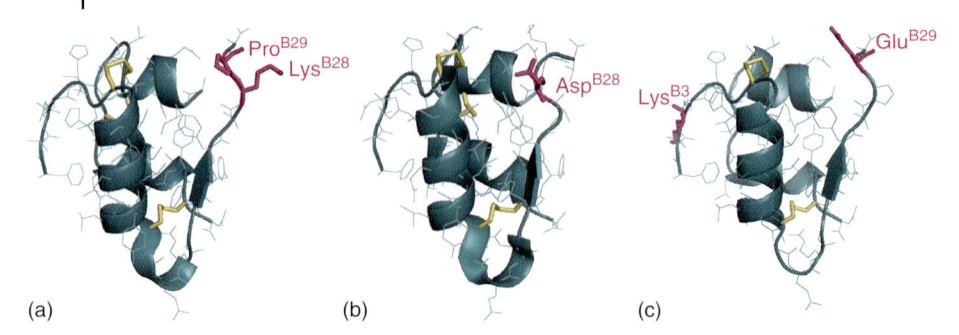

Figure 3.6 Commercially available rapid-acting insulin analogs for prandial use. (a) Insulin lispro; Lys^{B28}Pro^{B29}-human insulin, CAS number 133107-64-9, API in Humalog®, manufactured by Eli Lilly and Co. (b) Insulin aspart; Asp^{B28}-human insulin, CAS number 116094-23-6, API in NovoRapid®/NovoLog®, manufactured by Novo Nordisk A/S. (c) Insulin glulisine; Lys^{B3}-Glu^{B29}-human insulin, CAS number 160337-95-1; API in Apidra®, manufactured by Sanofi.

rapid dissociation of the hexamers into monomers [45, 47] (Figure 3.7). Consequently, the substantial concentration dilution ($\geq 10^5$) of insulin zinc hexamers is not necessary for the analog to dissociate from hexamers to monomers/dimers. This dissociation is a prerequisite for capillary absorption and its acceleration provides the patient with a more appropriate therapy for mealtime administration. In addition, Humalog® (insulin lispro) also improves control of postprandial hyperglycemia and reduces the frequency of severe hypoglycemic events by lessening the duration of action [48, 49].

The launch of Humalog® ushered in two additional rapid-acting insulin analogs to the market, insulin aspart, the API for the formulations NovoRapid®/NovoLog® (Novo Nordisk, Bagsværd, Denmark), and insulin glulisine, the API for the formulation Apidra® (Sanofi, Paris, France) (Figure 3.6). To enable multi-use capability, both insulin analogs are formulated under neutral pH conditions with phenolic preservative. The engineering strategy employed for insulin aspart, Asp^{B28}-human insulin, exploited the same criticality of Pro^{B28} to stabilization of the dimer interface as insulin lispro; however, the approach utilized the replacement of Pro^{B28} with a negatively-charged aspartic acid residue to eliminate critical van der Waals interactions involved in dimer stabilization [43, 45, 50]. Moreover, akin to Lys^{B28}Pro^{B29}-human insulin, Asp^{B28}-human insulin can be formulated as hexamers in the presence of zinc and preservatives [50] for improved formulation stability without affecting the rapid time-action following SC injection [51]. Thus, Novolog® SC behavior is similar to that outlined for Humalog® (Figure 3.7).

The protein engineering strategy of insulin glulisine, Lys^{B3}-Glu^{B29}-human insulin, exploited substitutions of both the Lys^{B29} with a negatively-charged glutamic acid as well as Asn^{B3} with a positively charged lysine. The glutamic acid substitution at B29, adjacent to the dimer self-association region, weakens but does not eliminate dimerization, imparting a degree of stability to the molecule under formulation conditions. Additionally, the charge swap favorably lowers

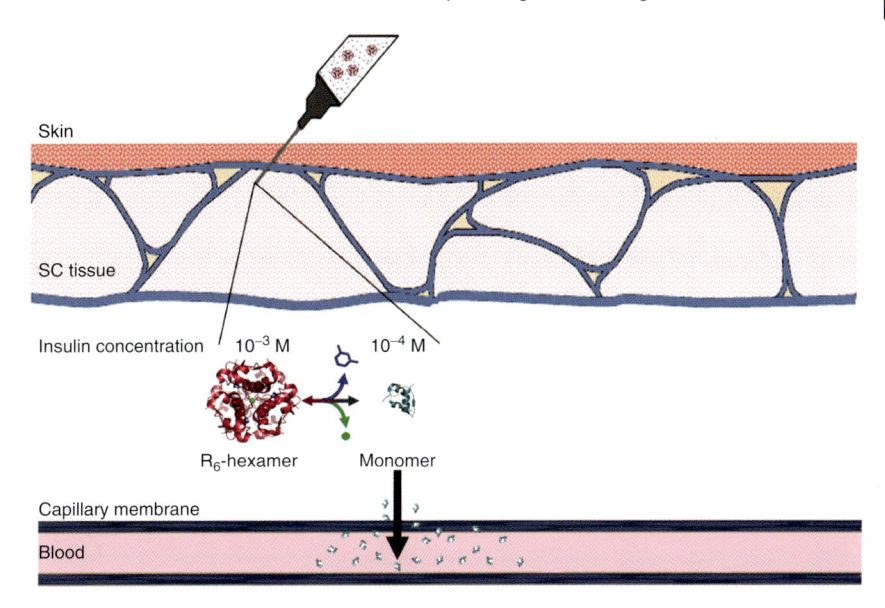

Figure 3.7 The dissociation of formulated insulin lispro hexamers in the SC space [47]. The surrounding SC tissue absorbs the hydrophobic phenolic preservative (blue), destabilizing the R_6-hexamer, allowing zinc (green) to diffuse away, and then the hexamer dissociates into monomers that can be readily absorbed [28].

the pI of the molecule making this analog slightly more soluble under physiological pH [43, 52]. The lysine substitution at Asn^{B3} enhances solution chemical stability by eliminating a deamidation site that is traditionally minimized by hexamerizing the molecule to an R_6-hexamer state [52]. Interestingly, unlike insulin lispro and insulin aspart, insulin glulisine is formulated as a dimeric complex in a zinc-free formulation. Consequently, the B1–B8 region is unlikely to occupy a phenolic-induced R-state that minimizes Asn^{B3} deamidation in other insulin formulations; therefore, substitution of Asn^{B3} with lysine would significantly improve the chemical stability profile of the analog. The engineering and formulation of insulin glulisine as a readily dissociable dimeric complex was presumably designed to eliminate any lag in absorption associated with hexameric dissociation of other monomeric analogs, insulin lispro, and insulin aspart (Figure 3.7). Pharmacological studies confirmed that the insulin glulisine demonstrates a rapid time-action profile similar to insulin lispro. However, formulated Apidra® demonstrates only a slightly faster PK profile and no difference in PD properties from Humalog® (insulin lispro) and Novolog® (insulin aspart) [53]. Consequently, the similarity in PK profiles suggests that the hexameric breakdown of insulin lispro and insulin aspart must be rapid relative to the rate-limiting step of capillary absorption (Figure 3.7).

In addition to MDI dosing regimens based on pen injector or syringe/vial prescription formats, Humalog®, Novolog®, and Apidra® are soluble formulations approved for use in external infusion pumps for continuous subcutaneous insulin infusion (CSII). As stated previously, the Humalog® and Novolog® formulations utilize a hexameric association state driven by the presence of the phenolic preservative and zinc in the presence of a phosphate buffer. This formulation minimizes the physical aggregation of insulin that can lead to clogging of the infusion sets. Apidra®, because of its dimeric nature, necessitates the inclusion of a surfactant, polysorbate 20, to minimize aggregation. In general, with regard to clinical efficacy, all three rapid-acting insulin analogs provide equivalent control. However, the hexameric formulations of insulin lispro and insulin aspart outperform insulin glulisine regarding the occurrence of infusion catheter occlusions. This performance difference suggests that hexamerization is more effective than surfactant at improving the physical stability of the insulin analog and thus, more useful at mediating insulin-mediated aggregation [54].

3.5
Long-Acting Insulin Analog Formulations (Basal Insulin)

As noted above, insulin secretion can be parsed into two phases, prandial/bolus and basal phases [23]. Prandial insulin secretion is variable and dictated by a response to glucose absorption from a meal whereas, the basal insulin secretion rate is more steady, at approximately 1 unit of insulin per hour, and, in part, directed at controlling hepatic glucose output [55]. Appropriate basal insulin levels are critical in diabetes therapy to regulate hepatic glucose output, maintain appropriate fasting blood glucose levels, and establish the proper maintenance of glucose homeostasis during diurnal cycling of the body and intermittent food intake. Consequently, basal insulin formulations attempt to provide a very protracted profile while trying to minimize peaks. Therefore, basal insulin analog formulations possess a very different profile when compared to prandial insulin formulation.

Since the 1930s, products utilizing crystallized forms of insulin in suspension formulations have been used to create intermediate- to long-acting basal insulin preparations. One popular intermediate-acting formulation for basal insulin use is Neutral Protamine Hagedorn (NPH) [4, 56]. The NPH formulation is prepared by co-crystallizing insulin with protamine [57], where the latter is a heterogeneous mixture of four closely related basic peptides (each approximately 30 amino acids in length and enriched with arginine) that are isolated from fish sperm [58]. NPH crystals are oblong and tetragonal in nature with volumes between 1 and 20 μm^3 [56, 59]. These crystalline suspension formulations produce an intermediate time-action that is exploited for basal insulin control. The protracted time-action achieved is a consequence of shifting the rate-limiting step from monomeric/dimeric absorption to the dissolution of insulin/protamine crystals. NPH formulations generally exhibit peak activity from 6 to 12 h with a duration of activity from 18 to 24 h.

In the 1990s, the "basal" utility of rapid-acting analog insulin lispro was exploited using this traditional insulin formulation strategy [4, 56, 59], wherein hexameric insulin lispro in the presence of phenolic preservatives and zinc, could be cocrystallized with protamine sulfate to produce NPH-like crystals of the rapid-acting analog. The NPH-like crystals of insulin lispro are also oblong tetragonal crystals, similar to those observed with NPH insulin formulations [56]. The NPH-like suspension of insulin lispro possesses the physicochemical properties relative to human insulin NPH [60]; however, to prepare the NPH-analogous tetragonal crystals with intermediate-acting time-action, significant modifications to the protamine/insulin crystallization procedure was required due to the reduced self-association properties of insulin lispro [61].

PK/PD studies reported for the insulin lispro protamine suspension (ILPS), also referred to as neutral protamine lispro (NPL), demonstrated that the pharmacology was analogous to human insulin NPH [61, 62]. Recent clinical trials of ILPS alone in type 2 diabetes mellitus (T2DM) and in combination with insulin lispro in T1DM have demonstrated clinical utility [63–65]. Moreover, the PK/PD profiles of ILPS in T2DM patients support utility as a once-daily therapy regimen [66]. Of further note, ILPS in T1DM patients demonstrated a more predictable response with reduced intrasubject variability compared to Lantus® (formulation explained in further detail later in the text) [67]. ILPS is available as an intermediate-acting "basal" formulation in several EU countries and Japan.

The ability of insulin lispro to form NPH-like crystals in the presence of protamine allowed for the preparation of homogeneous, biphasic mixture preparations containing crystalline protamine/insulin component and a soluble rapid-acting hexameric component generated either extemporaneously or as a manufactured premix formulation. Based on the work with insulin lispro, premixed formulations composed of insulin aspart with a protamine-retarded crystalline preparation of insulin aspart were subsequently prepared [68]. The pharmacological properties of the rapid-acting and intermediate-acting components for these mixtures, that is, Humalog® premix (75/25) and Novolog® premix (70/30), have been clinically demonstrated in these commercially stable formulations [69, 70]. These premixed formulations are commercially available in many countries.

Although the time action profiles of NPH-like crystalline components prepared from rapid-acting insulin analogs are most suitable for a MDI regimen with prandial insulin, in either a basal/bolus or premixed dosing schemes, the pharmacology is clearly a compromise because the profiles poorly mimic the nondiabetic physiologic profile of endogenous basal insulin secretion. The variable absorption, pronounced PD peaks, and duration of action of less than 24 h make these crystalline suspension formulations less than optimal as a once-daily basal insulin therapy [9]. To address the limitations of the NPH-based formulation strategies, insulin analogs and chemically-modified insulin derivatives were designed to provide a more physiologic basal insulin activity profile. Of the three basal insulin entities currently approved for commercial use, there is one insulin analog (insulin glargine) and two derivatized insulins (insulin detemir and insulin degludec) (Figure 3.8).

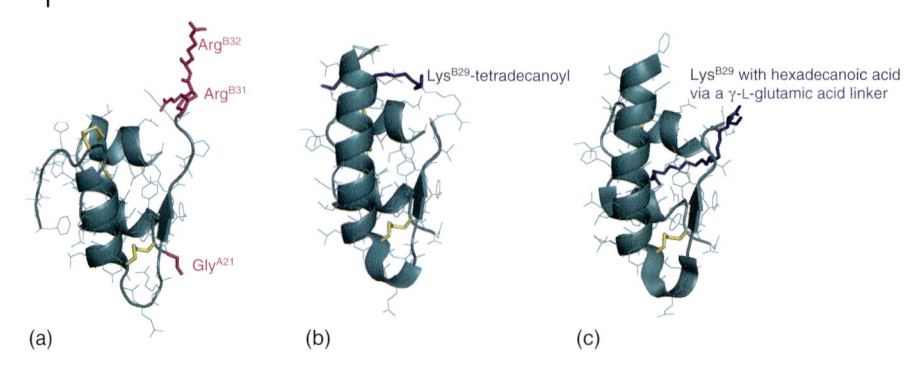

Figure 3.8 Long-acting insulin analogs and derivatives for use as basal insulin. (a) Insulin glargine; Gly^{A21}Arg^{B31}Arg^{B32}-human insulin, CAS number 160337-95-1, API in Lantus®, manufactured by Sanofi. (b) Insulin detemir; Lys^{B29}(N-tetradecanoyl) des(B30) human insulin, CAS number 169148-63-4, API in Levemir®, manufactured by Novo-Nordisk A/S. (c) Insulin degludec; Lys^{B29}-[N^6-[N-(15-carboxy-1-oxopentadecyl)-L-γ-glutamyl]-L-lysine], CAS number 844439-96-9, API in Tresiba®, manufactured by Novo-Nordisk A/S.

Insulin glargine is an insulin analog with potency similar to human insulin and was engineered to (i) shift the isoelectric point (pI) of the insulin from 5.4 to 6.7 and (ii) increase the chemically stability of the analog under acidic formulation conditions. The addition of two arginine residues, Arg^{B31} and Arg^{B32} at the C-terminus of the B-chain shifted the pI to 6.7, closer to physiological pH, making the analog far less soluble at physiological pH. Interestingly, these additional arginine residues are the first two amino acids of the natural C-peptide that is typically removed during the pancreatic production of insulin. In addition to engineering for pI, the Asn^{A21} residue was replaced with Gly^{A21}, yielding an insulin analog with improved chemical stability under acidic formulation conditions (pH 4.0). By eliminating the C-terminal asparagine that is susceptible to acid-mediated deamidation, the glargine analog maintains potency and has reduced high molecular weight polymer formation in the formulation [36, 71]. Thus, these protein-engineering changes enable a formulation-based time extension strategy wherein a soluble acidic formulation, Lantus® (Sanofi, Paris, France), can be prepared that eliminates the need for resuspension of crystals prior to dosing, a requirement of NPH-based formulations. The Lantus® formulation contains zinc presumably to augment pH-based insulin glargine precipitation that consequently slows the generation of readily absorbable monomers and dimers in the SC tissue (Figure 3.9). This time extension strategy produces a relatively constant rate of absorption into the bloodstream to provide a glucodynamic profile sufficient for once-daily administration in a preponderance of patients [72]. This acidic formulation of insulin glargine, as with other insulin formulations, is preserved with *m*-cresol to enable multi-injections from either a vial or a cartridge format.

As with all insulin preparations, the time course of the Lantus® formulation can vary from injection to injection, referred to as intrapatient variability, and

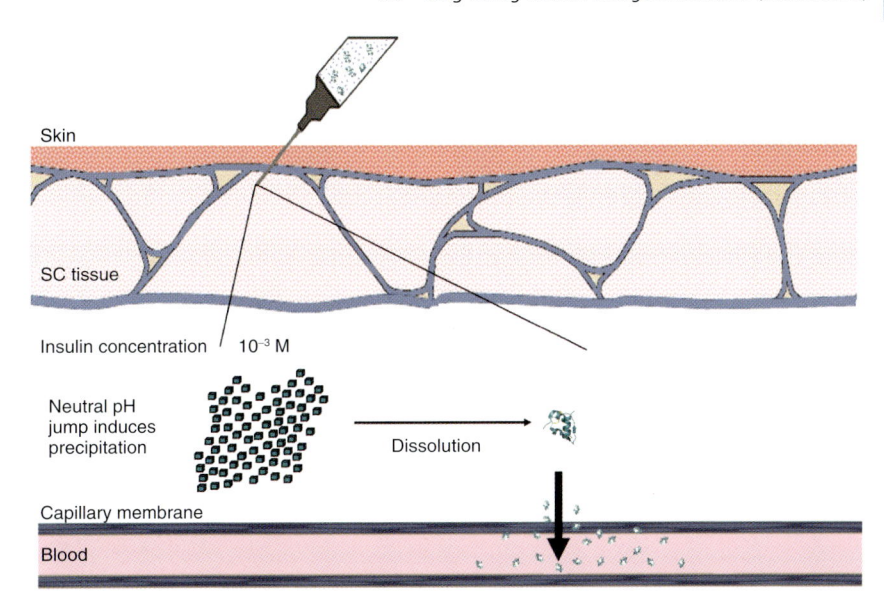

Figure 3.9 The SC precipitation and dissolution of formulated insulin glargine. The pH neutralization of the acidic Lantus® formulation induces the precipitation of insulin glargine as it transitions through its pI. The precipitated insulin glargine slowly dissolves releasing monomers that can be readily absorbed over an extended period of time, ~24 h [9].

from individual to individual, referred to as interpatient variability. One source of variability is the rate of absorption, which is dependent on numerous parameters, including but not limited to, blood supply, temperature, and the patient's physical activity, consistency of precipitation, or other absorption limiting events. Although insulin glargine (Lantus®) improved intrapatient PK and PD variability (referred to as coefficient of variation) relative to NPH-like formulation, the variability remains high; thus limiting the pharmaceutical optimization of basal insulin therapy [73]. In part to address issues with variability, an alternate time extension strategy exploiting fatty acid derivatization of des-Thr^{B30} human insulin was used to create insulin detemir, commercially available as Levemir® (Novo-Nordisk, Bagsværd, Denmark).

Insulin detemir achieves its protracted time action by a combination of intermolecular interactions and physiological binding events. Specifically, the elegant design of the Levemir® formulation enables the generation of higher-order hexameric association states in the SC tissue following injection and phenolic preservative absorption. The covalent linkage of the 14-carbon, myristoyl fatty acid to the ε-amino group of Lys^{B29} enables the binding of the monomeric species of insulin to albumin, a serum protein with a long serum half-life [74].

Originally, the proposed mechanism of time extension for insulin detemir was thought to be primarily driven by binding of the tetradecanoyl-acylated insulin to albumin; however, an elegant utilization of the T → R transition underlies a

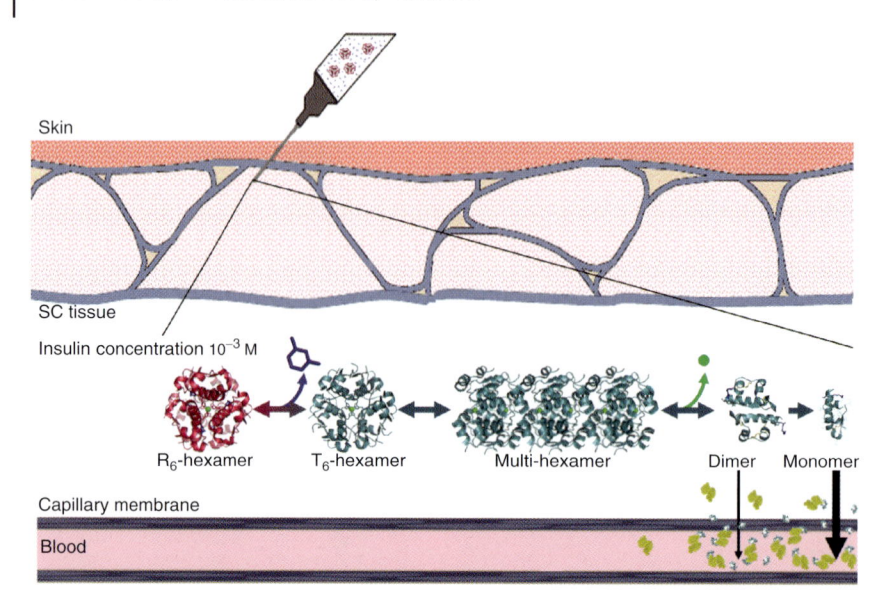

Figure 3.10 The SC association properties of insulin detemir. The Levemir® formulation stabilizes insulin detemir in a R_6-hexameric (red) complex that converts to a T_6-hexamer (teal) after the absorption of the phenolic preservative (blue ring). The T_6-hexamer has the propensity to form T_6-multihexamer complexes through interactions introduced by the covalent attachment of tetradecanoyl fatty acid. The creation of zinc-stabilized T_6-multihexamer complexes introduces an additional rate-limiting step in the dissolution into dimers and monomers capable of capillary absorption [74, 75]. In the plasma, insulin detemir is bound to albumin (yellow), which regulates the disposition and clearance of insulin detemir.

more complex mechanism. Post-injection, preservative absorption by the SC tissue shifts the hexameric conformation from a R_6-state in the formulation to a T_6-state. Experiments suggest the T_6-hexamer of insulin detemir prolongs absorption predominately through hexamer stability driven by dihexamerization [74, 75]. The multi-hexamerization formation appears to be a consequence of the symmetrical arrangement of fatty acid moieties around the outside of the hexamers, as shown by X-ray crystallographic studies [76] that stabilize the dihexamer species through hydrophobic interactions. Multi-hexamerization dissociation introduces another time-dependent event for the formation of albumin-binding monomers (Figure 3.10). Additional, but less prominent contributions to prolongation of absorption are attributed to albumin binding within the depot. Moreover, albumin binding may play a more critical role in controlling distribution and elimination.

Clinical studies demonstrated that Levemir® (insulin detemir) displays lower PK and PD variability relative to Humulin N® (NPH) and Lantus® (insulin glargine) [77–80]. Unfortunately, the duration of action for insulin detemir is generally considered to be <24 h [79, 81]; thus, the duration of action may not be sufficient to warrant the classification of Levemir® as a truly once-daily basal insulin formulation for all patients.

Lantus® (insulin glargine) and Levemir® (insulin detemir) both provide an improvement in basal insulin therapy through the introduction of a soluble formulation that simplifies administration; however, both products fail to achieve the dual goals desired in a once-daily basal insulin product, that is, full 24-h coverage in all patients with low intrapatient variability. Moreover, a true 24-h basal insulin could minimize nocturnal hypoglycemia. These deficiencies in the first generation basal insulin analogs continue to drive the exploration of improved basal insulin derivatives and formulations.

To address the deficiencies in insulin detemir, insulin glargine, and generally, limitations in suspension basal insulin formulations, Novo Nordisk developed a next generation ultra-long basal insulin, insulin degludec, which is formulated as Tresiba® and commercially marketed in the EU and Japan. Insulin degludec is an acylated insulin derivative that utilizes a des-ThrB30 human insulin acylated at the ε-amino group of LysB29 with hexadecandioic acid via a γ-L-glutamic acid linker. It demonstrates a protracted C_{max} (10–12 h) and terminal $t_{1/2}$ (~20 h) in T2DM patients [82] (Figure 3.8). Building on the concept of exploiting the T ↔ R conformational changes in hexameric acyl-modified insulins to stabilize hexamer/hexamer interactions and slow insulin payout from the SC depot, a phenomenon first observed with insulin detemir, insulin degludec uses the specific fatty acid and linker elements to increase hexameric association to multihexamers in the SC depot [83]. Specifically, insulin degludec forms stable T_3R_3 hexamers that form stable dihexamers in the formulation, Tresiba®. This formulation contains phenolic preservatives and higher zinc concentrations (6 mol Zn/hexamer) that convert to a T_6-state dihexamers capable of self-associating into strands of multihexamers in the SC site after the preservative is absorbed by the surrounding SC tissue. Although the exact nature of the stabilizing interactions in either the T_3R_3-dihexamer or the T_6-multihexamer states are unknown, it is hypothesized that the terminal carboxylic acid residue of the fatty di-acid chain coordinates to a zinc atom in a neighboring hexamer through a T_3 interface and subsequent hexameric self-association propagation. Unlike other insulin preparations, insulin degludec is stable in the presence of high zinc concentrations suggesting that the di-acid and glutamic linker create additional high-affinity zinc binding sites that prevent insulin degludec from precipitating due to nonspecific metal binding (Figure 3.11). The association into soluble multihexamers adds an additional rate-limiting step to the release of albumin binding insulin monomers in the SC tissue and maintains the advantage of a reproducible absorption kinetic profile from a soluble depot relative to a precipitation/dissolution-based depot [77, 84].

Clinically, insulin degludec demonstrated comparable glycemic control to insulin glargine with a reduced hypoglycemia profile in both T1DM patients [85] and T2DM patients [86]. Moreover, in the latter study, the protracted time action of Tresiba® (insulin degludec) enabled every other day dosing while maintaining efficacy.

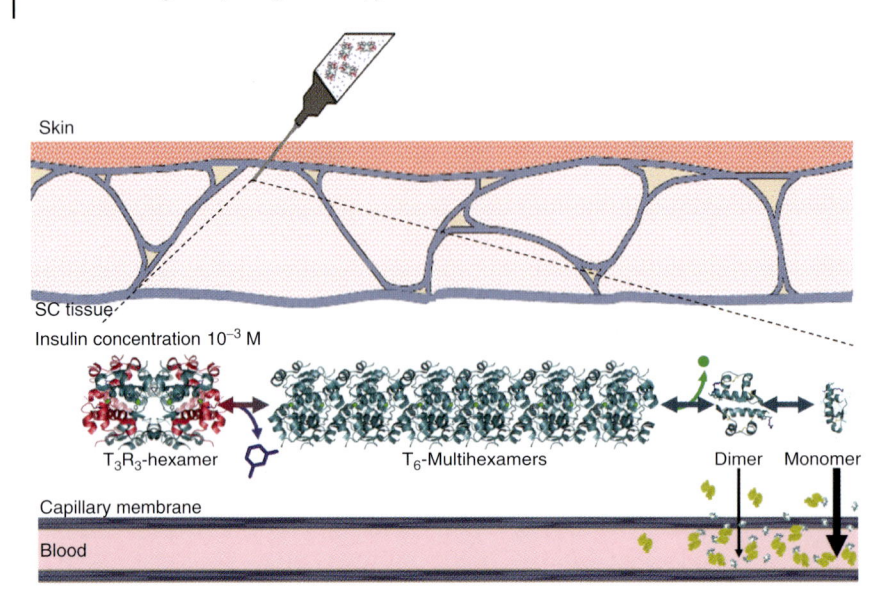

Figure 3.11 The SC properties of formulated insulin degludec. The Tresiba® formulation stabilizes insulin degludec in a T_3R_3-hexameric dimeric complex (teal/rouge) that converts to a T_6-multihexameric complex (teal) after the absorption of the phenolic preservative (blue ring). The T_6-hexamer has the propensity to form T_6-multihexamers in the presence of additional zinc through interactions introduced by the covalent attachment via a γ-L-glutamic acid linker of hexadecandioic acid. The creation of zinc-stabilized multihexamers introduces an additional rate-limiting step in the creation of dimers and monomers capable of capillary absorption [83]. In the plasma, insulin degludec is bound to albumin (yellow), which regulates the drug disposition and clearance of insulin degludec.

3.6
Conclusions and Future Considerations

In 2013, based on the Annual Reports of Eli Lilly and Co, Sanofi, and Novo-Nordisk A/S, insulin analog formulations containing insulin lispro, insulin aspart, insulin glulisine, insulin glargine, and insulin detemir account for ~83% (~$17.4B) of the total worldwide sales of insulin (~$20.8B). Basal insulin sales accounted for ~46% (~$9.6B) of the sales and rapid-acting insulin analog-containing formulations comprised ~37% (~$7.7B) of analog sales. Moreover, the rate of insulin analog sales growth is ~15% annually, due in part, to the improved PK/PD properties of these molecules, which improve patient outcomes and the patient experiences with these exogenous, patient-administered therapies.

Improvements in prandial insulin analogs and their associated formulations are expected to enable the development of an artificial pancreas, an external pump that rapidly and continuously controls blood glucose in conjunction with constant blood glucose monitoring. Currently, prandial insulins are too slow in their absorption phase and too prolonged in time action to be optimally used in an artificial pancreas [10]. Numerous approaches to increase absorption include the modification of the SC tissue to enable rapid insulin dispersion with a concomitant increase in capillary bed access to accelerate absorption [87] or disruption of the hexameric state of insulin and masking of surface charges to facilitate more rapid absorption of monomers [88].

Concerning basal insulin analogs, future engineering endeavors may rest in altering insulin disposition in the body after a SC injection to restore appropriate hepatic and peripheral activity, which is not optimal with current insulin preparations [89]. The ability of a basal insulin to limit peripheral activity while maintaining hepatic activity may more accurately reproduce insulin metabolism seen with endogenously secreted insulin [90–93].

List of Abbreviations

API	active pharmaceutical ingredient
CSII	continuous subcutaneous insulin infusion
DCCT	diabetes control and complications trial
DNA	deoxyribonucleic acid
DP	drug product
ILPS	insulin lispro protamine suspension
MDI	multiple daily injection
NPH	neutral protamine hagedorn
NPL	neutral protamine lispro
PD	pharmacodynamic
PK	pharmacokinetic
SC	subcutaneous
T1DM	type 1 diabetes mellitus
T2DM	type 2 diabetes mellitus
UKPDS	UK Prospective Diabetes Study

References

1. Bliss, M. (1982) *The Discovery of Insulin*, McClelland and Stewart Limited, Toronto, pp. 189–211.
2. Brange, J. (1987) *Galenics of Insulin*, Springer-Verlag, Berlin, pp. 17–39.
3. Schlichtkrull, J., Munck, O., and Jersild, M. (1965) Insulin rapitard and insulin actrapid. *Acta Med. Scand.*, **177**, 103–113.
4. Hagedorn, H.C. et al. (1984) Landmark article Jan 18, 1936: Protamine insulinate. By H.C. Hagedorn, B.N. Jensen, N.B. Krarup, and I. Wodstrup. *J. Am. Med. Assoc.*, **251** (3), 389–392.
5. Hallas-Moller, K. (1956) The lente insulins. *Diabetes*, **5** (1), 7–14.
6. Brange, J. (1987) *Galenics of Insulin*, Springer-Verlag, Berlin, pp. 1–5.
7. Galloway, J.A. (1988) in *Diabetes Mellitus* (eds J.H. Potvin, C.R. Shuman, and J.A. Galloway), Lilly Research Laboratories, Indianapolis, IN, pp. 105–133.
8. Bergenstal, R.M., Bailey, C.J., and Kendall, D.M. (2010) Type 2 diabetes: assessing the relative risks and benefits of glucose-lowering medications. *Am. J. Med.*, **123** (4), 374.e 9–374.e 18.

9. Bolli, G.B., Andreoli, A.M., and Lucidi, P. (2011) Optimizing the replacement of basal insulin in type 1 diabetes mellitus: no longer an elusive goal in the post-NPH era. *Diabetes Technol. Ther.*, **13** (Suppl. 1), S43–S52.
10. Heinemann, L. and Muchmore, D.B. (2012) Ultrafast-acting insulins: state of the art. *J. Diabetes Sci. Technol.*, **6** (4), 728–742.
11. Sanger, F. and Thompson, E.O. (1953) The amino-acid sequence in the glycyl chain of insulin. II. The investigation of peptides from enzymic hydrolysates. *Biochem. J.*, **53** (3), 366–374.
12. Sanger, F. and Thompson, E.O. (1953) The amino-acid sequence in the glycyl chain of insulin. I. The identification of lower peptides from partial hydrolysates. *Biochem. J.*, **53** (3), 353–366.
13. Sanger, F. and Tuppy, H. (1951) The amino-acid sequence in the phenylalanyl chain of insulin. 2. The investigation of peptides from enzymic hydrolysates. *Biochem. J.*, **49** (4), 481–490.
14. Sanger, F. and Tuppy, H. (1951) The amino-acid sequence in the phenylalanyl chain of insulin. I. The identification of lower peptides from partial hydrolysates. *Biochem. J.*, **49** (4), 463–481.
15. Conlon, J.M. (2001) Evolution of the insulin molecule: insights into structure-activity and phylogenetic relationships. *Peptides*, **22** (7), 1183–1193.
16. Blundell, T.L. and Wood, S.P. (1975) Is the evolution of insulin Darwinian or due to selectively neutral mutation? *Nature*, **257** (5523), 197–203.
17. Steiner, D.F. *et al.* (1967) Insulin biosynthesis: evidence for a precursor. *Science*, **157** (3789), 697–700.
18. Steiner, D.F. and Oyer, P.E. (1967) The biosynthesis of insulin and a probable precursor of insulin by a human islet cell adenoma. *Proc. Natl. Acad. Sci. U.S.A.*, **57** (2), 473–480.
19. Adams, M.J., Blundell, T.L., Dodson, E.J., Dodson, G.G., Vijayan, M., Baker, E.N., Harding, M.M., Hodgkin, D.C., Rimmer, B., and Sheat, S. (1969) Structure of rhombohedral 2 zinc insulin crystals. *Nature*, **224**, 491–495.
20. Blundell, T.L. *et al.* (1971) Atomic positions in rhombohedral 2-zinc insulin crystals. *Nature*, **231** (5304), 506–511.
21. Riggs, A.D. (1981) Bacterial production of human insulin. *Diabetes Care*, **4** (1), 64–68.
22. Schade, D.S.S., Santiago, J.V., Skyler, J.S., and Rizza, R.A. (1983) *Intensive Insulin Therapy*, Medical Examination Publishing, Princeton, NJ, p. 24.
23. Galloway, J.A. and Chance, R.E. (1994) Improving insulin therapy: achievements and challenges. *Horm. Metab. Res.*, **26** (12), 591–598.
24. Menting, J.G. *et al.* (2013) How insulin engages its primary binding site on the insulin receptor. *Nature*, **493** (7431), 241–245.
25. Baker, E.N. *et al.* (1988) The structure of 2Zn pig insulin crystals at 1.5 A resolution. *Philos. Trans. R. Soc. Lond. B Biol. Sci.*, **319** (1195), 369–456.
26. Ciszak, E. *et al.* (1995) Role of C-terminal B-chain residues in insulin assembly: the structure of hexameric LysB28ProB29-human insulin. *Structure*, **3** (6), 615–622.
27. Goldman, J. and Carpenter, F.H. (1974) Zinc binding, circular dichroism, and equilibrium sedimentation studies on insulin (bovine) and several of its derivatives. *Biochemistry*, **13** (22), 4566–4574.
28. Birnbaum, D.T. *et al.* (1997) Assembly and dissociation of human insulin and LysB28ProB29-insulin hexamers: a comparison study. *Pharm. Res.*, **14** (1), 25–36.
29. Brange, J. and Langkjoer, L. (1993) Insulin structure and stability. *Pharm. Biotechnol.*, **5**, 315–350.
30. Huus, K. *et al.* (2005) Thermal dissociation and unfolding of insulin. *Biochemistry*, **44** (33), 11171–11177.
31. Brange, J. *et al.* (1997) Toward understanding insulin fibrillation. *J. Pharm. Sci.*, **86** (5), 517–525.
32. Kjeldsen, T. and Pettersson, A.F. (2003) Relationship between self-association of insulin and its secretion efficiency in yeast. *Protein Expr. Purif.*, **27** (2), 331–337.
33. Haataja, L. *et al.* (2013) Proinsulin intermolecular interactions during secretory

trafficking in pancreatic beta cells. *J. Biol. Chem.*, **288** (3), 1896–1906.
34. Birnbaum, D.T. *et al.* (1996) Hierarchical modeling of phenolic ligand binding to 2Zn–insulin hexamers. *Biochemistry*, **35** (17), 5366–5378.
35. Derewenda, U. *et al.* (1989) Phenol stabilizes more helix in a new symmetrical zinc insulin hexamer. *Nature*, **338** (6216), 594–596.
36. Brange, J. and Langkjaer, L. (1992) Chemical stability of insulin. 3. Influence of excipients, formulation, and pH. *Acta Pharm. Nord.*, **4**, 149–158.
37. Brader, M.L. and Dunn, M.F. (1991) Insulin hexamers: new conformations and applications. *Trends Biochem. Sci.*, **16** (9), 341–345.
38. Brange, J. and Volund, A. (1999) Insulin analogs with improved pharmacokinetic profiles. *Adv. Drug Deliv. Rev.*, **35** (2-3), 307–335.
39. Charman, S.A. *et al.* (2001) Lymphatic absorption is a significant contributor to the subcutaneous bioavailability of insulin in a sheep model. *Pharm. Res.*, **18** (11), 1620–1626.
40. Kaminskas, L.M. and Porter, C.J. (2011) Targeting the lymphatics using dendritic polymers (dendrimers). *Adv. Drug Deliv. Rev.*, **63** (10-11), 890–900.
41. Porter, C.J. and Charman, S.A. (2000) Lymphatic transport of proteins after subcutaneous administration. *J. Pharm. Sci.*, **89** (3), 297–310.
42. Supersaxo, A., Hein, W.R., and Steffen, H. (1990) Effect of molecular weight on the lymphatic absorption of water-soluble compounds following subcutaneous administration. *Pharm. Res.*, **7** (2), 167–169.
43. Brange, J. *et al.* (1990) Monomeric insulins and their experimental and clinical implications. *Diabetes Care*, **13** (9), 923–954.
44. Brange, J. *et al.* (1988) Monomeric insulins obtained by protein engineering and their medical implications. *Nature*, **333** (6174), 679–682.
45. Brems, D.N. *et al.* (1992) Altering the association properties of insulin by amino acid replacement. *Protein Eng.*, **5** (6), 527–533.
46. Howey, D.C. *et al.* (1994) [Lys(B28), Pro(B29)]-human insulin. A rapidly absorbed analogue of human insulin. *Diabetes*, **43** (3), 396–402.
47. Bakaysa, D.L. *et al.* (1996) Physicochemical basis for the rapid time-action of LysB28ProB29-insulin: dissociation of a protein-ligand complex. *Protein Sci.*, **5** (12), 2521–2531.
48. Holleman, F. *et al.* (1997) Reduced frequency of severe hypoglycemia and coma in well-controlled IDDM patients treated with insulin lispro. The Benelux-UK Insulin Lispro Study Group. *Diabetes Care*, **20** (12), 1827–1832.
49. Anderson, J.H. Jr., *et al.* (1997) Mealtime treatment with insulin analog improves postprandial hyperglycemia and hypoglycemia in patients with non-insulin-dependent diabetes mellitus. Multicenter Insulin Lispro Study Group. *Arch. Intern. Med.*, **157** (11), 1249–1255.
50. Whittingham, J.L. *et al.* (1998) Interactions of phenol and m-cresol in the insulin hexamer, and their effect on the association properties of B28 pro → Asp insulin analogues. *Biochemistry*, **37** (33), 11516–11523.
51. Heinemann, L. *et al.* (1997) Comparison of the time-action profiles of U40- and U100-regular human insulin and the rapid-acting insulin analogue B28 Asp. *Exp. Clin. Endocrinol. Diabetes*, **105** (3), 140–144.
52. Becker, R.H. and Frick, A.D. (2008) Clinical pharmacokinetics and pharmacodynamics of insulin glulisine. *Clin. Pharmacokinet.*, **47** (1), 7–20.
53. Home, P.D. (2012) The pharmacokinetics and pharmacodynamics of rapid-acting insulin analogues and their clinical consequences. *Diabetes Obes. Metab.*, **14** (9), 780–788.
54. Bode, B.W. (2011) Comparison of pharmacokinetic properties, physicochemical stability, and pump compatibility of 3 rapid-acting insulin analogues-aspart, lispro, and glulisine. *Endocr. Pract.*, **17** (2), 271–280.
55. Waldhausl, W. *et al.* (1979) Insulin production rate following glucose ingestion

estimated by splanchnic C-peptide output in normal man. *Diabetologia*, **17** (4), 221–227.
56. Deckert, T. (1980) Intermediate-acting insulin preparations: NPH and lente. *Diabetes Care*, **3** (5), 623–626.
57. Balschmidt, P. *et al.* (1991) Structure of porcine insulin cocrystallized with clupeine Z. *Acta Crystallogr. B*, **47** (Pt. 6), 975–986.
58. Hoffmann, J.A., Chance, R.E., and Johnson, M.G. (1990) Purification and analysis of the major components of chum salmon protamine contained in insulin formulations using high-performance liquid chromatography. *Protein Expr. Purif.*, **1** (2), 127–133.
59. Krayenbühl, C. and Rosenberg, T. (1946) Crystalline protamine insulin. *Rep. Steno Hosp. (Kbh.)*, **1**, 60–73.
60. Yip, C.M. *et al.* (2000) Structural studies of a crystalline insulin analog complex with protamine by atomic force microscopy. *Biophys. J.*, **78** (1), 466–473.
61. DeFelippis, M.R. *et al.* (1998) Preparation and characterization of a cocrystalline suspension of [LysB28,ProB29]-human insulin analogue. *J. Pharm. Sci.*, **87** (2), 170–176.
62. Janssen, M.M. *et al.* (1997) Nighttime insulin kinetics and glycemic control in type 1 diabetes patients following administration of an intermediate-acting lispro preparation. *Diabetes Care*, **20** (12), 1870–1873.
63. Chacra, A.R. *et al.* (2010) Comparison of insulin lispro protamine suspension and insulin detemir in basal-bolus therapy in patients with Type 1 diabetes. *Diabet. Med.*, **27** (5), 563–569.
64. Fogelfeld, L. *et al.* (2010) A randomized, treat-to-target trial comparing insulin lispro protamine suspension and insulin detemir in insulin-naive patients with Type 2 diabetes. *Diabet. Med.*, **27** (2), 181–188.
65. Strojek, K. *et al.* (2010) Addition of insulin lispro protamine suspension or insulin glargine to oral type 2 diabetes regimens: a randomized trial. *Diabetes Obes. Metab.*, **12** (10), 916–922.
66. Hompesch, M. *et al.* (2009) Pharmacokinetics and pharmacodynamics of insulin lispro protamine suspension compared with insulin glargine and insulin detemir in type 2 diabetes. *Curr. Med. Res. Opin.*, **25** (11), 2679–2687.
67. Ocheltree, S.M. *et al.* (2010) Comparison of pharmacodynamic intrasubject variability of insulin lispro protamine suspension and insulin glargine in subjects with type 1 diabetes. *Eur. J. Endocrinol.*, **163** (2), 217–223.
68. Balschmidt, P. (1998) AspB28 insulin crystals. US Patent 5840680, filed Dec. 11, 1996 and Published Nov. 24, 1998.
69. Heise, T. *et al.* (1998) Time-action profiles of novel premixed preparations of insulin lispro and NPL insulin. *Diabetes Care*, **21** (5), 800–803.
70. Weyer, C., Heise, T., and Heinemann, L. (1997) Insulin aspart in a 30/70 premixed formulation. Pharmacodynamic properties of a rapid-acting insulin analog in stable mixture. *Diabetes Care*, **20** (10), 1612–1614.
71. Brange, J., Havelund, S., and Hougaard, P. (1992) Chemical stability of insulin. 2. Formation of higher molecular weight transformation products during storage of pharmaceutical preparations. *Pharm. Res.*, **9** (6), 727–734.
72. Wang, F., Carabino, J.M., and Vergara, C.M. (2003) Insulin glargine: a systematic review of a long-acting insulin analogue. *Clin. Ther.*, **25** (6), 1541–1577, discussion 1539–1540.
73. Vora, J. and Heise, T. (2013) Variability of glucose-lowering effect as a limiting factor in optimizing basal insulin therapy: a review. *Diabetes Obes. Metab.*, **15** (8), 701–712.
74. Havelund, S. *et al.* (2004) The mechanism of protraction of insulin detemir, a long-acting, acylated analog of human insulin. *Pharm. Res.*, **21** (8), 1498–1504.
75. Jonassen, I. *et al.* (2006) Biochemical and physiological properties of a novel series of long-acting insulin analogs obtained by acylation with cholic acid derivatives. *Pharm. Res.*, **23** (1), 49–55.
76. Whittingham, J.L. *et al.* (2004) Crystallographic and solution studies of

N-lithocholyl insulin: a new generation of prolonged-acting human insulins. *Biochemistry*, **43** (20), 5987–5995.
77. Heise, T. *et al.* (2004) Lower within-subject variability of insulin detemir in comparison to NPH insulin and insulin glargine in people with type 1 diabetes. *Diabetes*, **53** (6), 1614–1620.
78. Hermansen, K. *et al.* (2001) Comparison of the soluble basal insulin analog insulin detemir with NPH insulin: a randomized open crossover trial in type 1 diabetic subjects on basal-bolus therapy. *Diabetes Care*, **24** (2), 296–301.
79. Porcellati, F. *et al.* (2007) Comparison of pharmacokinetics and dynamics of the long-acting insulin analogs glargine and detemir at steady state in type 1 diabetes: a double-blind, randomized, crossover study. *Diabetes Care*, **30** (10), 2447–2452.
80. Vague, P. *et al.* (2003) Insulin detemir is associated with more predictable glycemic control and reduced risk of hypoglycemia than NPH insulin in patients with type 1 diabetes on a basal-bolus regimen with premeal insulin aspart. *Diabetes Care*, **26** (3), 590–596.
81. Lepore, M. *et al.* (2000) Pharmacokinetics and pharmacodynamics of subcutaneous injection of long-acting human insulin analog glargine, NPH insulin, and ultralente human insulin and continuous subcutaneous infusion of insulin lispro. *Diabetes*, **49** (12), 2142–2148.
82. Heise, T. *et al.* (2012) Ultra-long-acting insulin degludec has a flat and stable glucose-lowering effect in type 2 diabetes. *Diabetes Obes. Metab.*, **14** (10), 944–950.
83. Jonassen, I. *et al.* (2012) Design of the novel protraction mechanism of insulin degludec, an ultra-long-acting basal insulin. *Pharm. Res.*, **29** (8), 2104–2114.
84. Klein, O. *et al.* (2007) Albumin-bound basal insulin analogues (insulin detemir and NN344): comparable time-action profiles but less variability than insulin glargine in type 2 diabetes. *Diabetes Obes. Metab.*, **9** (3), 290–299.
85. Birkeland, K.I. *et al.* (2011) Insulin degludec in type 1 diabetes: a randomized controlled trial of a new-generation ultra-long-acting insulin compared with insulin glargine. *Diabetes Care*, **34** (3), 661–665.
86. Zinman, B. *et al.* (2011) Insulin degludec, an ultra-long-acting basal insulin, once a day or three times a week versus insulin glargine once a day in patients with type 2 diabetes: a 16-week, randomised, open-label, phase 2 trial. *Lancet*, **377** (9769), 924–931.
87. Muchmore, D.B. and Vaughn, D.E. (2012) Accelerating and improving the consistency of rapid-acting analog insulin absorption and action for both subcutaneous injection and continuous subcutaneous infusion using recombinant human hyaluronidase. *J. Diabetes Sci. Technol.*, **6** (4), 764–772.
88. Heinemann, L. *et al.* (2012) U-100, pH-Neutral formulation of VIAject((R)): faster onset of action than insulin lispro in patients with type 1 diabetes. *Diabetes Obes. Metab.*, **14** (3), 222–227.
89. Herring, R., Jones, R.H., and Russell-Jones, D.L. (2014) Hepatoselectivity and the evolution of insulin. *Diabetes Obes. Metab.*, **16** (1), 1–8.
90. Dimitriadis, G. *et al.* (2011) Insulin effects in muscle and adipose tissue. *Diabetes Res. Clin. Pract.*, **93** (Suppl. 1), S52–S59.
91. Moore, M.C. *et al.* (2014) Novel PEGylated basal Insulin LY2605541 Has a preferential hepatic effect on glucose metabolism. *Diabetes*, **63** (2), 494–504.
92. Shojaee-Moradie, F. *et al.* (2000) Novel hepatoselective insulin analog: studies with a covalently linked thyroxyl-insulin complex in humans. *Diabetes Care*, **23** (8), 1124–1129.
93. Shojaee-Moradie, F. *et al.* (1998) Novel hepatoselective insulin analogues: studies with covalently linked thyroxyl-insulin complexes. *Diabet. Med.*, **15** (11), 928–936.

John M. Beals received a PhD (1986) in chemistry, with specialization in biochemistry at the University of Notre Dame (Notre Dame, IN). After receiving his doctorate, he attended Cornell University (Ithaca, NY) on a National Institute of Health Postdoctoral Fellowship. Since 1990, he has worked at Eli Lilly and Company (Indianapolis, IN) focusing on the discovery, development, and manufacture of proteins for use in bioproduct therapies.

**Part III
Case Histories**

Successful Drug Discovery, First Edition. Edited by János Fischer and David P. Rotella.
© 2015 Wiley-VCH Verlag GmbH & Co. KGaA. Published 2015 by Wiley-VCH Verlag GmbH & Co. KGaA.

4
The Discovery of Stendra™ (Avanafil) for the Treatment of Erectile Dysfunction

Koichiro Yamada, Toshiaki Sakamoto, Kenji Omori, and Kohei Kikkawa

4.1
Introduction

Erectile dysfunction (ED) is defined as the inability to achieve or maintain an erection sufficient for a satisfactory sexual performance [1]. ED is a nonmalignant disorder; however, it erodes the quality of life with impairment of physical and psychosocial health. Recent accumulating evidence indicates that ED is a common problem associated with comorbidities of psychogenic and organic origin such as cardiovascular diseases, diabetes mellitus, prostate disease, metabolic syndrome, pelvic surgery, and use of various medications [2]. Epidemiological data have shown a high prevalence and incidence of ED worldwide, mainly because of the development of western lifestyles and rapidly aging populations [3]. Therefore, safe and effective medicines offer significant benefits.

The mechanism underlying penile erection has been clarified over the past two decades to indicate that the nitric oxide (NO)/cGMP (cyclic guanosine monophosphate) signaling pathway plays a significant role in the penis. Sexual stimulation induces the NO release from NANC (nonadrenergic, noncholinergic) nerve terminals located in the corpus cavernosum and from the endothelium (i) in response to the release of acetylcholine by parasympathetic endothelial nerve endings and (ii) by shear stress elicited by increased blood flow in the corporeal sinusoids [4]. NO then activates guanylyl cyclase, which causes cGMP production from guanosine triphosphate (GTP). Increased intracellular cGMP levels inhibit calcium entry into the cell to decrease intracellular calcium concentrations and cause relaxation of the vascular smooth muscles of the corpora cavernosa and increased blood flow. Then, subtunical veins and venules are compressed and local venous return is occluded, making the penis expand and creating an erection (Figure 4.1). A cGMP-degradation enzyme, cyclic nucleotide phosphodiesterase (PDE) 5 [PDE5], which is rich in the cavernosal smooth muscles, negatively regulates NO-dependent penile erection by hydrolysis of cGMP. The blockade of PDE5 increases cGMP levels and leads to a penile erection. Thus far, several orally available PDE5 inhibitors have been developed and approved as first-line therapy for ED on the basis of their high efficacy rates and favorable safety profiles. In the

Successful Drug Discovery, First Edition. Edited by János Fischer and David P. Rotella.
© 2015 Wiley-VCH Verlag GmbH & Co. KGaA. Published 2015 by Wiley-VCH Verlag GmbH & Co. KGaA.

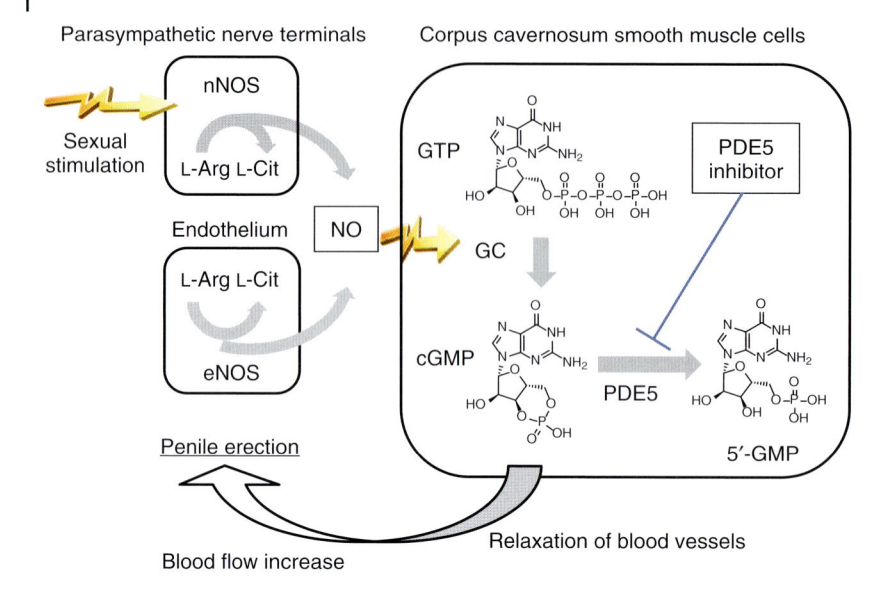

64 | *4 The Discovery of Stendra™ (Avanafil) for the Treatment of Erectile Dysfunction*

Figure 4.1 Mechanism of action of PDE5 inhibitors.

Sildenafil (**1**)

Vardenafil (**2**)

Tadalafil (**3**)

Figure 4.2 PDE5 inhibitors on the market.

United States, sildenafil (**1**, Viagra®) was launched in 1998 followed by vardenafil (**2**, Levitra®) and tadalafil (**3**, Cialis®) in 2003 [5–7] (Figure 4.2).

PDEs are a family of enzymes that hydrolyze intracellular cyclic nucleotides (cAMP (cyclic adenosine monophosphate) and/or cGMP). Based on their sequence similarity, enzymatic characteristics, and sensitivity to inhibitors or endogenous regulators, PDEs are classified into 11 subfamilies designated PDE 1–11 encoded by 21 genes [8]. Since each PDE has a specific role in the physiological regulation of the cells in various tissues, it is important to generate specific inhibitors toward a targeted PDE subfamily in order to avoid adverse effects resulting from the inhibition of other PDE subfamilies. Selective inhibitors of the cGMP specific PDE, PDE5, are optimal for the treatment of ED. From this point of view, sildenafil is reported to inhibit PDE6, which plays a critical role in photo signal transduction in the eye resulting in visual disruptions [9]. Sildenafil is also reported to inhibit PDE1, a calmodulin-regulated cAMP/cGMP-PDE, which is known to be the major cGMP-hydrolyzing PDE present in vascular smooth muscle cells. Therefore, dual blockade of PDE1 and PDE5 is expected to cause an excessive cGMP accumulation in the cardiovascular system and result in unwanted hypotension, especially when NO stimulating agents such as NTG (nitroglycerin) are administered. Tadalafil (**3**) is a PDE5 inhibitor, which shows a weaker inhibitory effect on PDE1 and PDE6 than either vardenafil and sildenafil, although it inhibits PDE11, which is a dual substrate PDE prominent in skeletal muscle and the prostate. Although the physiological relevance of PDE11 is unclear, the relationship between the symptoms of back pain and myalgia and the effect of PDE11 inhibition by tadalafil has been discussed [10]. Treatment with sildenafil and vardenafil has also been reported to cause back pain and myalgia but with less frequency [11, 12]. Other potential issues of the marketed PDE5 inhibitors for ED include their relatively slow onset of action (30–120 min), long half-life ($T_{1/2}$: 3.8–17.5 h), and long duration of action [13, 14].

4.2 Discovery of Avanafil

4.2.1 Differentiation Strategies to Develop a New Drug

While the first marketed PDE5 inhibitor, sildenafil (**1**), has been shown to be effective for the treatment of ED, it has several issues as described above. Our investigation toward the development of a new more optimal PDE5 inhibitor for the treatment of ED included incorporation of the following physicochemical characteristics: (i) higher PDE5 selectivity for improved tolerability with less unwanted side effects such as visual disturbance, excessive hypotension, myalgia, and so on; (ii) appropriately short duration of action compared with other PDE5 inhibitors; and (iii) faster onset of action. A novel and safe PDE5 inhibitor which satisfies these three requirements should be beneficial for ED patients compared with existing PDE5 inhibitors. Faster T_{onset} should be accomplished by rapid absorption in the gastrointestinal tract and/or higher tissue permeability of the compound. Shorter T_{offset} should be accomplished by the introduction of metabolically unstable functional groups into the molecule. Based on this idea, we initiated in July 1998 the synthesis of novel PDE5 inhibitors with an ideal pharmacokinetic (PK) profile, shown in blue (type a) in Figure 4.3.

4.2.2 Discovery of Isoquinoline Derivatives from Isoquinolinone Lead

In our company (MTPC), Ukita *et al.* had developed the isoquinolinone (**5**, T-1032) as a prototype PDE5 inhibitor, by optimizing the naphthalene lead (**4**) [15]. T-1032 showed potent PDE5 inhibitory activity, but demonstrated insufficient selectivity against bovine retina PDE6 (32-fold in the trypsin-activated form, and 650-fold in the light-activated form) as shown in Figure 4.4. T-1032

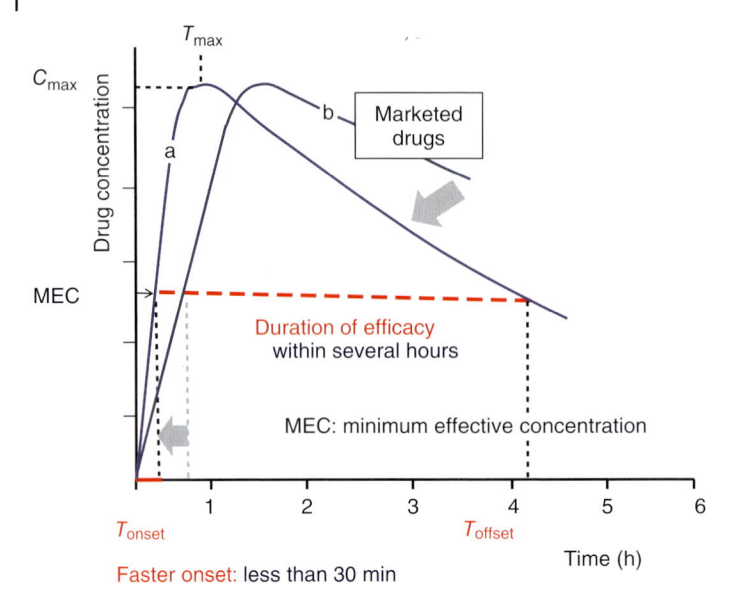

Figure 4.3 Our goal image for plasma drug concentration.

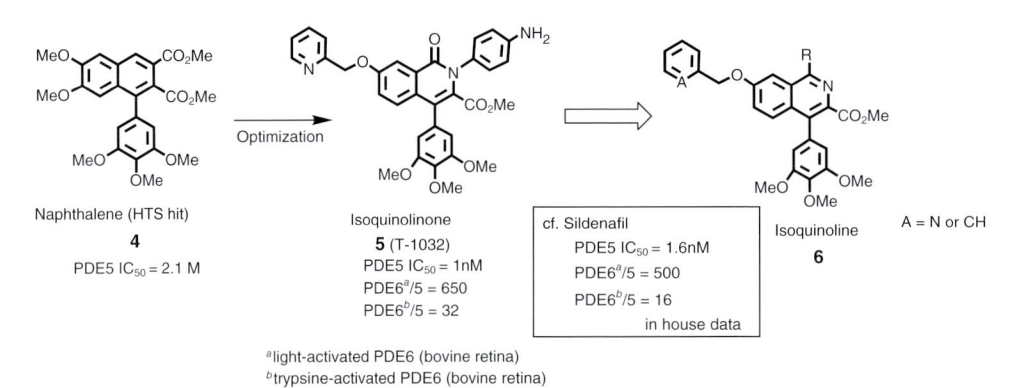

Figure 4.4 Synthetic strategy of isoquinolinones from isoquinoline lead.

and sildenafil showed similar selectivity against PDE6. In regard to the PDE6 inhibition assay, light (photon) causes a conformational change of the rhodopsin. The photoexcited rhodopsin activates a heterotrimeric G protein, transducin, and the resultant transducin α subunit with GTP binds to PDE6 holoenzyme containing inhibitory γ subunits. By this process, de-inhibition by the γ subunit occurs and the catalysis of cGMP to 5′-GMP is accelerated. Thus, light activation is assumed to form physiologically activated PDE6, and therefore, the assay with light-activated PDE6 was employed in the early period of the screening process. PDE6 preparation activated by treatment of trypsin, which is reported elsewhere, was used to compare an inhibitory profile of candidates on PDE6 with those of other representative PDE5 inhibitors [16].

4.2 Discovery of Avanafil

Table 4.1 SAR at the 1-position of the isoquinolines (**6a–h**).

	–NR¹R²	A	IC$_{50}$ (nM)	PDE6^a/5 selectivity		–NR¹R²	A	IC$_{50}$ (nM)	PDE6^a/5 selectivity
6a	(pyridin-2-ylmethylamino)	C	2.24	410	6e	(NHCH₂CH₂NMe₂)	C	5.46	1060
6b	(pyridin-4-ylmethylamino)	C	2.22	470	6f	(morpholine-CH₂-OH)	C	0.9	740
6c	(2-(pyridin-2-yl)ethylamino)	C	7.08	670	6g	(morpholine-CH₂-OH)	N	3.06	250
6d	(3-chloro-4-methoxybenzylamino)	C	3.46	>2890	6h	(4-hydroxypiperidine)	N	6.74	280

^{a}Light activated PDE6 (bovine retina).

Further research on T-1032 was terminated owing to its severe retinal toxicity in dogs at that time. Since this series of isoquinolinone derivatives showed insufficient selectivity against PDE6, we started designing a new type of PDE5 inhibitors using T-1032 as the lead compound. First, we examined whether the carbonyl group of isoquinolinone **5** plays a crucial role in the inhibitory activity of PDE5. For this purpose, the isoquinoline derivative **6**, which has no carbonyl group at the 1-position (Figure 4.4), was synthesized. As shown in Table 4.1, the isoquinolines (**6a–h**) were prepared from the reaction of the corresponding chloride **7** Table 4.1 with various amines. Among them, **6f** exhibited a potent PDE5 inhibitory activity (IC$_{50}$ = 0.9 nM) with moderate selectivity against PDE6 (IC$_{50}$ ratio: PDE6/PDE5 = 740). Isoquinoline (**6d**), substituted with a 3-chloro-4-methoxybenzylamino group at the 1-position, showed the highest selectivity against PDE6 (>2890-fold). We concluded that the carbonyl group in this scaffold is not crucial for PDE5 inhibitory activity.

4.2.3 Scaffold-Hopping Approaches from the Isoquinoline Leads

Since isoquinolines (**6**) showing potent PDE5 inhibitory activity were closely similar to isoquinolinones (**5**) in structure, it was plausible that undesired issues

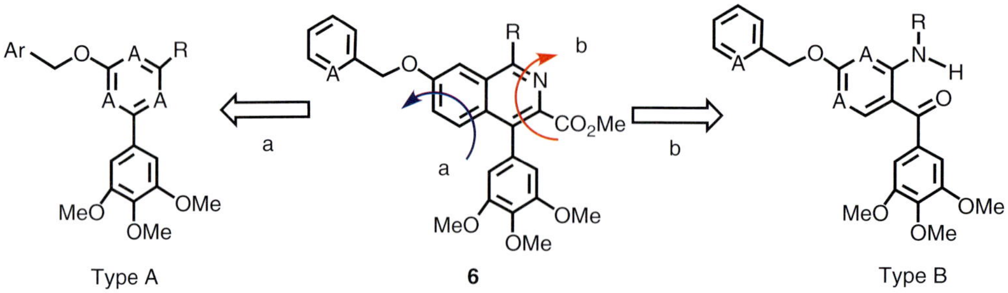

Figure 4.5 Transformation from bicyclic leads into monocyclic leads.

observed in the latter derivatives might also occur in the isoquinoline series (**6**). Therefore, we changed the scaffold of this series drastically. The ring-break framework modifications are commonly employed for creating new compounds and sometimes achieve more dramatic changes in potency, selectivity, bioavailability, and so on. We tried to cleave the bicyclic ring of the isoquinoline part into a monocyclic ring. Type A illustrates the disruption of the left side phenyl ring and type B opening of the pyridyl ring in the isoquinoline part (see Figure 4.5).

4.2.3.1 Monocyclic Type A Series: Tetrasubstituted Pyrimidine Derivatives

In the series of Type A, we synthesized the trisubstituted pyrimidine derivatives (**8, 10**), which showed a moderate inhibitory activity for PDE5 (see Figure 4.6). Since the derivatives (**8, 10**) possess a coplanar structure, we examined the stereochemical effects of the compound **9** on inhibitory activities for PDE5 and PDE6 by bromination. The *ortho*-bromosubstituted phenyl compound (**9**) was found to have superior potency and selectivity for PDE5. Therefore, we assumed that the perpendicular plane between the trimethoxyphenyl group and the pyrimidine nucleus is important for exhibiting PDE5 inhibitory activity. Based on this hypothesis, the tetrasubstituted pyrimidine (**11**) was synthesized. The introduction of a methoxycarbonyl group at the 5-position for twisting the plane potentiated the inhibitory activity for PDE5 ($IC_{50} = 0.7$ nM) and the relaxant effect on isolated rabbit corpus cavernosum ($EC_{30} = 11.1$ nM).

4.2.3.2 Monocyclic Type B Series: Trisubstituted Pyrimidine Derivatives

Next, the cleavage of the right side ring of isoquinoline derivatives was investigated. In this case, we designed the monocyclic but pseudo bicyclic compound by generating an intramolecular hydrogen bond, that is, one of the aryl rings was replaced by introducing a carbonyl group and an amine group to form the virtual aryl ring as illustrated in Figure 4.7. This ring-breaking technique was recently reported by the Novartis group in their scaffold morphing discovery of the anthranilic amide inhibitors of kinase domain receptor (KDR) [17]. In order to lower the lipophilicity and simplify the synthetic strategy, we designed the pyrimidine derivative (**12**) instead of the benzene analog. Consequently, we found

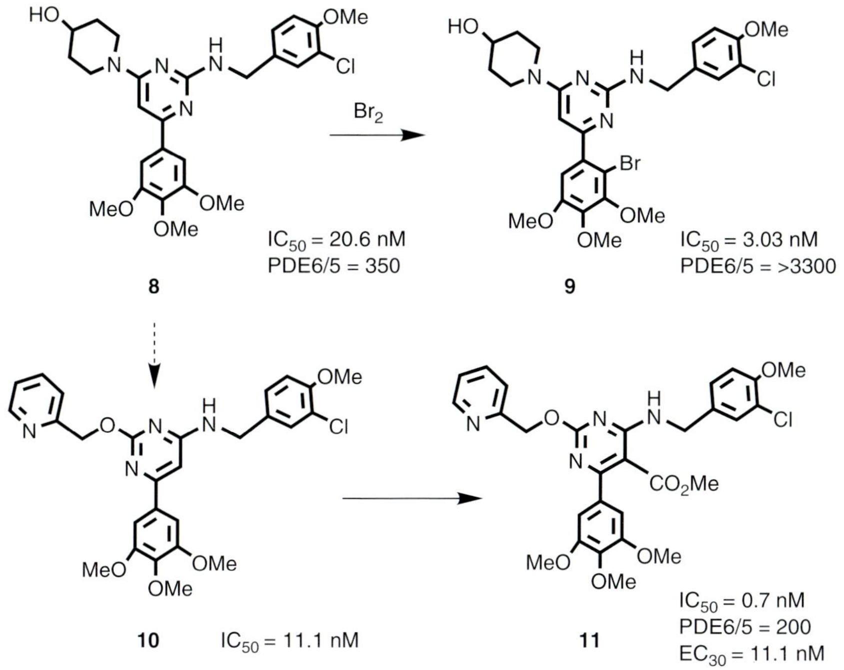

Figure 4.6 Monocyclic Type A series.

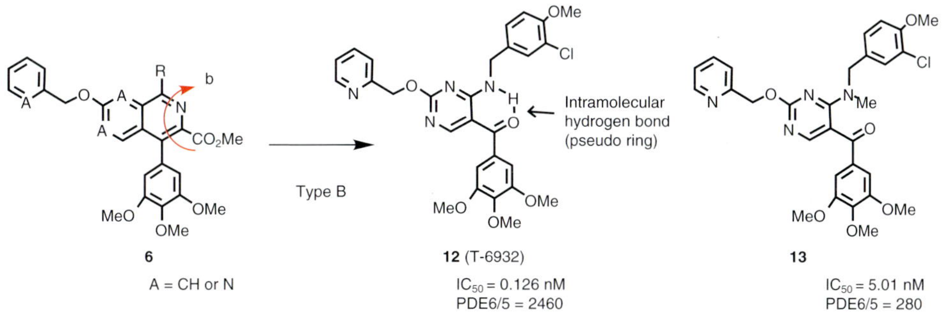

Figure 4.7 Monocyclic Type B series.

a pyrimidine derivative (**12**, T-6932) with an IC$_{50}$ of 0.126 nM and 2460-fold selectivity for PDE5 over PDE6 as a potential new lead.

The importance of the intramolecular hydrogen bond of T-6932 was supported by the reduced potency of the *N*-methyl derivative (**13**), which cannot form this hydrogen bond. Thus, the potent PDE5 inhibitory activity of T-6932 was attributed to a rigid conformation induced by the intramolecular hydrogen bond. The synthetic route to pyrimidine derivative **12** (T-6932) from the corresponding dichloride derivative (**14**) is shown in Scheme 4.1.

Although lower lipophilicity was expected in the pyrimidine derivatives compared with benzene analogs, T-6932 still demonstrated high lipophilicity

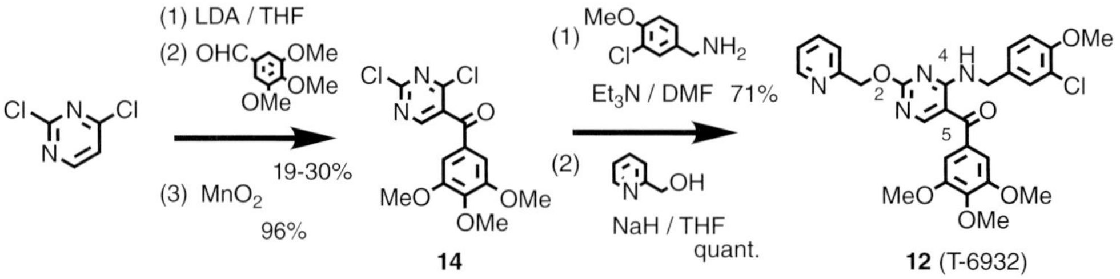

Scheme 4.1 Synthetic route to pyrimidine derivative 12 (T-6932).

(cLogP = 4.59) and showed a moderate relaxant effect on isolated rabbit corpus cavernosum (EC$_{30}$ = 53 nM). Modifications at the 2-, 4-, and 5-position of the pyrimidine moiety were explored to improve its profile, that is, reduced lipophilicity, increased solubility, and improved selectivity for PDE5. In addition, we hoped to introduce the properties of rapid onset and reduced metabolic stability into the newly synthesized compounds. We preferentially selected substituents bearing primary hydroxyl groups such as –CH$_2$OH, and alicyclic or cyclic amines to improve solubility (rapid-onset) and metabolic liability (short duration). The structure–activity relationships (SAR) at the 2-, 4-, and 5-positions of the pyrimidine derivatives (**15**) are described in the following sections.

4.2.3.3 SAR of Substitution at the 2-Position of the Pyrimidine Ring (15)

The 2-substituted-5-(3,4,5-trimethoxybenzoyl) pyrimidines (**15a–o**) were synthesized by the same procedure as used for the preparation of compound (**12**). Among the examined derivatives, the pyrimidine compounds (**15c, 15e, 15j, 15k**) bearing a hydroxyl group at the 2-position had a tendency to exhibit a high PDE5 inhibitory activity and a potent relaxant effect on isolated rabbit corpus cavernosum compared with those without a hydroxyl group. In particular, compound (**15n**), which has an (S)-prolinol group at the 2-position, showed high PDE5 inhibitory activity with an IC$_{50}$ of 0.205 nM and 5800-fold selectivity for PDE5 over PDE6.

4.2.3.4 SAR of Substitution at the 4-Position of the Pyrimidine Ring (16)

The 4-substituted-5-(3,4,5-trimethoxybenzoyl) pyrimidines (**16a–l**) were obtained by the same procedure as used for the preparation of compound (**12**). Introduction of benzylamino groups at the 4-position of the pyrimidine moiety had a tendency to exhibit high PDE5 inhibitory activity. As shown in Table 4.3, compound (**12**) incorporating a 3-chloro-4-methoxybenzylamino group at the 4-position showed the most potent inhibitory activity for PDE5. The influence of the substituents at the 3-position of 4-methoxybenzyl moiety on the inhibitory activity for PDE5 were in the following order: Cl > NO$_2$ > CN > F > OMe > H > NMe$_2$.

Introduction of an *ortho*-MeO group in the benzene ring (**16f**) or a methyl group at the benzylic position (**16l**) resulted in loss of PDE5 inhibitory activity. Considering the data from the above SAR study on the 4-position of the compounds

4.2 Discovery of Avanafil | 71

Table 4.2 SAR at the 2-position on the pyrimidine ring (15a–o).

No	R–	IC$_{50}$ (nM)	Selectivity PDE6^a/5	EC$_{30}$ (nM)	No	R–	IC$_{50}$ (nM)	Selectivity PDE6^a/5	EC$_{30}$ (nM)
12	T-6932	0.126	2460	53	15h		0.383	1620	>100
15a		6.62	870	>100	15i		2.31	2870	100
15b		35.9	NT	NT	15j		0.737	1670	39.7
15c		15.7	550	8.04	15k		1.75	2860	11
15d		2.62	2140	>100	15l		1.53	6530	>100
15e		2.69	>3710	28.6	15m		2.67	>3740	>100
15f		41.6	1010	>100	15n		0.205	5800	>100
15g		32.1	NT	NT	15o		0.691	2850	NT

IC$_{50}$ = half maximal effective concentration for inhibition of PDE5, EC$_{30}$ = 30% effective concentration for relaxation in the isolated rabbit corpus cavenosum precontracted with 5 μM phenylephrine, NT = not tested.
^{a} Inhibitory effect on PDE6 was measured with light-activated bovine PDE6.

Table 4.3 SAR at the 4-position on the pyrimidine ring (16a–m).

NO	R–	IC$_{50}$ (nM)	Selectivity PDE6^a/5	EC$_{30}$ (nM)	NO	R–	IC$_{50}$ (nM)	Selectivity PDE6^a/5	EC$_{30}$ (nM)
12	OMe, Cl (T-6932)	0.126	2460	53	16f	OMe, OMe	>100	NT	NT
13	Me, OMe, Cl	5.01	280	19.7	16g	OMe, F	7.29	540	>100
16a	OMe, Cl	1.46	1280	30.4	16h	OMe	27.5	390	>100
16b	OMe, NO$_2$	0.665	5680	7.59	16i	Cl, OMe	7.39	780	>100
16c	OMe, CN	3.79	>2630	11.9	16j	OMe, N	32.6	NT	NT
16d	OMe, OMe	14.7	200	35	16k	OMe, N	>100	NT	NT
16e	OMe, NMe$_2$	89.9	NT	NT	16l	OMe, Cl, Me	23.9	290	NT

IC$_{50}$ = half maximal effective concentration for inhibition of PDE5, EC$_{30}$ = 30% effective concentration for relaxation in the isolated rabbit corpus cavenosum precontracted with 5 μM phenylephrine, NT = not tested.
^{a} Inhibitory effect on PDE6 was measured with light-activated bovine PDE6.

(**12, 16**), the 3-chloro-4-methoxybenzylamino group was the best choice for the functional group at the 4-position of the pyrimidine derivatives.

4.2.3.5 SAR of Substitution at the 5-Position of the Pyrimidine Ring (19, 20)

In order to decrease lipophilicity, the influence of substitution of the trimethoxybenzoyl group at the 5-position of the pyrimidine moiety was investigated. The synthetic route of the compounds (**19**) is shown in Scheme 4.2. Regio-selective condensation of the chloride derivative (**17**) with benzylamines and subsequent oxidation of the 2-SMe group of **18** with *m*-CPBA (*meta*-chloroperoxybenzoic acid), followed by treatment with sodium 2-pyridinylmethoxide gave the corresponding ester. Hydrolysis of the ester and subsequent condensation with various amines afforded the corresponding amides (**19**).

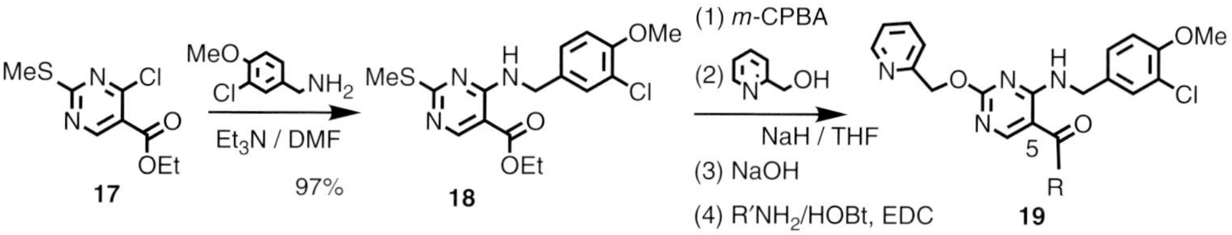

Scheme 4.2 Synthetic route to 5-substituted pyrimidines (**19a–m**).

As shown in Table 4.4, the presence of esters (**19a, 19b**) or carboxamides (**19c–m**) did not weaken the inhibitory activity for PDE5, and the physicochemical properties of the pyrimidine derivatives (**19**) was improved by introducing hydrophilic functional groups at this position.

The introduction of the (*S*)-prolinol group at the 2-position of the pyrimidine moiety (**15n**) conferred the highest selectivity against PDE6 and potent inhibitory activity for PDE5 on the derivatives as shown in Table 4.2. As a next step, we investigated the SAR at the 5-position of pyrimidine derivatives (**20**) bearing the (*S*)-prolinol group at the 2-position. The compounds (**20a–j**) were synthesized by a route similar to the preparation of **19**, and were evaluated for their inhibitory activity (see Table 4.5). Among them, the 2-pyrimidinylmethylcarbamoyl derivative (**20b**) (TA-1790; avanafil) showed high PDE5 inhibitory activity, high selectivity against PDE6 (2410-fold for the light-activated form), and also the most potent relaxant effect on isolated rabbit corpus cavernosum (EC_{30} = 2.05 nM).

Furthermore, avanafil (**20b**) demonstrated excellent *in vivo* effects, rapid onset of action, and short duration of efficacy, compared to sildenafil, and therefore, was further evaluated. The biological and pharmacological properties of avanafil are described in Section 4.3.

4.2.3.6 Core Structure Modifications of the Pyrimidine Nucleus of Avanafil

The importance of the pyrimidine nucleus of avanafil (**20b**) was investigated using the benzene (**21**), pyridine (**22–24**), pyridazine (**25**), pyrazine (**26**), triazine (**27**),

Table 4.4 SAR at the 5-position on the pyrimidine ring (**19a–m**).

NO	R-	IC$_{50}$ (nM)	Selectivity PDE6^a/5	EC$_{30}$ (nM)	NO	R-	IC$_{50}$ (nM)	Selectivity PDE6^a/5	EC$_{30}$ (nM)
12	(3,4,5-triOMe-phenyl)	0.126	2460	53	19g	NHCH$_2$CH$_2$OH	44.1	NT	NT
19a	–OEt	5.51	130	30	19h	NHCH$_2$CH$_2$CH$_2$OH	13	410	2.55
19b	OCH$_2$-(2-pyridyl)	2.1	1060	3.1	19i	NHCH$_2$CH$_2$CH$_2$OMe	15	240	7.69
19c	NHCH$_2$-(2-pyridyl)	2.72	480	8.59	19j	N(cyclohexyl-OH)	3.48	510	16.1
19d	NHCH$_2$CH$_2$-(2-pyridyl)	4.96	450	NT	19k	NHCH$_2$CH$_2$-morpholine	5.59	620	6.93
19e	N(Me)CH$_2$-(2-pyridyl)	84.8	NT	NT	19l	NHCH$_2$CH$_2$-(4-OH-piperidine)	1.28	NT	> 100
19f	NHCH$_2$-(2-pyrimidyl)	12.2	700	20.6	19m	N-(4-OH-piperidine)	>100	NT	NT

IC$_{50}$ = half maximal effective concentration for inhibition of PDE5, EC$_{30}$ = 30% effective concentration for relaxation in the isolated rabbit corpus cavenosum precontracted with 5 µM phenylephrine, NT = not tested.
^{a} Inhibitory effect on PDE6 was measured with light-activated bovine PDE6.

and thiazole (**28**), analogs which bear the three functional groups at the same position, that is, (S)-prolinol, 3-chloro-4-methoxy-benzylamine, and N-(2-pyrimidinylmethyl)carboxamide. Synthesis of these compounds was described elsewhere (see Ref. [18]). Their IC$_{50}$ values for inhibition of PDE5 and EC$_{30}$ values for the relaxant effect on isolated rabbit corpus cavernosum are shown in Table 4.6.

Among the core structure analogs of avanafil (**20b**), pyrazine (**26**), and triazine (**27**) showed potent inhibitory activity for PDE5, but avanafil (**20b**) bearing the pyrimidine nucleus was the most potent in the relaxant effect on isolated rabbit corpus cavernosum.

Table 4.5 SAR at the 5-position on the pyrimidine ring (**20a–j**).

NO	NR1R2	IC$_{50}$ (nM)	Selectivity PDE6^a/5	EC$_{30}$ (nM)	NO	NR1R2	IC$_{50}$ (nM)	Selectivity PDE6^a/5	EC$_{30}$ (nM)
20a		1.22	4180	22.1	20f		7.29	860	NT
20b	TA-1790	5.2	2410	2.05	20g		3.57	NT	>100
20c		3.03	2680	5	20h		6.07	2000	2.74
20d		5.07	NT	>100	20i		6.57	1470	4.86
20e		>100	NT	NT	20j		2.17	390	7

IC$_{50}$ = half maximal effective concentration for inhibition of PDE5, EC$_{30}$ = 30% effective concentration for relaxation in the isolated rabbit corpus cavenosum precontracted with 5 μM phenylephrine, NT = not tested.
^{a} Inhibitory effect on PDE6 was measured with light-activated bovine PDE6.

4.3 Pharmacological Features of Avanafil

4.3.1 PDE Inhibitory Profiles

The inhibitory potency and PDE selectivity of avanafil, sildenafil, vardenafil, and tadalafil against the 11 PDEs were compared (Table 4.7). Avanafil showed a potent and selective inhibitory effect on PDE5 isolated from canine lung. The IC$_{50}$ values of avanafil, sildenafil, vardenafil, and tadalafil for PDE5 isolated from canine lung were 5.2, 1.6, 0.084, and 4.0 nM, respectively. The selectivity ratio for avanafil's inhibitory effect on trypsin-activated PDE6/PDE5 was approximately 120, which was higher than those of sildenafil (16-fold) and vardenafil (21-fold). Sildenafil showed a selectivity ratio of inhibitory effect on PDE1/PDE5 of 375. In contrast,

Table 4.6 PDE5 inhibitions and relaxant effects of the core structure analogs of avanafil (20b).

No	A	IC$_{50}$ (nM)	EC$_{30}$ (nM)	No	A	IC$_{50}$ (nM)	EC$_{30}$ (nM)
20b	TA-1790	5.2	2.05	25		2.02	60.8
21		51	NT	26		3.38	13.4
22		103	NT	27		1.67	21.4
23		5.49	158	28		66	NT
24		4.22	NT				

IC$_{50}$ = half maximal effective concentration for inhibition of PDE5, EC$_{30}$ = 30% effective concentration for relaxation in the isolated rabbit corpus cavenosum precontracted with 5 μM phenylephrine, NT = not tested.

avanafil demonstrated a ratio of over 10 000. Although tadalafil showed the lowest inhibitory effect on PDE6 among the PDE5 inhibitors tested, it potently inhibited PDE11 with a selectivity ratio of 25 for PDE11/PDE5. Avanafil exhibited a selectivity ratio of >19 000 for PDE11/PDE5. Avanafil also exhibited a high selectivity for PDE5 over other PDEs (PDE1, PDE2, PDE3, PDE4, PDE7, PDE8, PDE9, and PDE10). Thus, avanafil is a potent PDE5-specific inhibitor with well-balanced PDE inhibitory profiles [19, 20].

4.3.2
In Vivo Pharmacology

4.3.2.1 Potentiation of Penile Tumescence in Dogs

The potentiating effect of intravenous avanafil or sildenafil (3–300 μg kg^{-1}) on pelvic nerve stimulation (pseudo sexual stimulation)-induced penile tumescence

Table 4.7 Inhibitory effects (IC$_{50}$) and selectivity (fold difference vs PDE5) of avanafil, sildenafil, vardenafil, and tadalafil toward 11 isozymes of PDE.

Drugs		PDE1 Rat heart	PDE2 Bovine adrenal gland	PDE3 Canine heart	PDE4 Canine lung	PDE5 Canine lung	PDE6[a] Bovine retina	PDE7B rHuman[b]	PDE8A rHuman[b]	PDE9A rHuman[b]	PDE10 Rat striatum	PDE11A rHuman[b]
Avanafil (TA-1790)	IC50 (nM)	53 000	51 000	>100 000	5700	5.2	630	27 000	12 000	>100 000	6200	>100 000
	Selectivity versus PDE5	10 192	9808	>19 231	1096	1	121	5192	2308	>19 231	1192	>19 231
Sildenafil	IC50 (nM)	600	63 000	26 000	5000	1.6	25	22 000	>100 000	3600	5400	7800
	Selectivity versus PDE5	375	39 375	16 250	3125	1	16	13 750	>62 500	2250	3375	4875
Vardenafil	IC50 (nM)	85	23 000	2200	1200	0.084	1.8	1500	84 000	1400	1500	500
	Selectivity versus PDE5	1012	273 810	26 190	14 286	1	21	17 857	1 000 000	16 667	17 857	5952
Tadalafil	IC50 (nM)	42 000	>100 000	>100 000	59 000	4	2200	>100 000	>100 000	>100 000	35 000	100
	Selectivity versus PDE5	10 500	>25 000	>25 000	14 750	1	550	>25 000	>25 000	>25 000	8750	25

IC$_{50}$ = half maximal inhibitory concentration; PDE = cyclic 3′,5′-nucleotide phosphodiesterase.
a) Inhibitory effect on PDE6 was measured with trypsine-activated bovine PDE6.
b) rHuman means human PDE enzyme prepared using recombinant DNA technique.

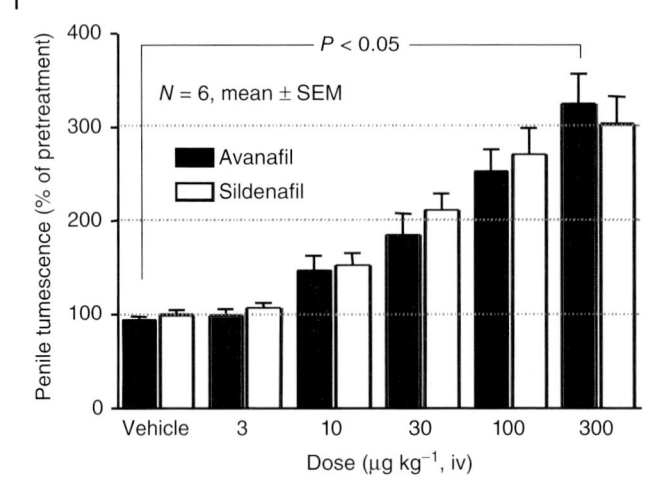

Figure 4.8 Influence of intravenous avanafil and sildenafil on the pelvic nerve stimulation-induced tumescence in anesthetized dogs. (Modified from Ref. [19].) Drugs were administered 5 min before stimulation. Induced tumescence in the absence of compounds represents 100%. Statistical analysis was performed using one-way ANOVA with randomized complete block design.

was evaluated in anesthetized dogs (Figure 4.8). Both avanafil and sildenafil caused a potentiation of the tumescence in a dose-dependent manner. The calculated $ED_{200\%}$ of avanafil and sildenafil on tumescence was 37.5 and 34.6 µg kg^{-1}, respectively. The plasma concentrations of avanafil and sildenafil 4 min after intravenous administration were 59.6 and 38.8 ng ml^{-1}, respectively.

In order to estimate the *in vivo* activity after gastrointestinal absorption, avanafil or sildenafil (100, 300, and 1000 µg kg^{-1}) was administered intraduodenally, and the potentiating effect on the tumescence was examined in anesthetized dogs (Figure 4.9a,c). Both avanafil and sildenafil caused a potentiation of penile tumescence. The time to peak response for avanafil and sildenafil was 10 and 30 min, respectively. The $ED_{200\%}$ of avanafil and sildenafil on tumescence was calculated as 151.7 and 79.0 µg kg^{-1}, respectively, at their peak times. Tumescence rigidity due to avanafil and sildenafil correlated with the corresponding plasma concentrations (Figure 4.9b,d). The calculated plasma $ED_{200\%}$ of avanafil and sildenafil was 24.7 and 6.6 ng ml^{-1}, respectively. Considering plasma protein binding of the compounds, the unbound concentration of avanafil and sildenafil at $ED_{200\%}$ was calculated as 3.5 and 2.0 nM, respectively [19].

These studies suggest that avanafil has a potentiating effect on tumescence in response to pelvic nerve stimulation, and that its *in vivo* pharmacological profile is unique because the rapid onset of action and relatively short duration of action correlated with the plasma drug concentration.

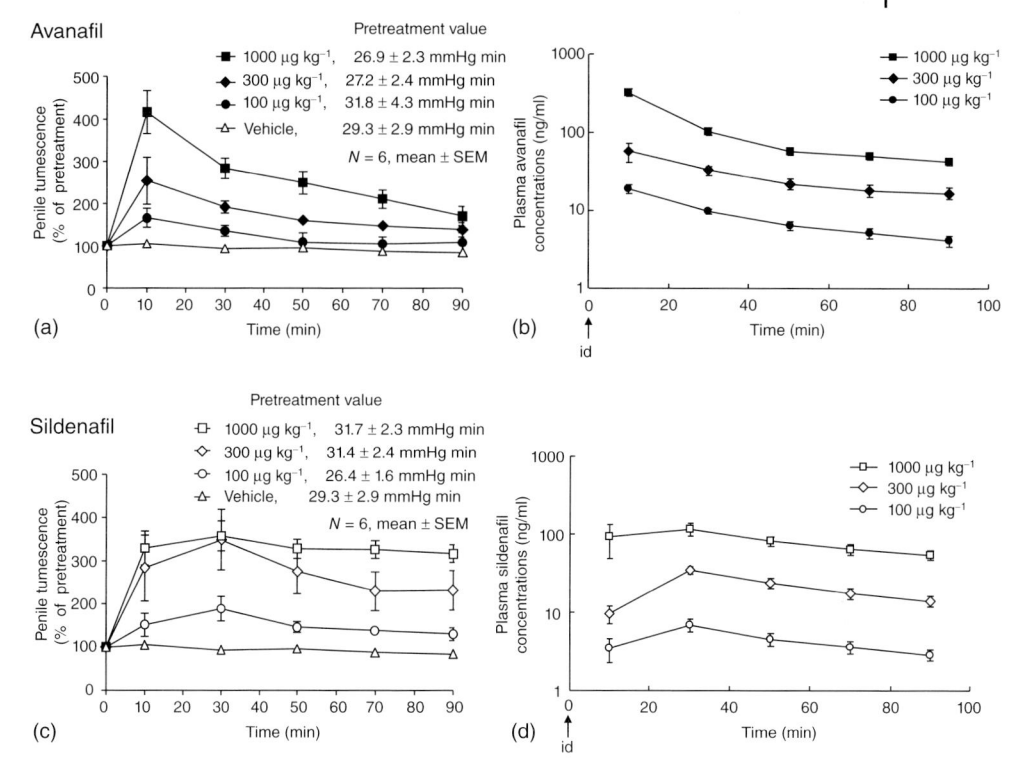

Figure 4.9 Influence of intraduodenal avanafil and sildenafil on the pelvic nerve stimulation-induced tumescence in anesthetized dogs. (Modified from Ref. [19].) Induced tumescence in absence of compounds represents 100%. (a, b) avanafil, (a, c) penile tumescence, (b, d) plasma concentration, and (c, d) sildenafil. Repeated measures ANOVA $p < 0.05$ versus vehicle.

4.3.2.2 Influence on Retinal Function in Dogs

The effect of avanafil and sildenafil on electroretinogram (ERG) was examined in anesthetized dogs. Intraduodenal administration of avanafil (10 and 30 mg kg^{-1}) or its vehicle did not affect the ERG waveform during the experimental period (Figure 4.10a,b). In contrast, administration of sildenafil (1–10 mg kg^{-1}) caused a change in ERG waveform, with the highest dose (10 mg kg^{-1}) strongly affecting the waveform shape (Figure 4.10c) [21].

As mentioned in Section 4.3.2.1, the pharmacologically active dose of avanafil after intraduodenal administration ranged from 0.1 to 1 mg kg^{-1}. Therefore, avanafil is expected to have a wide therapeutic window against the retinal side effects, probably because of its high selectivity for PDE5 over PDE6.

4.3.2.3 Influence on Hemodynamics in Dogs

As PDE5 inhibitors are classified as mild vasodilators for systemic and pulmonary vascular beds, the hemodynamic effects of avanafil and sildenafil were evaluated in anesthetized dogs. Interestingly, common and different hemodynamic changes

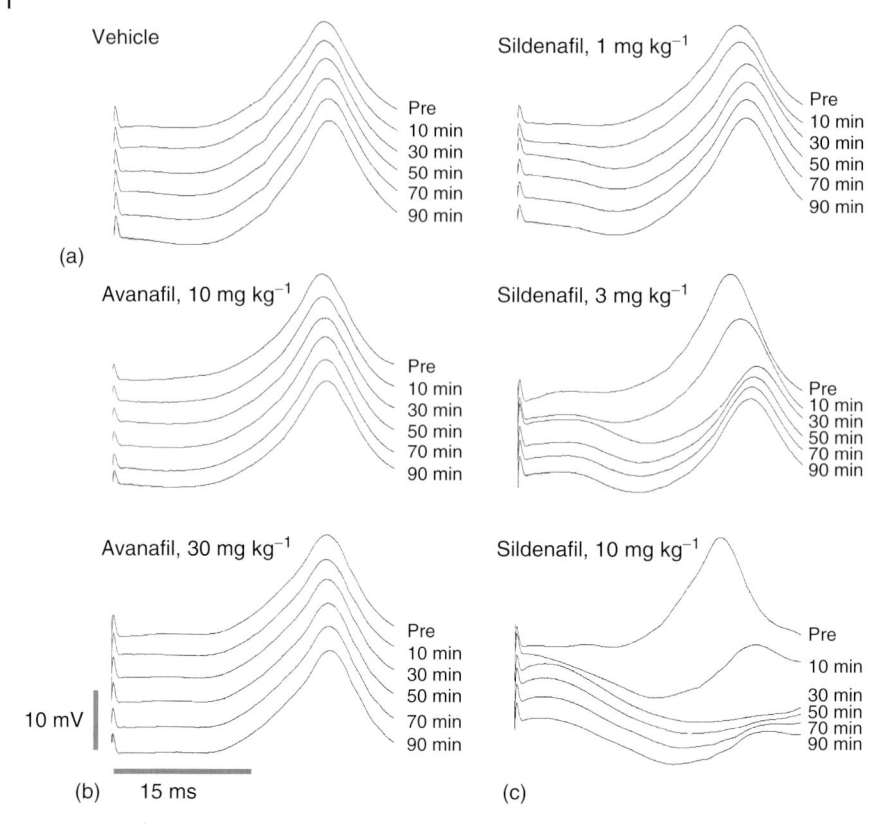

Figure 4.10 (a–c) Typical tracings of ERG waveform in anesthetized dogs treated with vehicle, avanafil or sildenafil. (Modified from Ref. [21]). Test compounds were administered intraduodenally. ERG was induced by a light-adapted 30 Hz flicker stimulation, and was recorded before (pre) and after administration. avanafil dose range was higher than that of sildenafil.

were observed between avanafil and sildenafil. Intravenous infusion of avanafil and sildenafil caused decreases in systemic arterial pressure, total peripheral resistance, and pulmonary arterial pressure (PAP) at the lowest dose of 1 µg kg^{-1} per minute. Sildenafil, but not avanafil, caused vasodilation in vascular beds of common carotid and vertebral arteries at the higher dose of 10 µg kg^{-1} per minute (data not shown). These hemodynamic data provide evidence that avanafil is a highly selective PDE5 inhibitor [21].

4.3.2.4 Influence on Nitroglycerin (NTG)-Induced Hypotension in Dogs

Since PDE5 inhibitors theoretically potentiate NO/cGMP signaling, concomitant use of nitrates is contraindicated in clinical situations. Therefore, the influence of avanafil on NTG-induced hypotension was examined in anesthetized dogs, in comparison with sildenafil. As we expected, intraduodenal administration of avanafil and sildenafil (0.1 and 1 mg kg^{-1}) potentiated NTG-induced hypotension

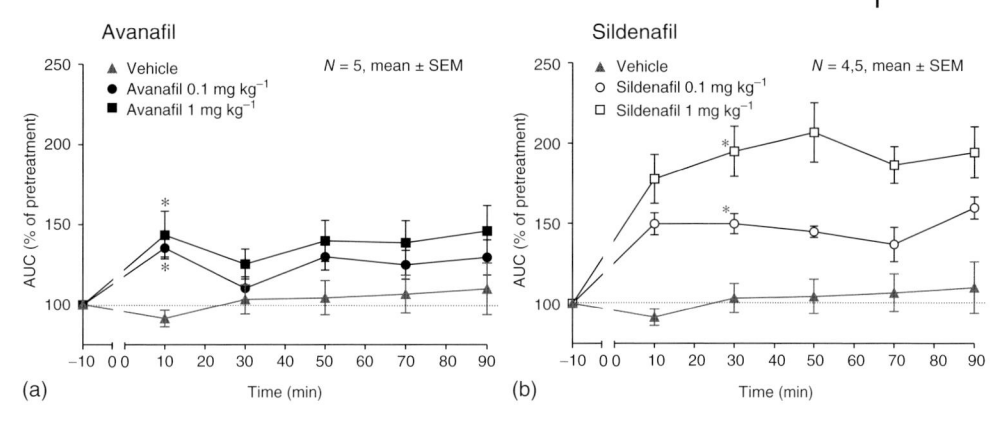

Figure 4.11 Potentiation of nitroglycerin-induced hypotension by avanafil and sildenafil in anesthetized dogs. ([Modified from Ref. [21]). Statistical analysis was performed at estimated time of peak plasma concentrations, including avanafil 10 min and sildenafil 30 min. Data were analyzed by Dunnett test. Asterisk indicates $p < 0.05$ versus vehicle, (a) avanafil and (b) sildenafil.

in the same dose range; however, the potentiating effect of 1 mg kg^{-1} avanafil was significantly weaker than that of 1 mg kg^{-1} sildenafil (Figure 4.11). The difference in NTG-induced hypotension between avanafil and sildenafil is likely due to the difference in the pharmacokinetic profiles of these compounds as discussed in Section 4.3.2.1 [21].

4.4
Clinical Studies of Avanafil

In a Phase I study assessing safety and tolerability, the mean time to reach the maximal drug concentration of avanafil after single oral administration (12.5–800 mg) ranged from 0.555 to 0.686 h, and the terminal plasma half-life ranged from 1.07 to 1.23 h. Avanafil showed no accumulation following multiple dosing [22]. In an additional phase I study of avanafil (50, 100, and 200 mg) in Korea, no serious adverse events were reported, and there were no clinically relevant changes in vital signs, ECG (electrocardiogram) recordings, physical examination findings, or color discrimination test results [23]. In addition, avanafil (200 mg) exhibited a weaker potentiation of NTG-induced hypotension than sildenafil (100 mg) [24] probably because of its short-acting hemodynamic profile, which was similar to the result of our animal study as shown in Section 4.3.2.4 (Table 4.8).

In Phase II and III studies (TA-05, TA-301, TA-302, TA-303, TA-314) in patients with mild to severe ED (total number patients: approximately 2000), oral treatment with avanafil (50, 100, 200 mg) resulted in significant improvements in the Sexual Encounter Profile (SEP) 3 and International Index of Erectile Function (IIEF) erectile function (EF) domain scores regardless of the presence of hypertension or diabetes. In these studies, avanafil showed a rapid onset of action without

Table 4.8 Clinical PK data of avanafil in comparison to marketed PDE5 inhibitors [14].

Drugs		T_{max} (h)	$T_{1/2}$ (h)
Avanafil	50 mg	0.686	1.07
	100 mg	0.555	1.23
	200 mg	0.593	1.19
Sildenafil	100 mg	1.16	3.82
Vardenafil	20 mg	0.660	3.94
Tadalafil	20 mg	2.0	17.5

severe adverse reactions [14, 20, 25]. The recent clinical study (TA-501) indicated that avanafil-treated patients achieved a statistically significant improvement over placebo, in the mean proportion of attempts that resulted in erections sufficient for successful intercourse, as early as 10 min for the 200 mg dose and 12 min for the 100 mg dose. These data strongly support the unique profile of rapid onset of action [26].

Avanafil has several metabolites after incubation with human microsomal fractions. In clinical trials, inactive forms of avanafil, M4 (PDE5; $IC_{50} = 51$ nM) and M16 (PDE5; $IC_{50} = 4100$ nM), were identified as major metabolites (M4: ~23%, M16: ~29%) in plasma. Avanafil is rapidly metabolized to M4 and M16 by CYP3A4 (cytochrome P450 3A4) at the 2-position of (S)-prolinol, and excreted into the feces. These metabolic profiles of avanafil may explain the relative short duration (Figure 4.12) [25].

To date, severe adverse reactions have not been reported, although the number of patients treated with avanafil is limited. The most common adverse reactions (greater than or equal to 2%) include headache, flushing, nasal congestion, nasopharyngitis, and back pain. These side-effects resolved after discontinuation. Coadministration with nitrates is a contraindication, which is similar to other PDE5 inhibitors.

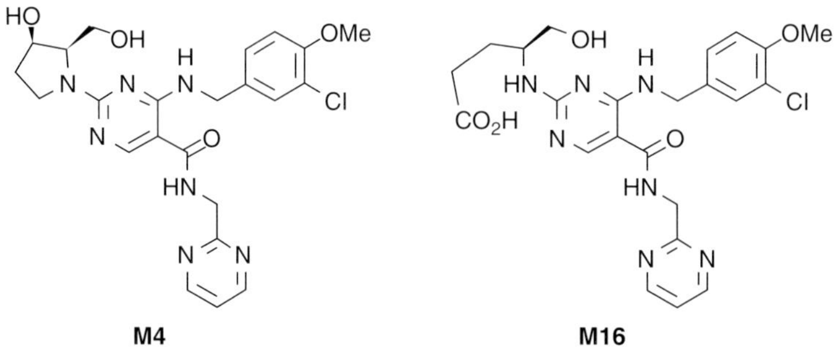

Figure 4.12 Two major circulating metabolites (M4 and M16) in human.

4.5
Conclusion

In summary, we succeeded in creating avanafil within 9 months after starting to seek a candidate compound having a new chemical scaffold from a prototype isoquinolinone inhibitor, T-1032. The goals of this project, that is, rapid onset of action, high selectivity for PDE5, enough safety based on short duration and PDE selectivity, were achieved by avanafil. Avanafil was licensed to Vivus Inc. (USA) and JW pharmaceuticals (Korea) in 2000 and 2006, respectively. After clinical trials, avanafil was approved by the Korea Food and Drug Administration in 2011 and launched in the Korean market as Zepeed®. Based on a comprehensive development program, avanafil was approved by the Food and Drug Administration in 2012 with a planned US launch under the tradename Stendra™. In 2013, avanafil was also approved under the trade name, Spedra™ by the European Medicines Agency. Avanafil is unique in properties and structurally different from sildenafil, vardenafil, and tadalafil, and therefore, avanafil may more appropriately be categorized as a second generation PDE5 inhibitor.

List of Abbreviations

nNOS	neuronal nitric oxide synthase
L-Arg	l-Arginine
L-Cit	l-Citrulline
NTG	nitroglycerin
ERG	electroretinogram
cAMP	cyclic adenosine monophosphate
cGMP	cyclic guanosine monophosphate
GTP	guanosine triphosphate
ATP	adenosine triphosphate
PK	pharmacokinetic

References

1. The National Institutes of Health Consensus Development Program (1992) Impotence, NIH Consensus Statement Online 1992 December 7–9, Vol. 10, pp. 1–31, *http://consensus.nih.gov/1992/1992impotence091html.htm* (accessed 9 August 2014).
2. Wespes, E., Eardley, I., Giuliano, F., Hatzichristou, D., Hatzimouratidis, K., Moncada, I., Salonia, A., and Vardi, Y. Guidelines on Male Sexual Dysfunction: Erectile Dysfunction and Premature Ejaculation, *http://www.uroweb.org* (accessed 9 August 2014).
3. Selvin, E., Burnett, A.L., and Platz, E.A. (2007) Prevalence and risk factors for erectile dysfunction in the U.S. *Am. J. Med.*, **120** (2), 151–157.
4. Albersen, M., Shindal, A., Mwamukonda, K., and Lue, T. (2010) The future is today: emerging drugs for the treatment of erectile dysfunction. *Expert Opin. Emerg. Drugs*, **15** (3), 467–480.
5. Viagra® (sildenafil citrate) (2003) *Summary of Product Specifications*, Pfizer Ltd, New York.

6. Levitra®, vardenafil hydrochloride (2003) *Summary of Product Specifications*, Bayer Healthcare.
7. Cialis® (tadalafil) (2003) *Summary of Product Specifications*, Lilly ICOS LLC, Indianapolis.
8. Omori, K. and Kotera, J. (2007) Overview of PDEs and their regulation. *Circ. Res.*, **100** (3), 309–327.
9. Marmor, M.F. and Kessler, R. (1999) Sildenafil (Viagra) and ophthalmology. *Surv. Ophthalmol.*, **44** (2), 153–162.
10. Brock, G., Glina, S., Moncada, I., Watts, S., Xu, L., Wolka, A., and Kopernicky, V. (2009) Likelihood of tadalafil-associated adverse events in integrated multiclinical trial database: classification tree analysis in men with erectile dysfunction. *Urology*, **73** (4), 756–761.
11. Bischoff, E. (2004) Potency, selectivity, and consequences of nonselectivity of PDE inhibition. *Int. J. Impot. Res.*, **16** (Suppl. 1), S11–S14.
12. Setter, S.M., Iltz, J.L., Fincham, J.E., Campbell, R.K., and Baker, D.E. (2005) Phosphodiesterase 5 inhibitors for erectile dysfunction. *Ann. Pharmacother.*, **39** (7–8), 1286–1295.
13. Shabsigh, R., Seftel, A.D., Rosen, R.C., Porst, H., Ahuja, S., Deeley, M.C., Garcia, C.S., and Giuliano, F. (2006) Review of time of onset and duration of clinical efficacy of phosphodiesterase type 5 inhibitors in the treatment of erectile dysfunction. *Urology*, **68** (4), 689–696.
14. Limin, M., Johnsen, N., Wayne, J., and Hellstrom, W.J.G. (2010) Avanafil, a new rapid-onset phosphodiesterase 5 inhibitor for the treatment of erectile dysfunction. *Expert Opin. Investig. Drugs*, **19** (11), 1427–1437.
15. Ukita, T., Nakamura, Y., Kubo, A., Yamamoto, Y., Moritani, Y., Saruta, K., Higashijima, T., Kotera, J., Takagi, M., Kikkawa, K., and Omori, K. (2001) Novel, potent, and selective phosphodiesterase 5 inhibitors: synthesis and biological activities of a series of 4-aryl-1-isoquinolinone derivatives. *J. Med. Chem.*, **44** (13), 2204–2218.
16. Kotera, J., Fujishige, K., Michibata, H., Yuasa, K., Kubo, A., Nakamura, Y., and Omori, K. (2000) Characterization and effects of methyl-2-(4-aminophenyl)-1,2-dihydro-1-oxo-7-(2-pyridinyl-methoxy)-4-(3,4,5-trimethoxyphenyl)-3-isoquinoline carboxylate sulfate (T-1032), a novel potent inhibitor of cGMP-binding cGMP-specific phosphodiesterase (PDE5). *Biochem. Pharmacol.*, **60**, 1333–1341.
17. Furet, P., Bold, G., Hofmann, F., Manley, P., Meyer, T., and Altmann, K.H. (2003) Identification of a new chemical class of potent angiogenesis inhibitors based on conformational considerations and database searching. *Bioorg. Med. Chem. Lett.*, **13** (18), 2967–2971.
18. Yamada, K., Matsuki, K., Omori, K., and Kikkawa, K. (2001) Cyclic compounds. PCT/JP2001/002034, WO Patent 2001083460A1.
19. Kotera, J., Mochida, H., Inoue, H., Noto, T., Fujishige, K., Sasaki, T., Kobayashi, T., Kojima, K., Yee, S., Yamada, Y., Kikkawa, K., and Omori, K. (2012) Avanafil, a potent and highly selective phosphodiesterase-5 inhibitor for erectile dysfunction. *J. Urol.*, **188** (2), 668–674.
20. Wang, R., Burnett, A.L., Heller, W.H., Omori, K., Kotera, J., Kikkawa, K., Yee, S., Day, W.W., DiDonato, K., and Peterson, C.A. (2012) Selectivity of avanafil, a PDE5 inhibitor for the treatment of erectile dysfunction: implications for clinical safety and improved tolerability. *J. Sex. Med.*, **9** (8), 2122–2129.
21. Mochida, H., Yano, K., Inoue, H., Yee, S., Noto, T., and Kikkawa, K. (2013) Avanafil, a highly selective phosphodiesterase type 5 inhibitor for erectile dysfunction, shows good safety profiles for retinal function and hemodynamics in anesthetized dogs. *J. Urol.*, **190** (2), 799–806.
22. Peterson, C. and Swearingen, D. (2006) Pharmacokinetics of avanafil, a new PDE5 inhibitor being developed for treating erectile dysfunction. *J. Sex. Med.*, **3** (Suppl. 3), 253–254.
23. Jung, J., Choi, S., Cho, S., Ghim, J., Hwang, A., Kim, U., Kim, B.S., Koguchi, A., Miyoshi, S., Okabe, H., Bae, K.S., and Lim, H.S. (2010) Tolerability and pharmacokinetics of avanafil, a phosphodiesterase type 5 inhibitor: a single- and

multiple-dose, double-blind, randomized, placebo-controlled, dose-escalation study in healthy Korean male volunteers. *Clin. Ther.*, **32** (6), 1178–1187.
24. Swearingen, D., Nehra, A., Morelos, S., and Peterson, C.A. (2013) Hemodynamic effect of avanafil and glyceryl trinitrate coadministration. *J. Interventions Clin. Prac.*, **26**, 1–10.
25. Kedia, G.T., Ückert, S., Assadi-Pour, F., Kuczyk, M.A., and Albrecht, K. (2013) Avanafil for the treatment of erectile dysfunction: initial data and clinical key properties. *Ther. Adv. Urol.*, **5** (1), 35–41.
26. Vivus, Inc. VIVUS Announces Study Results Showing STENDRA (Avanafil) is Effective for Sexual Activity Within 15 Minutes in Men With Erectile Dysfunction (ED). Press Release, *http://ir.vivus.com/releasedetail.cfm?ReleaseID=772324* (accessed 19 June 2013).

Koichiro Yamada was born in Fukui Prefecture, Japan, in 1951. He obtained his master's degree from Kanazawa University and joined Mitsubishi-Tanabe Pharma Corporation (MTPC), Medicinal Chemistry Research Lab. II, Saitama, Japan as a medicinal chemist in 1976. He got his PhD from the University of Tokyo (1982). He was a postdoctoral fellow at NIH in the United States (1990–1991). He joined the discovery team for PDE5 inhibitor as a leader and found avanafil. He was working as a manager, a senior principal research scientist, and an advisor for many research projects in MTPC until October 2013.

Toshiaki Sakamoto was born in Okayama Prefecture, Japan, in 1971. He received his master's degree of science from Waseda University in 1996. He joined Mitsubishi-Tanabe Pharma Corporation (MTPC) Medicinal Chemistry Research Lab. II, Saitama, Japan in 1996, and as a medicinal chemist has engaged in research to find new drug candidates. In 1998, he joined the project team of PDE5 inhibitor, and his synthetic studies contributed to the discovery of avanafil.

Kenji Omori was born in Kyoto in 1957. He earned his master's degree in 1981 from Kyoto University, and then, started his research in MTPC. In 1993, He received his PhD from Kyoto University, and worked in Pasteur Institute, Paris. After returning to MTPC, he immersed himself in cGMP signaling study including PDE5 project. Several novel PDEs were reported by him. In 2013, he moved to Nagoya University, Japan, where he is Designated Professor, Graduate School of Pharmaceutical Sciences, to expand his cGMP research and drug discovery.

Kohei Kikkawa was born in Tochigi Prefecture in 1959. He received his master's degree of science in 1983 and PhD in 1996 from Chiba University. He has been a Principal Research Scientist, MTPC, Pharmacology Research Laboratories II, Saitama, Japan and built a substantial career as a cardiovascular pharmacology researcher in MTPC since 1983. He was a visiting scientist at the Johns Hopkins University, Molecular Cardiology Medicine, in Baltimore, MD (1995–1997). He immersed himself in the research of PDE5 inhibitor as a project leader.

5
Dapagliflozin, A Selective SGLT2 Inhibitor for Treatment of Diabetes

William N. Washburn

5.1
Introduction

Type 2 diabetes mellitus (T2DM) has become a major worldwide health issue. Currently, 371 million patients are estimated to have developed T2DM; however, this number is projected to rise to 552 million by 2030 with increasing adoption of a sedentary lifestyle and a Western diet [1]. In its earliest stage, peripheral insulin resistance and improper first phase insulin release from the pancreas combine to require secretion of ever-increasing amounts of insulin to maintain normal postprandial glucose levels. Eventually the insulin required exceeds the pancreatic capacity resulting in the onset of hyperglycemia, the hallmark of T2DM. Over time, the burden of tissue proteins being modified by reaction with the elevated glucose manifests as complications of diabetes resulting in neuropathy, retinopathy, nephropathy, microvascular, and macrovascular complications, which give rise to an increased incidence of blindness, gangrene, renal failure, strokes, and cardiovascular events. T2DM is a progressive disease reflecting the exhaustion and death of increasing numbers of pancreatic beta cells over time. As a consequence, the initial medication will eventually fail to maintain the patient's blood sugar level in compliance, thereby necessitating addition of a second in combination. The need for additional new modes of treatment of T2DM is further underscored by the inability of current agents to maintain HbA1c levels <7.0 for ~40% of patients [2, 3].

For a normal healthy individual, five organs – liver, pancreas, skeletal muscle, adipose deposits, and kidney – are primarily responsible for maintaining glucose homeostasis. Current widely utilized treatments target various combinations of all but the kidney to (i) increase glucose cellular uptake, (ii) increase glucose metabolism, or (iii) decrease hepatic glucose output [4]. SGLT2 (sodium dependent glucose cotransporter) inhibitors are attractive because they represent for the first time a means to utilize the kidney to alleviate hyperglycemia [5].

Successful Drug Discovery, First Edition. Edited by János Fischer and David P. Rotella.
© 2015 Wiley-VCH Verlag GmbH & Co. KGaA. Published 2015 by Wiley-VCH Verlag GmbH & Co. KGaA.

5.2
Role of SGLT2 Transporters in Renal Function

The kidneys of a healthy individual filter 180 g of glucose daily from blood yet virtually every milligram is recovered from the glomerular filtrate passing down the S1–S3 segments of the proximal tubules. This recovery is mediated by two sodium dependent glucose cotransporters SGLT1 and SGLT2 that reside in the lumen-facing apical membranes of the epithelial cell surface [6]. These SGLT transporters couple the transport of glucose against a concentration gradient with the transport of Na^+ down a concentration gradient using the free energy gained from Na^+ transport to compensate for that required for glucose transport [7]. SGLT2, a low affinity high capacity glucose transporter, which is expressed only in the S1 and S2 segments of the proximal tubule, is responsible for recovery of as much as 90% of the filtered glucose [8, 9]. SGLT1, a high affinity low capacity glucose transporter residing in the S3 segment, recovers any remaining glucose. In addition to the renal tubule, SGLT1 is also expressed in the small intestine and the heart [10]. The role of SGLT1 in the heart is not known; however, studies have revealed that SGLT1 is required for absorption of glucose and galactose from the small intestine.

Clinical studies have revealed that the few individuals with nonfunctional SGLT2 were healthy in all aspects except for being profoundly glucosuric – spilling as much as 140 g of glucose daily in their urine [11, 12]. In contrast individuals with nonfunctional SGLT1 transporters presented with glucose–galactose malabsorption, which was manifested by severe diarrhea [13]. Given the stark contrast between these two phenotypes, we at Bristol-Myers Squibb elected to pursue selective SGLT2 inhibitors especially because all evidence suggested that these agents could promote removal of significant quantities of glucose safely in an insulin independent manner with minimal side effects [14]. Furthermore, it was anticipated that the risk of hyperglycemia would be low because these agents would not impact any of the hepatic counter-regulatory mechanisms. Moreover, unlike most of the anti-diabetic agents, they would not promote weight gain and possibly could cause weight loss depending on the magnitude of the uncompensated caloric expenditure.

The anticipated perturbation on renal function of a healthy individual is illustrated in Figure 5.1. The rate of glucose filtration increases linearly with glucose concentration in blood. The recovery rate is essentially coincident with the filtration rate; however the recovery capacity is finite. When the glucose flux exceeds the SGLT transport capacity of the kidney (Tm), the recovery rate can no longer increase. The recovery capacity is well above what is required for normal glycemic levels of 100 mg dl^{-1}. Consequently glycemic concentrations of healthy individuals do not reach levels that exceed recovery capacity, whereas those of uncontrolled diabetics do, resulting in progressively more glucosuria as the blood glucose levels approach Tm. A SGLT2 inhibitor reversibly lowers the glucose threshold value from ~180 to 200 mg dl^{-1} to a much lower value depending on the extent of inhibition thereby providing an insulin independent mechanism to reduce blood sugar

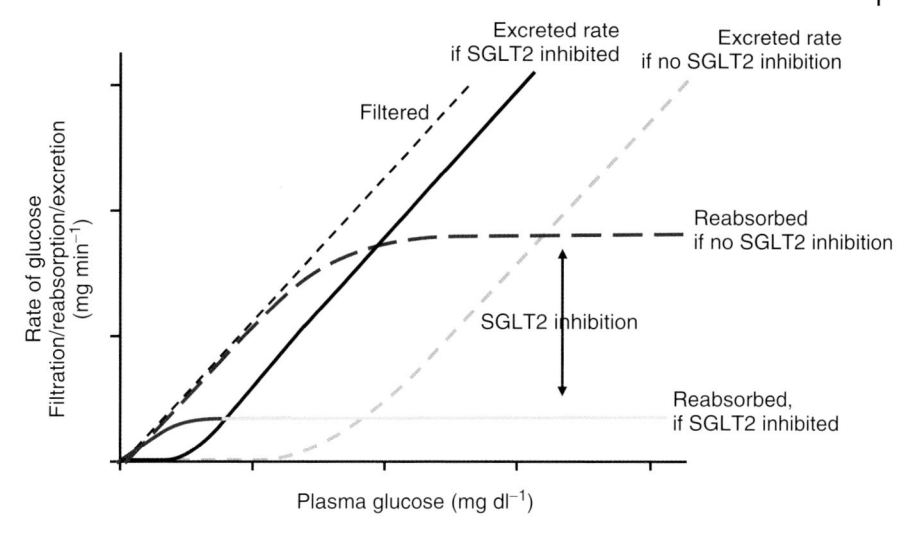

Figure 5.1 Renal processing of glucose: dependence of rates of filtration, recovery, and excretion on plasma glucose concentration in absence and presence of an SGLT2 inhibitor [15].

levels. If a selective SGLT2 inhibitor was employed, SGLT1 mediated transport would ensure that some glucose recovery would continue.

5.3
O-Glucoside SGLT2 Inhibitors

The impetus for the pharmaceutical effort to develop SGLT2 inhibitors was the understanding that SGLT2 inhibition was responsible for the glucosuria induced by the O-glucoside phlorizin [16]. However, the absorption, distribution, metabolism, and excretion (ADME) properties of phlorizin rendered it unsuitable as a drug [17]. When Bristol-Myers Squibb initiated a program to identify selective SGLT2 inhibitors, the immediate goal was to identify an inhibitor that would be amenable to once daily administration. Two literature disclosures – phenolic O-glucosides **2** reported by the Tanabe group and a Wyeth disclosure that administration of pyrazalones **3** produced glucosuria – served to initiate the program [18, 19]. During the first year three classes of *ortho* substituted O-glucosides were explored. These findings enabled the transition to *meta* substituted C-aryl glucosides that then became the sole focus of the program.

The Tanabe group had reported that modification of phlorizin **1** to generate a closely related series of dihydrochalcone O-glucosides represented by **2a** exhibited potential as orally administered SGLT2 inhibitors. Although these more lipophilic phlorizin analogs were more drug-like, the low oral bioavailability of **2a** was similar to that of phlorizin because, before absorption, the O-glucoside bond was readily cleaved by the glucosidases present in the intestinal tract to generate an inactive aglycone. Capitalizing on the finding that glucosidase substrate recognition requires a free glucose C6 hydroxyl, the Tanabe group selectively acylated the primary hydroxyl of **2a** to generate the methyl carbonate prodrug **2b**. Subsequent structure-activity relationship (SAR) evolution led to T-1095A **4a** which, when administered to rats in food as the prodrug T-1095 **4b**, produced a robust glucosuric response that in sub-chronic studies with diabetic rodents restored normal glucose metabolism [20]. Although oral bioavailability improved to 75%, clearance of the active agent was rapid in part due to cleavage by other glucosidases present in liver, kidney, and so on (Figure 5.2).

5.3.1
Hydroxybenzamide O-Glucosides

The Bristol-Myers Squibb group initially investigated whether **5a**, the O-glucoside hydroxybenzamide counterpart of **2a**, offered any advantage with respect to potency, selectivity, and susceptibility to glucosidase mediated cleavage. Compounds were initially characterized for SGLT2 and SGLT1 *in vitro* potency using Chinese hamster ovary (CHO) cells transfected with human SGLT1 and SGLT2 transporters to determine the EC_{50}s for inhibition of SGLT2 and SGLT1 mediated α-methylglucoside uptake [21]. Selectivity versus SGLT1 was the ratio of SGLT1/SGLT2 EC_{50}s. When appropriate, O-glucosides of interest were screened for susceptibility to glucosidase cleavage by incubation with rat gut homogenate, rat kidney homogenate, and cynomolgus monkey liver homogenate. In addition oxidative metabolic stability in rat liver homogenate was also measured. Acute studies using Swiss Webster mice and STZ rats (Sprague-Dawley rats made diabetic by administration of streptozotocin, a selective toxin for the pancreatic beta cells) provided an initial assessment of *in vivo* potency. To better emulate the T2DM condition, advanced *in vivo* studies employed Zucker Diabetic Fatty (ZDF) rats which are genetically predisposed to become diabetic.

The hydroxybenzamide O-glucosides were initially prepared by glucosylation of aglycone **6** using conditions reported by Tanabe [18]. Phenolic glucosylation of the secondary benzamide **6** proceeded in modest yield upon heating with acetobromoglucose in refluxing toluene containing a stirred suspension of cadmium carbonate (Scheme 5.1). After hydrolysis, evaluation of **5a** revealed that relative to

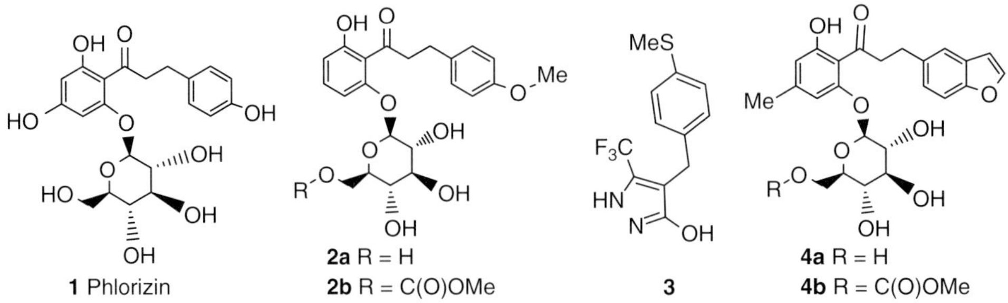

Figure 5.2 Glucosuric agents disclosed before 1999.

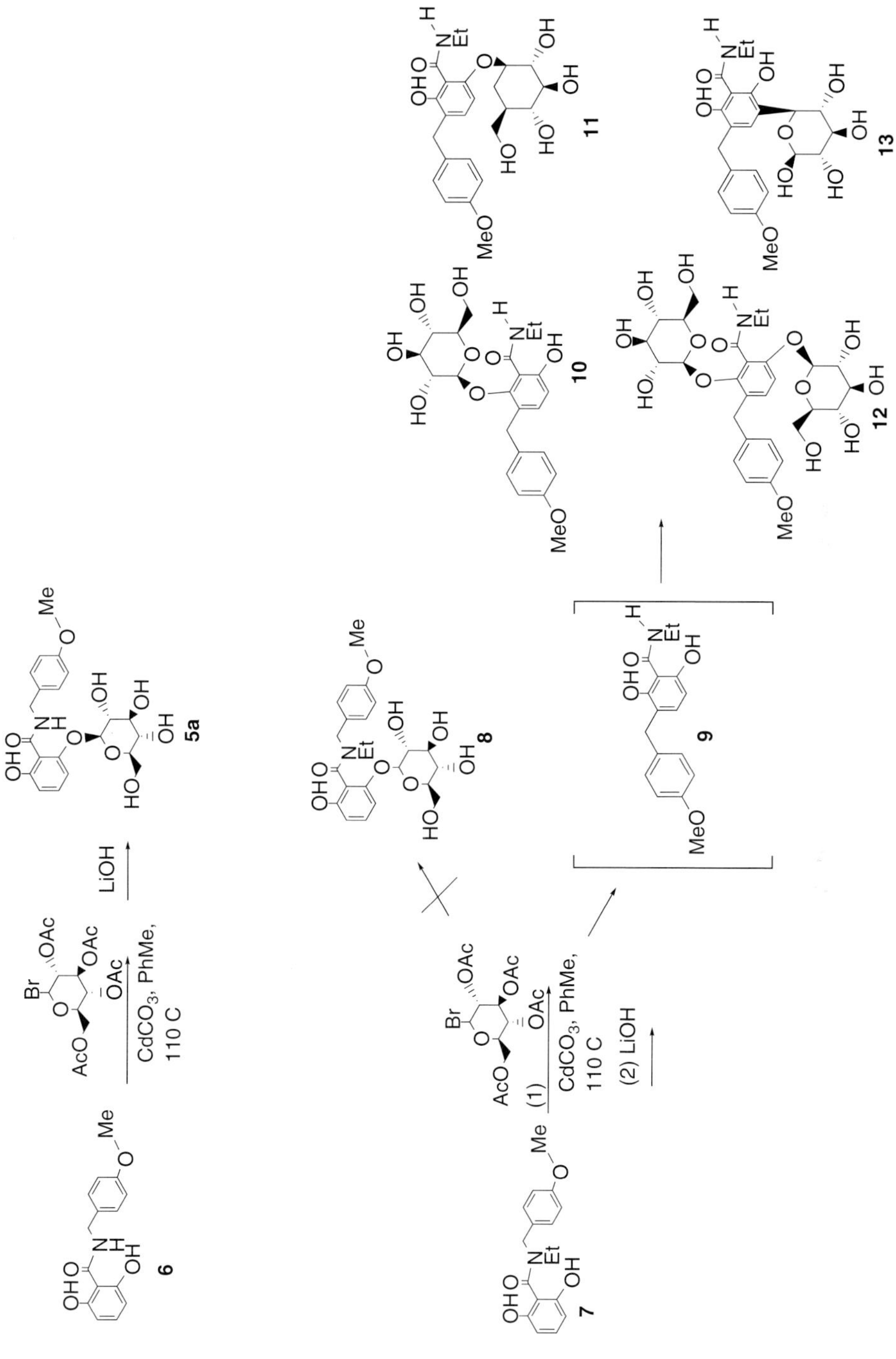

Scheme 5.1 Glycosylated products obtained with 2,6-dihydroxybenzamides.

2a SGLT2 potency had decreased eightfold but selectivity versus SGLT1 appeared somewhat greater (Table 5.1). Unlike **2a**, **5a** was stable to glucosidase mediated cleavage when incubated for 10 min with tissue homogenates prepared from rat intestinal tract, rat kidney, and cynomolgus monkey liver (Table 5.2).

Encouraged by this finding, the tertiary phenolic benzamide **7** was subjected to the same reaction conditions to ascertain if the O-glucoside **8** would exhibit improved potency/selectivity. Analysis of the chromatographic fractions revealed no evidence of formation of the desired product **8**; however, four products, isolated in ~1% yield, arose from glucosylation of aglycone **9**. Subsequent control studies revealed **7** was unstable when heated with catalytic HBr and was converted to **9** in poor yield. O-Glucosylation of **9** was the predominant reaction path generating two isomeric O-glucosides **10** and **11** as well as the bis-O-glucoside **12** (Scheme 5.1). *In vitro* profiling established that these three O-glucosides offered no advantage over **2a** because SGLT2 EC_{50} for each was >5000 nM. The C-glucoside structure assignment for the fourth product **13** was

Table 5.1 SGLT SAR for O-glucosides of *ortho* hydroxybenzamides.

Compound No.	X	R	SGLT2 EC_{50} (nM)	SGLT1 EC_{50} (nM)
2a	—	—	90	860
5a	6'-OH	4-OMe	750	>10 000
5b	6'-OH	H	2200	>10 000
5c	6'-OH	2-OMe	>10 000	>10 000
5d	6'-OH	3-OMe	1300	>10 000
5e	6'-OH	4-Me	2200	~10 000
5f	6'-OH	4-Cl	1100	>10 000
5g	6'-OH	4-Et	330	>10 000
5h	6'-OH	4-*t*-Bu	4500	>10 000
5i	6'-OH	4-NMe$_2$	1800	—
5j	6'-OH	4-Ph	680	>10 000
5k	6'-OH	4-Aza	>10 000	—
5l	H	4-OMe	720	>10 000
5m	6'-OMe	4-OMe	>10 000	—
5n	3'-OMe	4-OMe	>10 000	—
5o	4'-OMe	4-OMe	500	—
5p	5'-OMe	4-OMe	>10 000	—

Table 5.2 Susceptibility of selected β-O-glucosides to glucosidase cleavage.

Compound	Glucosidase rat gut (nMol min^{-1} mg^{-1})	Glucosidase rat kidney (nMol min^{-1} mg^{-1})	Glucosidase cyno liver (nMol min^{-1} mg^{-1})	Rat hepatic microsomal oxidation
2a	0.3	0.01	0.04	—
5a	Stable	Stable	Stable	—
5l	Stable	Stable	Stable	—
15	0.8	1.6	1.6	—
18a	Stable	0.3	1.9	—
18f	Stable	0.6	1.9	2.8
18q	Stable	0.9	1.0	—
18s	Stable	Stable	Stable	0.02
18v	ND	0.01	0.03	0.14
18w	ND	0.02	0.2	>0.8
18x	Stable	Stable	Stable	>2.2

unexpected. The fact that **13** exhibited modest SGLT2 potency, 1300 nM SGLT2 EC$_{50}$ with 30-fold selectivity versus SGLT1 was somewhat surprising as examples of *ortho* C-glucosides disclosed earlier exhibited very weak activity [22]. At the time the relative contribution of the *meta* C-glucoside versus the atypical polar substituents to the activity of **13** was unknown. Moreover, synthetic efforts were unsuccessful to generate sufficient quantities of **13** required for the *in vitro/in vivo* characterization to resolve this issue.

Further pursuit of **13** was suspended while the SAR of O-glucosides of phenolic benzamides was delineated. Two alternative routes were employed to prepare these O-glucosides. For screening a library of amine components, a 1-Hydroxy-7-azabenzotriazole/1-Ethyl-3-(3-dimethylaminopropyl)carbodiimide (HOAt/EDC) mediated coupling of the glucosylated hydroxyl benzoic acid **14** with amines provided an efficient means to survey a variety of benzamides. Using this route the tertiary amide **7** was prepared and found to lack SGLT2 activity. For preparative scale, O-glucosylation of the appropriate aglycone analogs of **6** with acetobromoglucose proceeded cleanly at 5 °C in ~55% when the reaction was run in presence of Ag$_2$O using 2,6-lutidine or quinoline as the solvent.

The resulting relatively flat SAR is summarized in Table 5.1. Only a secondary benzamide spacer is compatible with SGLT2 activity; the tertiary amide counterpart is not tolerated. A methoxyl probe was utilized to assess the response to substitution of each ring. For the distal ring the activity loss of **5c** relative to the parent **5b** established that *ortho* substitution was not tolerated whereas *meta* substitution **5d** provided little benefit but was tolerated. Although small *para* substituents typically increased SGLT2 affinity versus the parent **5b** two- to threefold, a 7× increase over **5b** to 330 nM EC$_{50}$ conferred by a *para* ethyl **5g** was the best achieved with this series. With respect to the benzamide ring, the near equivalence of **5a** and **5l** suggested that the C6'-hydroxyl is not important; moreover larger C6' substituents, see **5m**, appear incompatible. Elsewhere a methoxyl was tolerated at

C4' **5o** but abolished all activity at C3' **5n** or C5' **5p**. Given the poor prognosis for increasing potency of these benzamide phenolic O-glucosides and the failure to observe significant glucosuria after oral administration to Swiss Webster mice, further investigation ceased. The lasting impact from this approach was the unexpected formation and activity of compound **13**, which proved to be an essential lead that led to the discovery of C-glucosides such as dapagliflozin and analogs.

5.3.2
Benzylpyrazolone O-Glucosides

The report from Wyeth that benzylpyrazolones such as **3** induced glucosuria in mice following oral administration prompted further investigation at Bristol-Myers Squibb [19]. This activity was suggestive of a SGLT inhibitor; however, the structure was atypical of any disclosed SGLT inhibitor. Oral administration of **3** produced glucosuria only in mice but not in rats. Moreover, **3** was devoid of activity in either the SGLT1 or SGLT2 *in vitro* assays. We hypothesized that the active agent was a glycosylated metabolite of **3**. Characterization revealed the O-glucoside **15** to be a potent SGLT2 inhibitor (22 nM SGLT2 EC_{50} with 15-fold selectivity versus SGLT1) whereas the glucuronide **16** was devoid of any activity versus both transporters. When profiled versus the panel of glucosidases, **15** was an excellent substrate that was rapidly cleaved by all (see Table 5.2). Oral administration of **15** to rodents produced the same species-dependent response observed with **3** – glucosuria in mice but not in rats. Analysis of both blood and urine following oral administration of **3** or **15** to mice and rats revealed only glucuronide **16** in rat urine but a mixture of **15** and **16** in mouse urine. Moreover, the concentration of **15** in rat blood was low and rapidly cleared. Given the susceptibility of **15** to glucosidase cleavage, rapid conversion to **3** by both rodents was not surprising; however, mice are known to both glucosylate and glucuronidate xenobiotics whereas rats only glucuronidate thereby accounting for the species difference in glucosuric response. Because glucosylation is a rarely utilized metabolic pathway in humans, these results discouraged further pursuit of the pyrazalone O-glucosides as SGLT2 inhibitors at Bristol-Myers Squibb (Figure 5.3).

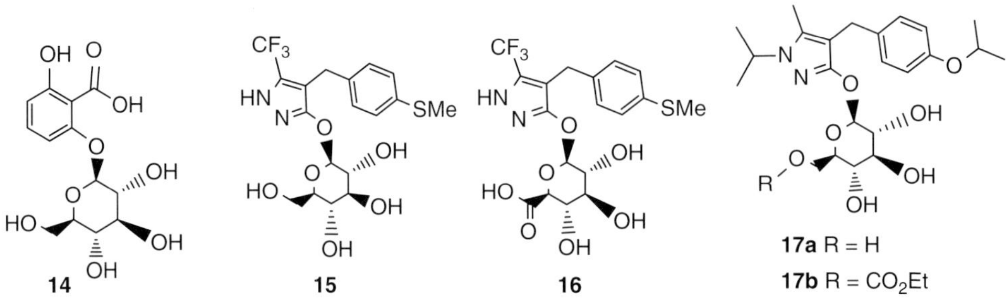

Figure 5.3 Pyrazolone O-glucosides.

Unbeknownst to the Bristol-Myers Squibb team, and several years before the effort at Bristol-Myers Squibb, the chemists at Kissei had reached a similar conclusion regarding the mechanism by which glucosuria was induced by these pyrazolones [23]. Moreover, their persistence culminated with identification of the pyrazolone O-glucoside SGLT2 inhibitor remogliflozin **17a** which entered clinical trials as remogliflozin etabonate **17b** [24].

The finding that the O-glucoside **15** and not the pyrazolone **3** was the glucosuric agent provided reassurance that the presence of a glucose moiety to serve as a targeting vector was an essential component of the active pharmacophore. Comparison of **2a** and **15** revealed structural similarities. In both cases the active agent was an O-glucoside of a hydroxylated vicinally substituted central planar ring linked via a spacer to a distal aryl ring. Because a similar spatial orientation of the distal ring could be achieved with either a one or three atom tether, these elements appeared to define a common pharmacophore required for SGLT2 inhibition. To confirm this hypothetical pharmacophore, the two structures were merged to generate O-glucosides of *ortho* benzylphenols **18**.

5.3.3
o-Benzylphenol O-Glucosides

Characterization of **18a** and analogs confirmed this expectation. The SAR, summarized in Table 5.3, was most encouraging as *in vitro* potency (<10 nM SGLT2 EC_{50}) and selectivity (>100-fold) were superior to that of any reported series of O-glucosides. *Para* substitution of the distal ring with a methoxyl **18d** or methyl **18f** or chlorine **18l** enhanced SGLT2 potency four- to fivefold over that of the parent **18a**. In contrast, a *meta* methoxyl **18c** or *meta* methyl **18e** decreased potency three- and sevenfold, respectively. The 10–15-fold loss of potency upon *ortho* substitution, regardless of methoxy **18b**, hydroxyl **18h**, or methyl **18g**, was suggestive of an unfavorable steric interaction. Both the steric and polar properties of a *para* substituent modulate SGLT2 potency. Two lipophilic substituents, *p*-ethyl **18i** or *p*-phenyl **18q**, produced 40-fold affinity increase to 1–2 nM EC_{50}. The binding pocket must be deep as potency is maintained even for large extended substituents such as styrenyl **18r**, although incorporation of branched groups such as *i*-propyl **18j** or *t*-butyl **18k** became progressively disfavored. Para hydroxylation **18s** did not reduce affinity relative to hydrogen; however, larger polar (**18m** or **18n**) or especially charged *para* substituents (compare **18o**–**18p**) decreased potency as much as 20-fold. No increase in SGLT2 potency was observed upon methylating any of the four open positions of the central aryl ring of **18a**. Methylation of C3′ **18t** or C5′ **18v** did not alter SGLT2 potency significantly; however, SGLT1 affinity was increased nine- and fourfold, respectively, resulting in a commensurate loss of selectivity. Methylation at C4′ **18u** or C6′ **18w** reduced SGLT2 potency 20–50-fold; larger substituents abolished all activity.

Initial glucosuric studies in mice with the lead compound **18f** appeared encouraging as robust efficacy was maintained and was not dependent on the administration route. The ED_{50}s measured for the decrease in plasma glucose

Table 5.3 SGLT SAR for O-glucosides of *ortho* benzylphenols.

Compound No.	X	R	SGLT2 EC$_{50}$ (nM)	SGLT1 EC$_{50}$ (nM)	Selectivity versus SGLT1
18a	H	H	41	2400	60
18b	H	2-OMe	570	>10 000	—
18c	H	3-OMe	136	>10 000	—
18d	H	4-OMe	10	>10 000	—
18e	H	3-Me	263	>10 000	—
18f	H	4-Me	8	2400	300
18g	H	2,4-Me$_2$	71	>10 000	—
18h	H	2-OH	590	>10 000	—
18i	H	4-Et	2	585	290
18j	H	4-*i*-Pr	7	>10 000	—
18K	H	4-*t*-Bu	65	>10 000	—
18l	H	4-Cl	20	>10 000	—
18m	H	4-MeSO$_2$	114	—	—
18n	H	4-OCH$_2$CONEt$_2$	885	—	—
18o	H	4-OCH$_2$CO$_2$H	1020	>10 000	—
18p	H	4-OCH$_2$CO$_2$Me	220	—	—
18q	H	4-Ph	1	1700	1700
18r	H	4-CH=CHPh	18	2700	150
18s	H	4-OH	52	—	—
18t	3-Me	H	63	270	4
18u	4-Me	H	~2000	—	—
18v	5-Me	H	33	500	15
18w	6-Me	H	710	>10 000	—
18x	3-Me	4-Me	10	250	25
18y	6-Me	4-Me	49	>10 000	>200

(0.3 and 1 mg kg^{-1}, respectively) for Swiss Webster mice given a glucose challenge before iv and oral administration of **18f** were within threefold of the ED$_{50}$s for glucosuria (0.15 and 0.4 mg kg^{-1}, respectively). In addition, the glucosuric response of *db/db* mice to a 2 mg kg^{-1} oral dose of **18f** was sufficient to reduce plasma glucose after 60 min. Results were very encouraging from a 4 week study with 3-month old *ob/ob* mice eating food containing 0.05% **18f** (estimated daily dosage of 20 mg kg^{-1} day^{-1}). The maximum decrease of 50–55% for fasting plasma glucose values relative to the control group (i.e., a drop from 560 to 280–290 mg dl^{-1}) observed on week 2 was maintained at week 4.

Screening of selected members of these *o*-benzylphenolic O-glucosides including **18a**, **18f**, and **18q** versus the panel of glucosidases revealed that for each the stability observed during incubation with rat intestinal homogenates did not extend to liver and kidney homogenates (Table 5.2). Concerns that **18f** might be hydrolyzed and reglycosylated in mice after absorption were confirmed when administration of the inactive *o*-benzylphenol aglycone **19** produced a comparable dose-dependent glucosuric response in mice. This finding established that the robust murine glucosuric response of **18f** reflected the ability of mice to repeatedly reglucosylate **19** just as was observed with pyrazolone **3**. Subsequent studies revealed that **18f** produced glucosuric responses in rats and monkeys that were respectively 10% and 1% of that measured for mice. We attributed these species-dependent differences to the inability of reglucosylation to occur in rats and monkeys. These findings greatly diminished our interest in **18f** beacuse we anticipated that in humans **18f** would be metabolized in a manner similar to that observed in rats and monkeys resulting in a weak glucosuric response.

This species-dependent susceptibility of **18f** and analogs to glucosidase mediated degradation was a major impediment to advancement of these O-glucosides. This realization prompted (i) replacement of mice with rats for *in vivo* screening and (ii) an effort to stericly retard enzymatic hydrolysis by introduction of substituents at C3′ or C6′ of the central aryl ring. The hope was that structural modifications would extend the exposure duration in humans such that these agents could be employed as short-acting agents to attenuate the postprandial glucose excursions. As expected, C3′ methylation of **18f** to generate **18x** maintained the 10 nM SGLT2 potency of **18f** while reducing SGLT1 selectivity 25-fold. In contrast **18y** arising from C6′ methylation of **18f** exhibited reduced SGLT2 potency (49 nM) but maintained >200-fold SGLT1 selectivity. In both cases the glucoside hydrolysis rates for the two tissue glucosidases were reduced by >20-fold. A similar decrease in susceptibility to glucosidases was observed for a number of counterparts of **18x** and **18y** containing a distal ring bearing alternative preferred *para* substituents. Oral administration of these hydrolysis resistant O-glucosides at 1 mg kg^{-1} to STZ rats caused an enhanced glucosuric response with longer duration relative to that from **18f** resulting in a 20–35% reduction in plasma glucose levels at 4 h post dose.

5.4
m-Diarylmethane C-Glucosides

The evolution to *o*-benzylphenolic O-glucosides **18** not only provided a promising new series of SGLT2 inhibitors for evaluation but also supplied the SAR understanding necessary to utilize the *meta* substituted C-glucoside side-product **13**. The expectation was that SGLT2 potency for **13** and **18** should exhibit a similar

SAR dependence because the two diarylmethane-containing glucosides are structurally similar. Without the impetus provided by **13**, no rationale would exist for preparation of the *meta* C-glucosides **20** because the weak SGLT2 activity of the few examples of *ortho* C-glucosides were not encouraging. This realization that the polar substituents of **13** should strongly disfavor SGLT2 potency redirected the synthetic effort to preparation of **20a**, the *meta* C-glucoside counterpart of **18f**. The direction of the program changed with the finding that **20a** was a potent SGLT2 inhibitor (EC$_{50}$ 22 nM) exhibiting high selectivity versus SGLT1 (>600×) and no inhibition versus facilitative glucose transporter (GLUT) 1 and GLUT 4 at 10 µM. Although both chemotypes were pursued concurrently, within a few months all effort became focused on the new C-glucoside series **20** due to their greater *in vitro* potency and selectivity and especially the ability to elicit greater plasma glucose reductions for longer duration in STZ rats than even the glucosidase resistant O-glucosides **18x** and **18y**.

5.4.1
Synthetic Route

The current synthetic route portrayed in Scheme 5.2 was the result of many process improvements over the initial route [25, 26]. In the original route benzyl ether protecting groups were employed throughout the synthesis and were removed by hydrogenolysis in the final step. As a consequence all intermediates were syrups and the final C-aryl glucoside an amorphous solid. In the current route trimethylsilyl ethers are employed during the organolithium addition. Before reduction of the crude O-methyl glucoside, the material is peracetylated. Under the specified conditions the Et$_3$SiH reduction yields almost exclusively the crystalline tetraacetoxy beta C-glucoside thereby enabling purification before alkaline hydrolysis to generate the final product as an amorphous solid. In contrast when benzyl ether protecting groups were employed, the α/β ratio was

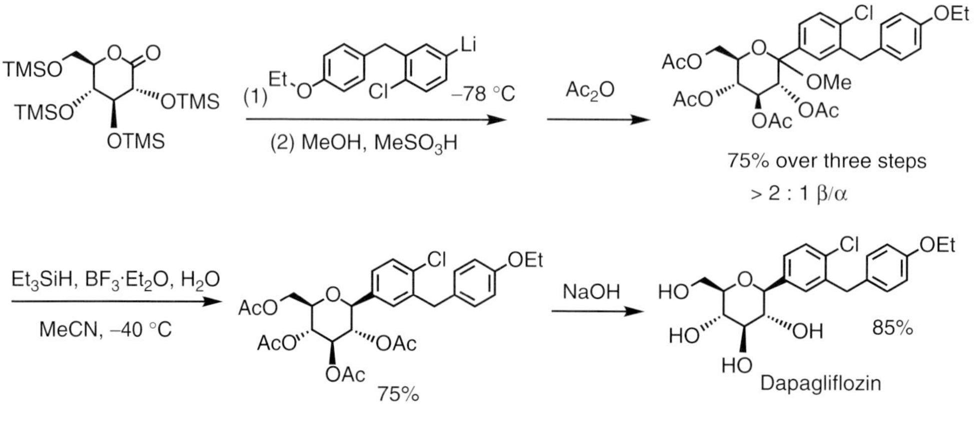

Scheme 5.2 Synthetic route employed for C-glucosides **30a–30af**.

much less favorable [27]. To date, all attempts to induce crystallization of these C-aryl glucosides have failed. However, after many months the process group found that these C-glucosides will cocrystallize in the presence of L-phenylalanine or L-proline or in some cases with diols to form stable 1 : 1 complexes which can be purified to homogeneity by repeated crystallizations.

5.4.2
Early SAR of C-Glucoside Based SGLT2 Inhibitors

As expected, the SAR shown in Table 5.4 for distal ring substitution of the *meta* C-aryl glucosides **20** was similar to that of *o*-benzylphenolic O-glucosides. Relative to the unsubstituted parent **20b**, SGLT2 potency increased 10–20-fold for small *para* alkyl substituents such as methyl **20a**, ethyl **20c**, and propyl **20d**; potency was

Table 5.4 SGLT SAR for C-glucosides of distal substituted diarylmethanes.

Compound No.	R	SGLT2 EC$_{50}$ (nM)	SGLT1 EC$_{50}$ (nM)
20a	4-Me	22	>10 000
20b	H	190	>10 000
20c	4-Et	10	>10 000
20d	4-Pr	14	—
20e	4-Bu	420	—
20f	4-Ph	110	—
20g	4-OMe	12	>10 000
20h	4-SMe	6	>10 000
20i	4-OCHF$_2$	25	>10 000
20j	4-OCF$_3$	35	>10 000
20k	4-C(O)Me	27	>10 000
20l	4-CH(OH)Me	62	>10 000
20m	4-CO$_2$H	340	—
20n	4-CF$_3$	64	>10 000
20o	4-S(O)Me	410	—
20p	4-S(O)$_2$Et	220	—
20q	4-NHSO$_2$Me	270	—
20r	4-OH	73	>10 000
20s	3-OMe	230	—
20t	3-Me	510	—
20u	2-Et	4000	—

comparable or slightly reduced relative to hydrogen for a larger alkyl substituents such as butyl **20e** or phenyl **20f**. Other small lipophilic groups such methoxyl **20g** or methylthio **20h** conferred high potency of ~10 nM; however, EC$_{50}$ for the fluorinated ethers HCF$_2$O **20i** or CF$_3$O **20j** increased to 25 and 35 nM, respectively. Oxidation of the alkyl moiety was not beneficial; the EC$_{50}$ progressively increased from 27 nM for acetyl **20k** to 62 nM for hydroxymethyl **20l** to 340 nM for CO$_2$H **20m**. Incorporation of polar groups such as sulfoxide **20o**, sulfone **20p**, or sulfonamide **20q** reduced potency ~twofold; however, *para* hydroxylation **20r** increased affinity threefold. *Meta* substitution of this ring was not advantageous but was tolerated. Potency relative to hydrogen was unchanged for methoxyl **20s** and decreased threefold for methyl **20t**. The 20-fold decrease in potency observed for the *o*-ethyl **20u** indicated that *ortho* substitution was unfavorable. The preference for beta linked aglycone is very strong for these C-glucosides as manifested by a 20-fold decrease in potency observed for the α-anomer **21c** compared to its β-anomer **20c**.

In summary, for this series a *para* substitution is strongly preferred for the distal ring. The SGLT2 binding pocket will accommodate any *para* substituent regardless of size, charge, or polarity without significant loss of potency; however, lipophilic groups of a defined size comparable to ethyl typically produce potency increases of 10–20-fold. All of these analogs of **20** exhibited such weak SGLT1 affinity (EC$_{50}$ > 10 000 nM in each instance) that SGLT1 selectivity was >1000. When compared to the *o*-benzylphenol glucosides **18**, both SGLT2 potency and SGLT1 selectivity was 10-fold greater for **20**. Compound **20c** was particularly noteworthy as the *in vivo* efficacy that **20c** exhibited in all the rodent models was superior to that of any O-glucoside or other analog of **20**. Oral administration of **20c** at 0.1 and 1 mg kg^{-1} reduced glycemic levels of STZ rats 17% and 57%, respectively, at 5 h post dose. Compound **20c** was eventually replaced as the lead compound once the benefits conferred by central ring substitution became apparent.

To ascertain if only the spatial presentation of the sugar and benzyl moieties of **20** was conducive to high SGLT2 potency, alternative presentations were systematically evaluated [25]. The assumption was that high affinity SGLT2 antagonists required the distal aryl ring to bear a lipophilic substituent properly oriented so that the substituent would extend into a favorable binding pocket. Therefore, to fully assess the consequences of eliminating or extending the methylene spacer, three derivatives of each chemotype were prepared in which the distal ring was either unsubstituted or was substituted with a *m*-methyl or *p*-methyl. If the distal phenyl was unsubstituted, binding affinity was reduced ~threefold when compared to **20b** regardless whether the spacer was deleted as in **21a** or elongated to either two **22a** or three methylenes **23a** (Table 5.5). These findings suggest that the SGLT2 binding site can accommodate several orientations for an unsubstituted phenyl but none confer high affinity. In all cases the *meta* methyl or a *para* methyl was not favorably oriented compared to **20a** resulting in steric clashes reducing affinity by 44–55-fold and 13–29-fold, respectively. The fact that a similar spacer induced response had been previously observed for benzylphenolic O-glucoside

5.4 m-Diarylmethane C-Glucosides

Table 5.5 Modulation of SGLT2 potency by spatial orientation of the C-glucoside aglycone.

Compound	A	R	SGLT2 EC$_{50}$ (nM)	Reference glucoside	Fold decrease in SGLT2 EC$_{50}$ from reference glucoside
21a	Bond	H	620	20b	3
21b	Bond	3-Me	1200	20a	55
21c	Bond	4-Me	286	20a	13
22a	CH$_2$CH$_2$	H	710	20b	4
22b	CH$_2$CH$_2$	3-Me	970	20a	44
22c	CH$_2$CH$_2$	4-Me	430	20a	20
23a	CH$_2$CH$_2$CH$_2$	H	480	20b	3
23b	CH$_2$CH$_2$CH$_2$	3-Me	1200	20a	55
23c	CH$_2$CH$_2$CH$_2$	4-Me	630	20a	29
24	S	4-Me	69	20a	3
25	O	4-Me	540	20a	22
26a	—	H	600	20b	3
26b	—	4-Et	>4000	20c	>400
27a	—	H	1250	20b	6
27b	—	4-Et	2300	20c	230
28a	—	H	1100	18a	27
28b	—	4-Me	500	18f	62

analogs **18** underscored the need to properly orient the substituted distal ring to achieve high affinity. This conclusion is further bolstered by the finding that replacement of the methylene spacer of **20b** with a sulfur atom **24** reduced binding affinity threefold whereas an oxygen spacer **25** induced a 25-fold reduction.

We also investigated whether the high affinity of C-arylglucosides such as **20c** was primarily due to the *meta* substituted diarylmethane glucoside or to proper orientation of the *para* ethyl substituent enabling this substituent to fit well into a binding pocket. Comparison of the unsubstituted *ortho* and *para* glucosides **26a** and **27a** to **20b** revealed *ortho* substitution reduced affinity threefold versus sixfold for *para*. However, when the distal phenyl contained a *para* ethyl substituent, the decrease in affinity for the *ortho* and *para* C-glucosides **26b** and **27b** relative to **20c** became >200-fold. We conclude that the SGLT2 binding site is rather promiscuous readily accommodating C-glucosides of two phenyls joined by a variety of linkers. However, only *meta* C-glucosides of diarylmethanes orient the distal ring *para* substituent to achieve a maximal binding interaction with a binding pocket.

Comparison of the two *ortho* benzyl glucosides – the unsubstituted **28a** and the *para* methyl **28b** – to their *o*-phenolic O-glucoside counterparts **18a** and **18f** revealed a 25–60-fold potency loss upon conversion of the *o*-phenolic O-glucosides to the benzyl counterpart. The lack of a substituent effect suggests that the aryl rings could not assume a conformation conducive to high affinity binding. This hypothesis is supported by a computation study which revealed that **18a** and **20b** have an energetically accessible conformation in common that is not available to **28a**. These findings suggest that an oxygen anomeric effect is the dominant influence determining the conformations accessible to the o-benzylphenolic O-glucosides **18**. Removal of this effect upon isosteric replacement of an oxygen with a methylene resulted in the aryl rings of **28** no longer being oriented to maximize SGLT2 binding energy.

5.4.3
Identification of Dapagliflozin

The impact on potency and selectivity upon substitution at the four open central ring positions of **20c** was assessed with a methyl probe (Table 5.6). C2′ methylation **29** decreased SGL2 potency 10-fold. In contrast C4′ methylation **30a** increased both SGLT2 and SGLT1 potency ~15-fold relative to that measured for **20c**. Methylation at C5′ **31** or C6′ **32** produced minor SGLT2 potency increases of two- and fivefold respectively without altering SGLT1 affinity. Given that substitution of C4′ and C4 produced the most favorable responses, the program focus became delineation of the SAR for various lipophilic substituents at those two positions of **30**. The choice of substituents was guided by the experimental observation that maximum *in vivo* efficacy was obtained with SGLT2 inhibitors which were ~90–95% bound to serum proteins. Moreover for the SGLT2 inhibitors comprising chemotypes **18** and **20**, there appeared to be a linear relationship between % free drug level and *c*Log*P* such that 5–10% of the drug was unbound if *c*Log*P* was ~2.0.

Analogs of **30** were designed with *c*Log*P* ranging from 1.5 to 2.5 by utilizing different combinations of substituents at C4 and C4′ in hopes of maximizing *in vivo* efficacy at 5 h post dose. This selection was subject to several constraints:

5.4 m-Diarylmethane C-Glucosides

Table 5.6 SGLT SAR for central ring methylation of C-aryl glucoside **20c**.

Compound	R	SGLT2 EC$_{50}$ (nM)	SGLT1 EC$_{50}$ (nM)	% decrease in blood glucose of STZ rats at 5 h post 0.1 mg kg^{-1} p.o. dose after vehicle correction
29	2-Me	98	>8000	—
30a	4-Me	0.6	600	41
31	5-Me	5.6	>4000	17
32	6-Me	2.7	>8000	28

To maximize *in vitro* SGLT2 potency the preferred C4 substituents for the distal ring were methyl, chloro, fluoro, ethoxy, ethyl, methoxy, or methylthio. SGLT2 potency was greatest if fluoro, methyl, or chloro were attached at C4′; employment of groups larger than methoxy or ethyl greatly reduced SGLT2 activity. Although SGLT1 potency did not appear influenced by these C4 distal ring substituents, the C4′ substituent markedly modulated SGLT1 potency and, as a consequence, selectivity versus SGLT1. If C4′ were hydrogen or fluorine, SGLT1 binding was least favored resulting in ∼2000-fold selectivity versus SGLT1; if methyl or chlorine, the SGLT1 potency increase caused selectivity to drop to ∼700–1000-fold; if methoxy or ethyl, selectivity decreased to ∼20-fold.

With these constraints ∼30 compounds were identified for which SGLT2 potency was <5 nM (Table 5.7). The fact that the SGLT2 potency for all of these compounds plateaued at ∼1–3 nM suggested that the capacity of the cell based assay to distinguish among these highly potent compounds had been exceeded. Lead compound selection was based on the decrease in blood glucose levels measured at 5 h after oral administration of 0.1 mg kg^{-1} drug to food restricted diabetic STZ rats (fasting glucose 580 mg dl^{-1}). These values, after correction for the glycemic decrease of the vehicle treated cohort, are summarized in Table 5.7. Efficacy achieved with most far exceeded the 17% glycemic reduction obtained with 0.1 mg kg^{-1} of **20c**. Modest *in vivo* responses due to low bioavailability were consistently obtained if the distal ring was substituted with a *para* methyl or *para* chloro such as **30i** and **30j**. These were removed before data analysis. For the remaining compounds with SGLT2 EC$_{50}$ < 5 nM, a plot of cLogP versus % decrease in blood glucose at 5 h suggests efficacy tracks cLogP until plateauing for cLogP values of ∼2 and then appears to decline as the agents became more lipophilic (Figure 5.4).

Table 5.7 In vitro and in vivo SAR for C4 and C4′ disubstituted diarylmethane C-glucoside SGLT2 inhibitors **30a–30af**.

Compound	R	cLogP	SGLT2 EC$_{50}$ (nM)	SGLT1 EC$_{50}$ (nM)	% decrease in blood glucose of STZ rats at 5 h post 0.1 mg kg^{-1} p.o. dose after vehicle correction
X = Me					
30a	Et	2.4	0.6	600	41
30b	OMe	1.3	1	540	35
30c	SO$_2$Me	−0.3	14	>10 000	—
30d	OH	0.7	2	766	31
30e	H	1.3	9	1300	19
30f	OCHF$_2$	1.7	1.4	1700	42
30g	SMe	1.9	1	790	47
30h	F	1.5	15	—	11
30i	Cl	2.1	2	—	25
30j	Me	1.8	2.5	—	17
30k	OEt	1.8	2	890	37
30l	Ac	0.8	3	1500	32
30m	n-Pr	3.0	1.1	—	24
X = Cl					
30n	OMe	1.5	0.8	725	38
30o	SMe	2.1	1.3	530	39
30p	SO$_2$Me	0.0	31	>10 000	9
30q	OH	0.9	3	1450	25
30r	Ac	1.0	4	1000	34
30s	OEt	2.0	1.1	1500	47
30t	CN	1.0	13	6000	19
30u	Me	2.0	2.1	—	32
30v	Et	2.6	1.2	—	47
30w	n-Pr	3.1	3.2	—	10
30x	Cl	2.2	1.6	—	10
X = F					
30y	Et	2.1	3	>10 000	35
30z	OMe	1	4	>10 000	28
30aa	SMe	1.6	1.8	>10 000	29
30ab	OH	0.6	19	>10 000	21
30ac	Et	2.1	3	>10 000	35
30ad	OEt	1.5	5	>10 000	37
30ae	Me	1.6	2.7	—	20
30af	Cl	1.9	2.1	—	14

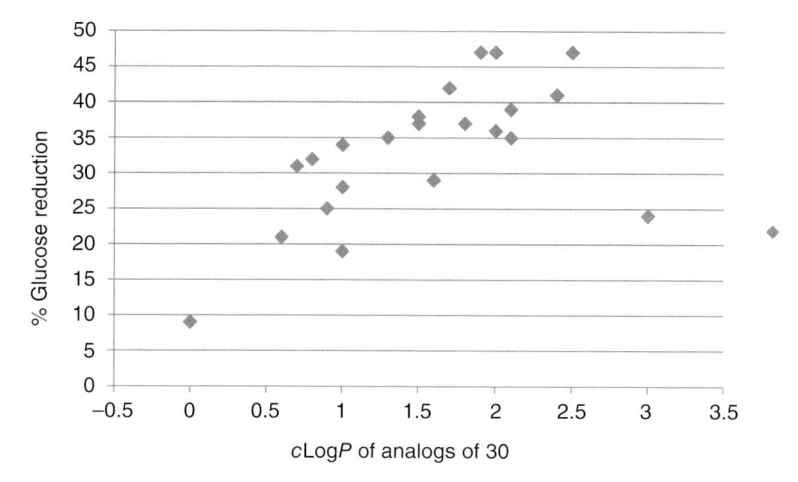

Figure 5.4 Correlation of cLogP with % reduction of blood glucose after correction for the vehicle response at 5 h after oral administration of 0.1 mg kg^{-1} dose of SGLT2 inhibitors **30a–30af** to food restricted STZ rat.

Profiles of three compounds **30g**, **30s**, and **30v** in the STZ rat assay were particularly promising as each not only reduced blood glucose levels 47% when administered at 0.1 mg kg^{-1} in the STZ assay but also exhibited dose-dependent efficacy at doses as low as 0.01 mg kg^{-1}. Upon evaluation in 24 h acute studies using ZDF rats, **30g**, **30s**, and **30v** produced comparable robust glycemic reductions that were maintained throughout a subsequent 2 week subchronic study. **30s** ultimately replaced the more advanced lead **30g** after **30g** was found to exhibit low but unacceptable oxidative instability when formulated with typical pharmaceutical excipients. Compound **30s** began clinical development as BMS-512148; however, it later became known by the tradename dapagliflozin [28]. The back-up for dapagliflozin **30v** entered clinical trials as BMS-655956. The uneventful progression of dapagliflozin through Phase 2 and Phase 3 studies caused progression of **30v** to be halted after completion of a single ascending dose (SAD) study.

5.5
Profiling Studies with Dapagliflozin

Pharmacological findings obtained using modestly glucosuric ZDF diabetic rats supported advancement of dapagliflozin **30s**. Further *in vitro* characterization of dapagliflozin revealed it to be a reversible potent SGLT2 inhibitor (Ki = 0.55 nM) with high selectivity versus all other SGLT transporters [29]. Dose-dependent glucosuria was observed following oral administration of **30s** at 0.01, 0.1, 1, and 10 mg kg^{-1} to ZDF rats that had been fed but were food restricted during the 6 h study. The amount of glucose spilt in urine per 400 g of body weight increased from 0.24 g for vehicle treated control to 2.2, 4.1, 4.2, and 4.2, respectively, accompanied by as much as a 10-fold increase in urine volume. Fortunately similar increases in urine output were not observed in clinical trials possibly because, unlike rats, the osmotic strength of human urine is sufficiently low that the additional glucose could be accommodated without a significant compensatory volume increase. Plasma glycemic levels of the ZDF rats receiving 1 and 10 mg kg^{-1} of dapa were effectively normalized at 6 h post dose; in addition the 0.1 mg kg^{-1} dose nearly achieved normalization (Figure 5.5). At 24 h glycemic reductions ranging from 20% to 60% were maintained for doses except the 0.01 mg kg^{-1} dose. A subsequent 2 week ZDF study with **30s** administered qd to ZDF rats for 2 weeks at 0.01, 0.1, and 1 mg kg^{-1} confirmed that dose-dependent plasma glucose reductions were maintained throughout the study under fed or fasting conditions (Figure 5.6). The animals exhibited no evidence of hypoglycemia even when severely fasted. Upon completion, clamp studies established that the glucose disposal rate had increased and that hepatic glucose output had decreased consistent with amelioration of the profound insulin resistance before treatment. Pancreatic staining revealed an increase in both insulin content and beta cell mass suggestive of beta cell preservation. Despite being profoundly glucosuric, no weight loss was observed during this study, a finding attributed to the increased food consumption of these ad lib fed hyperphagic animals.

In a subsequent study, the ability of **30s** to cause weight loss for ad lib fed diet induced obese (DIO) Sprague-Dawley rats was assessed over 4 weeks when administered po at 0.5, 1, and 5 mg kg^{-1} qd [30]. An additional cohort administered 5 mg kg^{-1} was pair fed to the vehicle control. All ad lib fed treated cohorts increased their water and food intake; however, the caloric increase did not fully compensate for the calories lost as glucosuria (Table 5.8). The decrease in respiratory quotient (RQ) indicated a metabolic shift incurred entailing increased reliance on fatty acids as an alternative energy source thereby accounting for the weight loss being primarily due to loss of fat. It is noteworthy that comparison

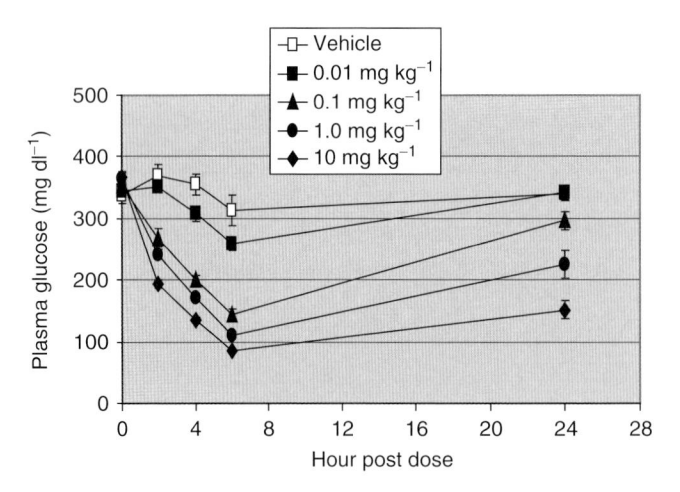

Figure 5.5 Dose-dependent reductions in plasma glucose of ZDF rats in an acute 24 h study after a single oral dose of dapagliflozin **30s**.

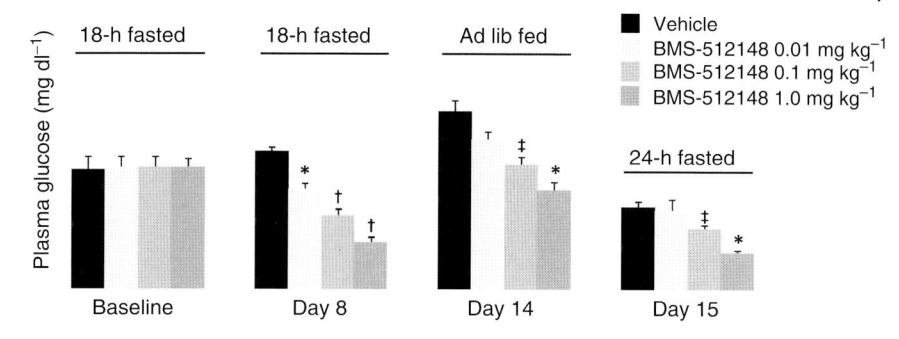

Figure 5.6 Dose-dependent reductions in plasma glucose of ZDF rats during a 2 week study with dapagliflozin **30s** administered daily.

Table 5.8 Four week weight loss study with dapagliflozin and diet induced obese (DIO) Sprague-Dawley rats.

Properties	Vehicle	Dapagliflozin treated cohorts (mg kg^{-1})			
		0.5	1.0	5	5 (pair fed to vehicle)
Decrease in body weight from day 1 (%)	0.5	4.3	4.1	6.1	9.9
Increase in total H$_2$O intake versus vehicle (%)	—	50	77	94	82
Increase in total kcal consumed versus vehicle (%)	—	7	14	14	0
Decrease in adipose mass from day 1 (%)	2.5	4.1	8.8	7.3	12.6
RQ (day 15)	0.96	0.8	0.8	0.81	0.78

of the two cohorts receiving 5 mg kg^{-1} of **30s** revealed the ad lib cohort only increased food intake sufficient to compensate for a third of the weight loss of the pair fed cohort. This finding bolstered expectations that T2DM treated with dapagliflozin would also lose weight.

The ADME profile determined for **30s** was highly supportive of clinical development [31]. The plasma protein binding of dapagliflozin ranged from 91% in humans to 95% in rat. All evidence indicated that dapagliflozin did not inhibit or upregulate P-450 enzymes. Upon incubation with mouse, rat, dog, monkey, and human hepatocytes, dapagliflozin was qualitatively converted to a similar mixture of Phase 1 (O-deethylation and hydroxylation to convert the diarylmethane moiety to a diarylcarbinol) and Phase 2 (glucuronidation of the C3 sugar hydroxyl) metabolites. Despite being a P-glycoprotein 1 (pgp) substrate, oral bioavailability was high due to the high permeability. The pharmacokinetic parameters measured in rats, dogs, and monkeys predicted that oral exposure, clearance, and elimination half-life in humans would be conducive to a single daily administration.

Clinical studies confirmed these predictions. At least 75% of a 50 mg oral dose was absorbed within 24 h; T_{max} was achieved in <1 h. Urinary excretion was the major elimination route comprising >75% of the dose primarily as the C3 sugar hydroxyl glucuronide conjugate generated by UGT1A9 [32]. The finding that the urine contained only 1.6% of the dose as unchanged dapa was unexpected. Dapa was a low clearance compound for which the terminal half-life was 13.8 h.

5.6
Clinical Studies with Dapagliflozin

Expectations based on early clinical studies suggesting that dapagliflozin would be a promising novel approach for safely treating T2DM have been substantiated by the drug's performance in more than 19 Phase 2/3 clinical trials involving >8500 patients [33]. These studies, many of which were placebo-controlled, established the efficacy and safety of dapagliflozin for treatment of T2DM either as monotherapy or in conjunction with other anti-diabetic agents. When administered once daily as a 5 or 10 mg tablet, T2DM patients achieved improved glycemic control manifested by HbA1c reductions of ≥0.5% with minimal increased risk of hypoglycemia. Additional benefits were a 2–3% reduction in body weight and a trend for a modest 3–5 mm reduction in systolic blood pressure. Dapagliflozin is contraindicated for patients with moderately or severely impaired renal function because efficacy is dependent on a normal or near-normal glomerular filtration rate. This medication is well tolerated although patients receiving dapagliflozin experienced a modest increase in the incidence of genital tract infections and a minor increase in urinary infections. The Food and Drug Administration (FDA), the European Commission, and health authorities of 40 countries have approved use of dapagliflozin as monotherapy or add-on therapy with other anti-diabetic agents including insulin. Most recently the European Commission approved a combination product formulated with dapagliflozin and metformin for treatment of T2DM.

5.7
Summary

Exploration of *meta* C-glucosides of appropriately substituted diarylmethanes identified a class of potent selective SGLT2 inhibitors. Subsequent evaluation of a promising subset resulted in the emergence of one compound dapagliflozin that exhibited a compelling safety and efficacy profile. Dapagliflozin represented a break from the traditional paradigm of insulin dependent anti-diabetic agents. Early disclosures of dapagliflozin properties halted on-going clinical evaluations of O-glucoside-based SGLT2 inhibitors and taught the medicinal chemistry community the essential structural features required for highly potent SGLT2 inhibitors. More than 20 groups initiated research efforts to identify proprietary counterparts [14]. Currently C-glucosides from eight other companies are in or have completed Phase 2/3 clinical trials; one other SGLT2 inhibitor, canagliflozin, has also been approved by the FDA and the European Commission. All indications are that selective SGLT2 inhibitors will play an important role in treatment of T2DM and improve quality of life for patients worldwide.

List of Abbreviations

ADME	absorption, distribution, metabolism, and excretion of a compound
CHO	Chinese hamster ovary
FDA	food and drug administration
GLUT	facilitative glucose transporter
HOAt/EDC	1-Hydroxy-7-azabenzotriazole/1-Ethyl-3-(3-dimethylaminopropyl)carbodiimide
pgp	P-glycoprotein 1
RQ	respiratory quotient
SAD	single ascending dose
SAR	structure-activity relationship
SGLT	sodium dependent glucose co-transporter
STZ rats	rats made diabetic by treatment with streptozotocin earlier
T2DM	type 2 diabetes mellitus
Tm	renal transport maximum
ZDF rats	rats genetically predisposed to become diabetic with age, a T2DM model

References

1. International Diabetes Federation. Diabetes Atlas (2012) http://www.idf.org/diabetesatlas/5e/Update2012 (accessed 8 August 2014).
2. de Pablos-Velasco, P., Bradley, C., Eschwège, E., Gönder-Frederick, L.A., Parhofer, K.G., Vandenberghe, H., and Simon, D. (2010) The PANORAMA pan-European survey: glycaemic control and treatment patterns in patients with type 2 diabetes. *Diabetologia*, **53** (Suppl. 1), 1012-P.
3. Chao, E.C. and Henry, R.R. (2010) SGLT2 inhibition--a novel strategy for diabetes treatment. *Nat. Rev. Drug Discov.*, **9** (7), 551–559.
4. Washburn, W.N. (2009) Development of renal glucose reabsorption inhibitors: a new mechanism for the pharmacotherapy of diabetes mellitus type 2. *J. Med. Chem.*, **52** (7), 1785–1794.
5. Abdul-Ghani, M.A. and DeFronzo, R.A. (2008) Inhibition of renal glucose reabsorption: a novel strategy for achieving glucose control in type 2 diabetes mellitus. *Endocr. Pract.*, **14** (6), 782–790.
6. Gerich, J.E. (2010) Role of the kidney in normal glucose homeostasis and in the hyperglycemia of diabetes mellitus: therapeutic implications. *Diabet. Med.*, **27** (2), 136–142.
7. Wright, E.M., Loo, D.D.F., and Hirayama, B.A. (2011) Biology of human sodium glucose transporters. *Physiol. Rev.*, **91** (2), 733–794.
8. Sabolic, I., Vrhovac, I., Eror, D.B., Gerasimova, M., Rose, M., Breljak, D.,

Ljubojevic, M., Brzica, H., Sebastiani, A., Thal, S.C., Sauvant, C., Kipp, H., Vallon, V., and Koepsell, H. (2012) Expression of Na+-D-glucose cotransporter SGLT2 in rodents is kidney-specific and exhibits sex and species differences. *Am. J. Physiol. Cell Physiol.*, **302** (8), C1174–C1188.

9. Chen, J., Williams, S., Ho, S., Loraine, H., Hagan, D., Whaley, J.M., and Feder, J.N. (2010) Quantitative PCR tissue expression profiling of the human SGLT2 gene and related family members. *Diabetes Ther.*, **1** (2), 57–92.

10. Sabolic, I., Skarica, M., Gorboulev, V., Ljubojević, M., Balen, D., Herak-Kramberger, C.M., and Koepsell, H. (2006) Rat renal glucose transporter SGLT1 exhibits zonal distribution and androgen-dependent gender differences. *Am. J. Physiol. Renal Physiol.*, **290** (4), F913–F926.

11. Calado, J., Santer, R., and Rueff, J. (2011) Effect of kidney disease on glucose handling (including genetic defects). *Kidney Int.*, **79**, S7–S13.

12. Santer, R. and Calado, J. (2010) Familial renal glucosuria and SGLT2: from a mendelian trait to a therapeutic target. *Clin. J. Am. Soc. Nephrol.*, **5** (1), 133–141.

13. Wright, E.M., Hirayama, B.A., and Loo, D.E. (2007) Active sugar transport in health and disease. *J. Intern. Med.*, **261** (1), 32–43.

14. Washburn, W.N. (2012) in *New Therapeutic Strategies for Type 2 Diabetes: Small Molecule Approaches*, 1st edn (ed R.M. Jones), Royal Chemical Society, Cambridge, pp. 29–87.

15. Adapted from Silverman, M. and Turner, R.J. (1992) in *Handbook of Physiology: Renal Physiology*, 1st edn (ed E.E. Windhager), Oxford University Press, Oxford, pp. 2017–2038.

16. Ehrenkranz, J.R., Lewis, N.G., Kahn, C.R., and Roth, J. (2005) Phlorizin: a review. *Diabetes Metab. Res. Rev.*, **21** (1), 31–38.

17. Rossetti, L., Shulman, G.I., Zawalich, W., and DeFronzo, R.A. (1987) Effect of chronic hyperglycemia on in vivo insulin secretion in partially pancreatectomized rats. *J. Clin. Invest.*, **80** (4), 1037–1044.

18. Tsujihara, K., Hongu, M., Saito, K., Inamasu, M., Arakawa, K., Oku, A., and Matsumoto, M. (1996) Na(+)-glucose cotransporter inhibitors as antidiabetics. I. Synthesis and pharmacological properties of 4′-dehydroxyphlorizin derivatives based on a new concept. *Chem. Pharm. Bull. (Tokyo)*, **44** (6), 1174–1180.

19. Kees, K.L., Fitzgerald, J.J., Steiner, K.E., Mattes, J.F., Mihan, B., Tosi, T., Mondoro, D., and McCaleb, M.L. (1996) New potent antihyperglycemic agents in db/db mice: synthesis and structure-activity relationship studies of (4-substituted benzyl) (trifluoromethyl)pyrazoles and -pyrazolones. *J. Med. Chem.*, **39** (20), 3920–3928.

20. Oku, A., Ueta, K., Arakawa, K., Ishihara, T., Nawano, M., Kuronuma, Y., Matsumoto, M., Saito, A., Tsujihara, K., Anai, M., Asano, T., Kanai, Y., and Endou, H. (1999) T-1095, an inhibitor of renal Na+-glucose cotransporters, may provide a novel approach to treating diabetes. *Diabetes*, **48** (9), 1794–1800.

21. Han, S., Hagan, D.L., Taylor, J.R., Xin, L., Meng, W., Biller, S.A., Wetterau, J.R., Washburn, W.N., and Whaley, J.M. (2008) Dapagliflozin, a selective SGLT2 inhibitor, improves glucose homeostasis in normal and diabetic rats. *Diabetes*, **57** (6), 1723–1729.

22. Link, J.T. and Sorensen, B.K. (2000) A method for preparing C-glycosides related to phlorizin. *Tetrahedron Lett.*, **41** (48), 9213–9217.

23. Ohsumi, K., Matsueda, H., Hatanaka, T., Hirama, R., Umemura, T., Oonuki, A., Ishida, N., Kageyama, Y., Maezono, K., and Kondo, N. (2003) Pyrazole-O-glucosides as novel Na(+)-glucose cotransporter (SGLT) inhibitors. *Bioorg. Med. Chem. Lett.*, **13** (14), 2269–2272.

24. Fujimori, Y., Katsuno, K., Nakashima, I., Ishikawa-Takemura, Y., Fujikura, H., and Isaji, M. (2008) Remogliflozin etabonate, in a novel category of selective low-affinity sodium glucose cotransporter (SGLT2) inhibitors, exhibits antidiabetic efficacy in rodent models. *J. Pharmacol. Exp. Ther.*, **327** (1), 268–276.

25. Deshpande, P.P., Ellsworth, B.A., Buono, F.G., Pullockaran, A., Singh, J., Kissick, T.P., Huang, M.-H., Lobinger, H., Denzel,

T., and Mueller, R.H. (2007) Remarkable β-1-C-arylglucosides:stereoselctive reduction of acetyl-protected methyl 1-C-aryl-glucosides without acetoxy-group participation. *J. Org. Chem.*, **72** (25), 9746–9749.

26. Deshpande, P.P., Singh, J., Pullockaran, A., Kissick, T., Ellsworth, B.A., Gougoutas, J.Z., Dimarco, J., Fakes, M., Reyes, M., Lai, C., Lobinger, H., Denzel, T., Ermann, P., Crispino, G., Randazzo, M., Gao, Z., Randazzo, R., Lindrud, M., Rosso, V., Buono, F., Doubleday, W.W., Leung, S., Richberg, P., Hughes, D., Washburn, W.N., Meng, W., Volk, K.J., and Mueller, R.H. (2012) A practical stereoselective synthesis and novel cocrystallizations of an amphipatic SGLT-2 inhibitor. *Org. Process Res. Dev.*, **16** (4), 577–585.

27. Ellsworth, B.E., Meng, W., Patel, M., Girotra, R.N., Wu, G., Sher, P., Hagan, D., Obermeier, M., Humphreys, W.G., Robertson, J.G., Wang, A., Han, S., Waldron, T., Morgan, N.N., Whaley, J.M., and Washburn, W.N. (2008) Aglycone exploration of C-arylglucoside inhibitors of renal sodium dependent glucose transporter SGLT2. *Bioorg. Med. Chem. Lett.*, **18** (17), 4770–4773.

28. Meng, W., Ellsworth, B.A., Nirschl, A.A., McCann, P.J., Patel, M., Girotra, R.N., Wu, G., Sher, P.M., Morrison, E.P., Biller, S.A., Zahler, R., Deshpande, P.P., Pullockaran, A., Hagan, D.L., Morgan, N., Taylor, J.R., Obermeier, M.T., Humphreys, W.G., Khanna, A., Discenza, L., Robertson, J.G., Wang, A., Han, S., Wetterau, J.R., Janovitz, E.B., Flint, O.P., Whaley, J.M., and Washburn, W.N. (2008) Discovery of dapagliflozin: a potent, selective renal sodium-dependent glucose cotransporter 2 (SGLT2) inhibitor for the treatment of type 2 diabetes. *J. Med. Chem.*, **51** (5), 1145–1149.

29. Uveges, A., Hagan, D., Onorato, J, and Whaley, J.M. (2011) Dapagliflozin selectively inhibits human SGLT2 versus SGLT1, SMIT1, SGLT4, and SGLT6. Presented at the American Diabetes Association 71st Scientific Sessions, San Diego, CA, June 24–28, 2011, Poster P-987.

30. Devenny, J.J., Godonis, H.E., Harvey, S.J., Rooney, S., Cullen, M.J., and Pelleymounter, M.A. (2012) Weight loss induced by chronic dapagliflozin treatment is attenuated by compensatory hyperphagia in diet-induced obese (DIO) rats. *Obesity*, **20** (8), 1645–1652.

31. Obermeier, M.T., Yao, M., Khanna, A., Koplowitz, B., Zhu, M., Li, W., Komoroski, B., Kasichayanula, S., Discenza, L., Washburn, W., Meng, W., Ellsworth, B.A., Whaley, J.M., and Humphreys, W.G. (2010) In vitro characterization and pharmacokinetics of dapagliflozin (BMS-512148), a potent sodium-glucose cotransporter type II (SGLT2) inhibitor, in animals and humans. *Drug Metab. Dispos.*, **38** (3), 405–414.

32. Kasichayanula, S., Liu, X., Lacreta, F., Griffen, S.C., and Boulton, D.W. (2014) Clinical pharmacokinetics and pharmacodynamics of dapagliflozin, a selective inhibitor of sodium-glucose co-transporter type 2. *Clin. Pharmacokinet.*, **53** (1), 17–27.

33. Washburn, W.N. and Poucher, S.M. (2013) Differentiating SGLT2 inhibitors in development for the treatment of type 2 diabetes mellitus. *Expert Opin. Investig. Drugs*, **22** (4), 463–486.

William N. Washburn graduated from Princeton University in 1967 with a major in Chemistry. He subsequently received his PhD in Organic Chemistry in 1971 from Columbia University studying nonbenzenoid aromatic systems with Professor Ronald Breslow. After completion of a postdoctoral appointment at Harvard University with Professor E. J. Corey, he became an Assistant Professor in the Chemistry Department at the University of California at Berkeley addressing physical organic and bio-organic problems. He was a member of the Research Laboratories of Eastman Kodak for 12 years before accepting a position with Bristol-Myers Squibb in 1991 where he was a Senior Research Fellow prior to his retirement in January 2014. Since joining Bristol-Myers Squibb, Dr Washburn's role, until his retirement, had been leading drug discovery programs pertaining to obesity and Type 2 Diabetes; specific examples are beta 3 adrenergic agonists, selective thyroid agonists, sodium glucose co-transporter inhibitors, TGR5 agonists, and melanin concentrating hormone receptor antagonists.

6
Elvitegravir, A New HIV-1 Integrase Inhibitor for Antiretroviral Therapy

Hisashi Shinkai

6.1
Introduction

Acquired immunodeficiency syndrome (AIDS) was first reported as a fatal disease in 1981. After AIDS was identified as being caused by a human immunodeficiency virus (HIV) in 1983 [1], the first antiretroviral drug zidovudine (AZT) was introduced in 1987 [2]. In 2014, anti-HIV drugs approved by the US Food and Drug Administration (FDA) can be categorized into seven groups: nucleoside reverse transcriptase inhibitors (seven drugs), nucleotide reverse transcriptase inhibitors (one drug), non-nucleoside reverse transcriptase inhibitors (five drugs), protease inhibitors (10 drugs), fusion inhibitors (one drug), co-receptor inhibitors (one drug), and integrase inhibitors (three drugs). The anti-HIV drugs, which are used in combination, dramatically reduced HIV-associated morbidity and mortality and turned a fatal HIV infection into a manageable chronic disease [3–5]. However, the use of these drugs for long periods has been relatively limited by their toxicity [6] and drug resistance development [7, 8]. New anti-HIV drugs with a good safety, an acceptable resistance profile, or novel mechanisms of action are thus needed. This review will summarize the discovery of elvitegravir as a novel monoketo acid (MKA) class of HIV-1 integrase inhibitors. Elvitegravir 150 mg is a once-daily HIV integrase inhibitor with potent antiviral activity and favorable safety that has been co-formulated with a cytochrome P450 CYP3A inhibitor (cobicistat 150 mg) plus the nucleoside/nucleotide reverse transcriptase inhibitor back bone (emtricitabine 200 mg/tenofovir disoproxil fumarate 300 mg), which was approved by FDA in 2012 and by the European Medicines Agency (EMA) in 2013. Elvitegravir 85 and 150 mg as stand-alone agents were also approved by EMA in 2013 and FDA in 2014.

Successful Drug Discovery, First Edition. Edited by János Fischer and David P. Rotella.
© 2015 Wiley-VCH Verlag GmbH & Co. KGaA. Published 2015 by Wiley-VCH Verlag GmbH & Co. KGaA.

6.2
Discovery of Elvitegravir

6.2.1
HIV-1 Integrase and Diketo Acid Inhibitors

HIV-1 integrase plays a central role in the insertion of viral DNA into the genome of host cells [9]. This virally encoded enzyme is thus essential for viral replication, and represents a crucial target for antiretroviral drugs [10]. HIV-1 integrase consists of three distinct structural domains (the zinc binding N-terminal, the catalytic core, and the DNA binding C-terminal) [11]. Both the N- and C-terminal domains are required for the 3′-processing and strand transfer steps [12–18], while the core domain alone can only carry out a disintegration reaction (Figure 6.1) [13, 14, 19–22]. DNA ending with the sequence CAGT is also required for functional activity in HIV-1 integrase [23]. Therefore, the catalytic core domain by itself is not able to assume the precise structure required for the functional enzyme. Full-length integrase complexed with DNA is essential to achieve the exact conformation of a functional integrase. HIV-1 integrase performs a two-step reaction, that is, hydrolysis removal of the terminal dinucleotide from each 3′-end of the viral DNA (3′-processing) and subsequent transesterification to join the 3′-end of the viral DNA to host DNA (strand transfer) [11, 24]. This catalytic reaction requires divalent metal ions, such as Mg^{2+} and Mn^{2+} [25]. The catalytic core domain of the enzyme contains two aspartates and one glutamate (the DDE motif consists of aspartic D64, D116, and glutamic E152), which are essential for enzymatic activity [11]. This catalytic triad potentially binds two divalent metal ions [25]. And a dual metal-ion mechanism is proposed as a chemically reasonable mechanism of action for polynucleotidyl transferases that is a large family of DNA-processing enzymes including HIV-1 integrase [26–30]. Two divalent cations (Mg^{2+} or Mn^{2+}) apparently coordinate the DDE motif of HIV-1 integrase with the phosphodiester backbone of the viral donor cDNA and the chromosomal acceptor DNA during the integration steps [31, 32].

Among numerous attempts to develop integrase inhibitors, the diketo acid class of compounds has been most aggressively developed because of the marked antiretroviral activities exhibited [10]. Diketo acid inhibitors selectively interrupt the strand transfer step, and in cell-based assays inhibit integration without affecting earlier phases of the HIV-1 replication cycle [33–37]. In addition, the binding of the diketo acid inhibitors was only detected when full-length integrase was assembled onto the viral DNA ends [33]. To explain these observations, it is proposed that the diketo acid inhibitors can bind only with the interface of the full-length integrase-DNA-divalent metal complex after 3′-processing [10]. Their ability to bind selectively to the enzyme complexed with the viral DNA and compete with the host DNA substrate may explain their selectivity for the strand transfer reaction [33, 37]. The binding and activity of diketo acid HIV-1 integrase inhibitors are divalent metal cation dependent [33, 34, 38, 39]. The

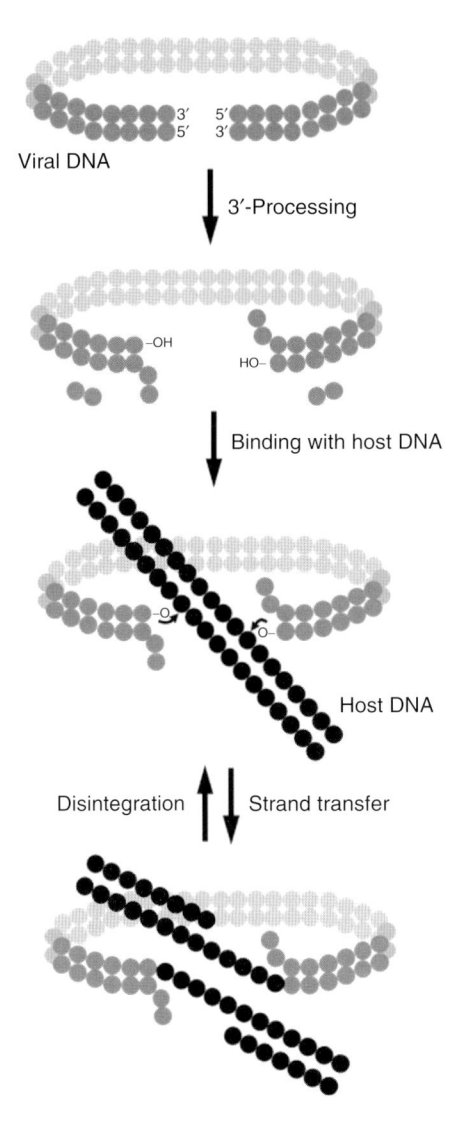

Figure 6.1 The integrase catalytic reactions.

diketo acid moiety consists of γ-ketone, enolizable α-ketone, and carboxylic acid. Its keto–enol acid tautomer is known to have the ability to simultaneously form stable five- and six-membered chelate rings with two divalent metal ions (Figure 6.2) [31, 40].

Thus, the interaction of the dual metal-chelating pharmacophore in the inhibitors with the divalent metal ions in the active site of the enzyme is considered to be a key factor in the inhibition of HIV-1 integrase. Based on the hypothesis that the diketo acid moiety as the dual metal-chelating pharmacophore was essential for the inhibitory activity of this series of integrase inhibitors [33], the structures

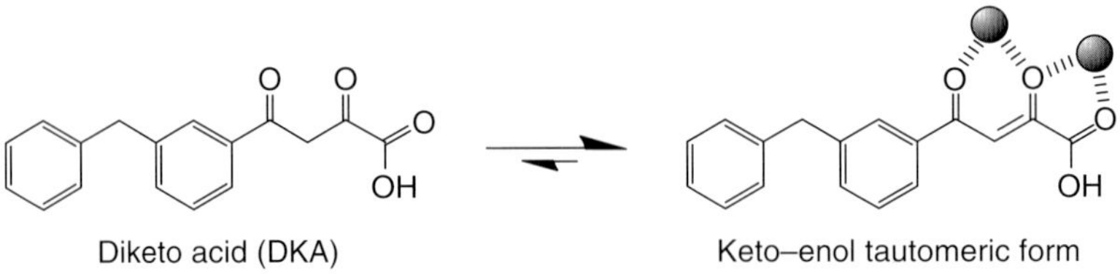

Figure 6.2 Diketo acid and its keto–enol tautomeric form.

of diketo triazole (S1360) [41] diketo tetrazole (5CITEP) [42], 6-carbonyl-1,2-catechol (catechol) [43], 7-carbamoyl-8-hydroxy-(1,6)-naphthyridine (L-870810) [43, 44], and 4-carbamoyl-5-hydroxy-6-pyrimidinone (raltegravir) [45] were synthesized as bioisosteres of the diketo acid pharmacophore (Figure 6.3). The carboxylic acid in the diketo acid motif can be replaced with not only well-known bioisosteres of a carboxylic acid group, such as triazole, tetrazole, and phenol [43, 46] but also by a basic pyridine [43]. This means that the carboxylic acid in the diketo acid motif can be replaced with not only acidic bioisosteres but also by a basic heterocycle bearing a lone pair donor atom [24]. The heteroaromatic nitrogen in the pyridine ring is believed to mimic the corresponding carboxyl oxygen in the diketo acid motif as a Lewis base equivalent [43]. Phenolic hydroxyl groups as alternatives to α-enol in the enolized diketo acid motif confirmed that the biologically active conformation is not the diketo acid form but the coplanar keto–enol acid form [24, 43–45, 47, 48]. Moreover, the carboxylic acid moiety of the diketo acid motif can be replaced with a neutral carbonyl group as a lone pair donor [49]. The 4-carbamoyl-5-hydroxy-6-pyrimidinone derivative raltegravir is the first HIV-1 integrase inhibitor, which was approved by FDA in 2007. Raltegravir-based regimen contains raltegravir 400 mg twice daily in combination with other inhibitors such as emtricitabine 200 mg plus tenofovir disoproxil fumarate 300 mg once-daily [50].

6.2.2
Monoketo Acid Integrase Inhibitors and Elvitegravir

All bioisosteres of the diketo acid motif have the three functional groups that mimic a ketone, enol, and carboxyl oxygen, and assume a coplanar conformation. The motif can simultaneously form stable five- and six-membered chelate rings with two divalent metal ions at the DDE motif of HIV-1 integrase [40]. Because the metal-chelating functions seemed relevant to the mechanism of inhibition [33], we first designed quinolone-3-glyoxylic acid (diketo acid) as a nonenolizable diketo acid that can assume a planer conformation without forming enol (Figure 6.4).

The central ketone of this motif cannot form enol, but still has three functional groups that potentially coordinate with the two metal ions. However, the nonenolizable diketo acid did not exhibit integrase inhibitory activity. This indicates the importance of the central ionizable enol in the diketo acid motif. In contrast, we found 4-quinolone-3-carboxylic acid (MKA1) does exhibit integrase inhibitory

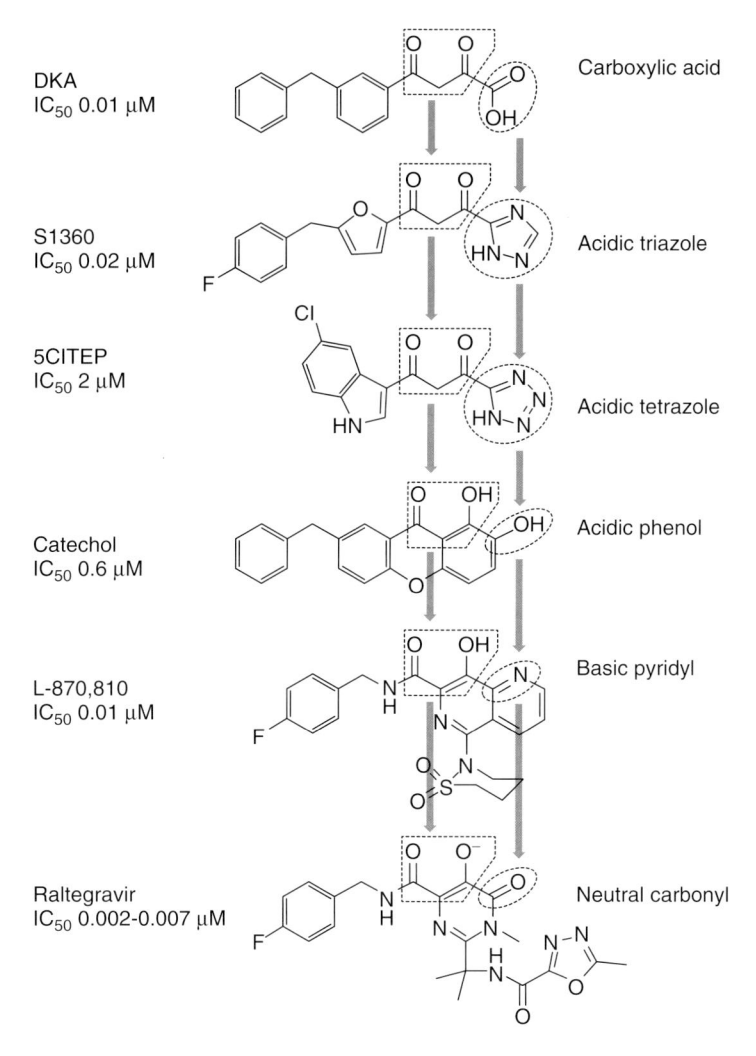

Figure 6.3 Bioisosteres of the diketo acid.

activity [34]. This novel quinolone integrase inhibitor has only two functional groups, a carboxylic acid and a β-ketone, which are coplanar. This result shows that the coplanar MKA motif in 4-quinolone-3-carboxylic acid can serve as an alternative to the diketo acid motif, even though the downsized MKA motif is unlikely to fully coordinate with the two divalent metal ions (Figure 6.4). The carboxylic acid in the MKA motif would be an alternative to the central ionizable enol in the diketo acid moiety. Although the finding regarding MKA suggests that full functions of the diketo acid motif may not be essential for inhibitory activity, the metal-chelating functions are still important for their inhibitory action. In fact, the MKA1 is far less potent than diketo acid because of its weaker metal-chelating ability (Figure 6.4). However, the weak metal-chelating ability of

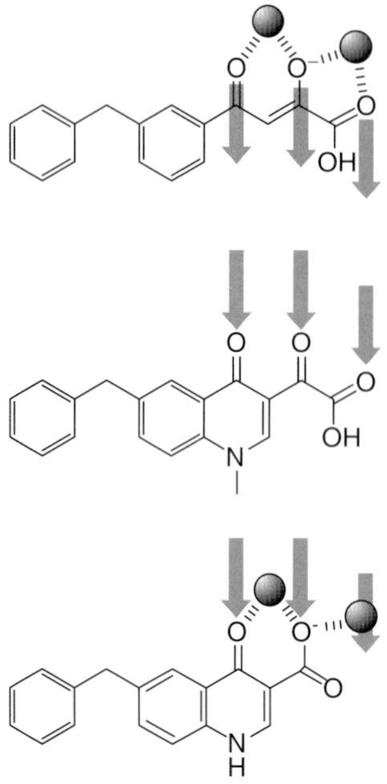

Figure 6.4 Enolized diketo acid, nonenolizable diketo acid, and monoketo acid.

the MKA motif was preferable from a safety standpoint. HIV-1 integrase belongs to a large family of DNA-processing enzymes (polynucleotidyl transferases), which contain the same arrangement of three catalytically essential carboxylates [10, 26, 51–53]. Because the catalytic core domain with two metal ions are highly conserved in all integrases and polynucleotidyl transferases, the strong metal-chelating ability of the diketo acids have a risk of inhibiting the family enzymes. In fact, the diketo acid compounds often exhibit adverse hepatic effects [54]. Therefore, we chose the MKA1 as an initial lead compound and started modifying structures around the core MKA moiety to lead to more potent compounds despite the comparatively low starting potency (IC$_{50}$ of 1.6 μM). The structural optimization process for the MKA integrase inhibitors is shown in Figure 6.5 [34, 55].

First, the effect of introducing substituents into the distal benzene ring was examined. Introduction of various groups at the *para* position of the benzene ring caused a loss of integrase inhibitory activity. In contrast, introduction of chloro or fluoro group at the *ortho* or *meta* position caused an increase in integrase inhibitory activity. Notably, introduction of 2-fluoro and 3-chloro substituents into the distal benzene ring (MKA2) led to a significant improvement (approximately a 36-fold increase) of its inhibition of strand transfer (IC$_{50}$ 44 nM) and its antiviral activity (EC$_{50}$ 810 nM).

Next, the effect of introducing substituents at the 1-position of the quinolone ring was examined. Introduction of an ethyl group or an isopropyl group at the 1-position of quinolone ring caused an increase in both integrase inhibitory

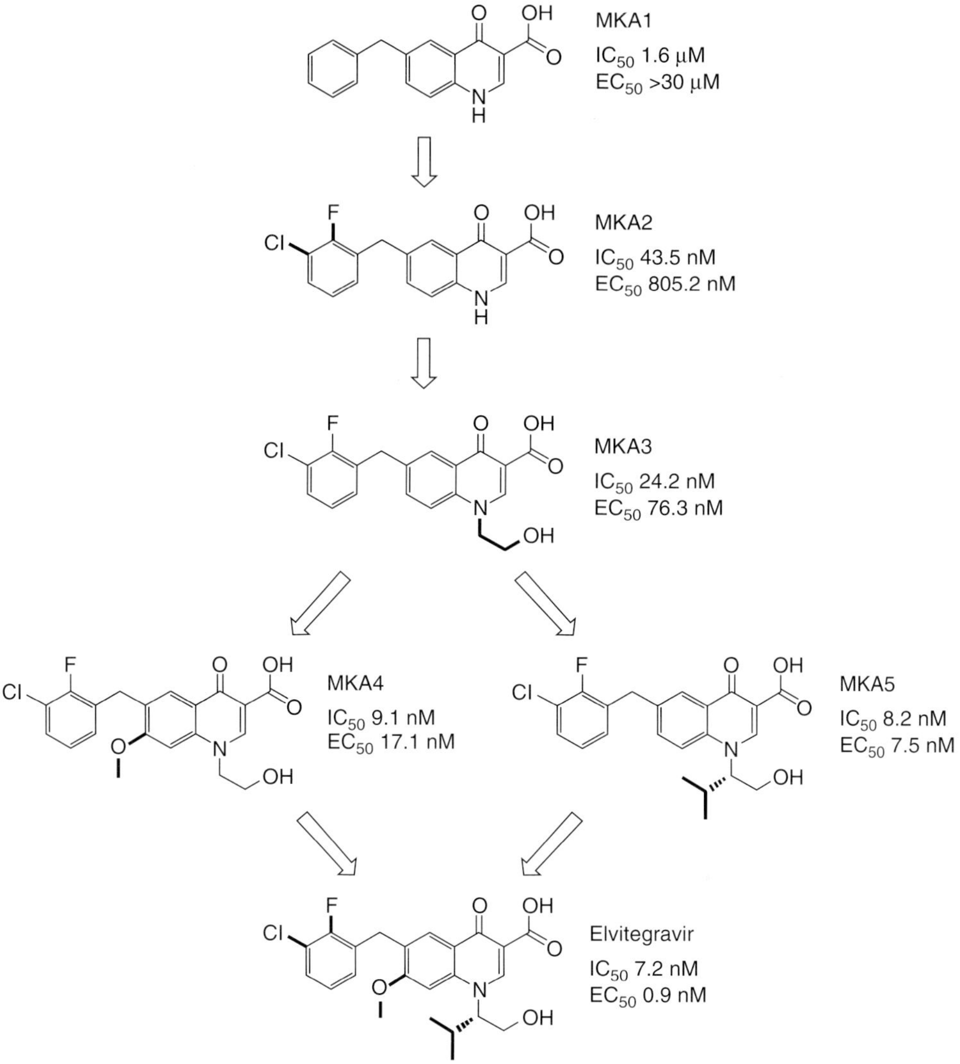

Figure 6.5 Structural optimization process for monoketo acid integrase inhibitors.

activity and antiviral activity. However, introduction of the other alkyl groups was less effective. Introduction of various polar groups at the 1-position of the quinolone ring was also examined. Among the compounds, 1-hydroxyethyl compound (MKA3) was 1.8-fold more potent at inhibiting strand transfer (IC_{50} 24 nM) and displayed an approximately 11-fold stronger antiviral activity (EC_{50} 76 nM) than MKA1.

Next, the effect of introducing substituents into the central benzene ring was examined. Although introduction of a fluoro group at the 5- or 8-position of the quinolone ring led to a decrease in the inhibition of strand transfer,

7-fluoroquinolone did not decrease its integrase inhibitory activity or antiviral activity. As the substitution at the 7-position of the quinolone ring was only tolerable, various groups were introduced. Among these, 7-methoxy compound (MKA4) showed a significant improvement of inhibition of strand transfer (IC$_{50}$ 9.1 nM) and antiviral activity (EC$_{50}$ 17.1 nM).

The independent introduction of a small alkyl group or a hydroxyethyl group at the 1-position of the quinolone ring led to a significant improvement of activity, so the effect of the combination of the small alkyl groups and the hydroxyethyl group was examined. A combination compound bearing a methyl group at the 1S-position of the hydroxyethyl moiety was 1.6-fold more potent at inhibiting strand transfer (IC$_{50}$ = 14.8 nM) and displayed an approximately 2.7-fold stronger antiviral activity (EC$_{50}$ = 27.7 nM) than MKA3. On the other hand, its enantiomer, was less potent at inhibiting both strand transfer (IC$_{50}$ = 38.3 nM) and HIV-1 replication (EC$_{50}$ = 115.5 nM) than MKA3. Thus, several alkyl groups, such as ethyl, *n*-propyl, isopropyl, *tert*-butyl, cyclohexyl, and phenyl group, were introduced at the 1S-position of the hydroxyethyl moiety. Among these, MKA5, bearing an isopropyl group at the 1S-position of the hydroxyethyl moiety, was approximately threefold more potent at inhibiting strand transfer (IC$_{50}$ 8.2 nM), and approximately 10-fold stronger at inhibiting HIV replication (EC$_{50}$ 7.5 nM) than MKA3. In contrast, the enantiomer of MKA5 was far less potent at inhibiting HIV-1 replication (EC$_{50}$ = 115.5 nM) than MKA3.

The introduction of both a methoxy group at the 7-position of the quinolone ring and an isopropyl group at the 1S-position of the hydroxyethyl moiety (elvitegravir) led to a synergistic improvement in antiviral activity (EC$_{50}$ 0.9 nM), although no additive or synergistic improvement in the inhibition of HIV-1 integrase (IC$_{50}$ 7.2 nM) was observed. This may be due to the condition of the strand transfer assay using 5 nM of target DNA, which can limit the potency of inhibitors. The enantiomer of elvitegravir was confirmed to be far less potent at inhibiting HIV-1 replication (EC50 = 108.8 nM) than MKA4.

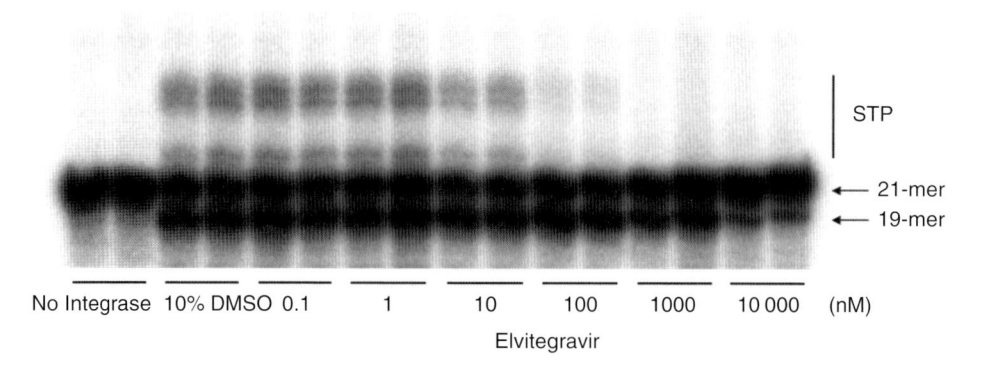

Figure 6.6 Inhibition of 3′-processing and strand transfer by elvitegravir. Gel electrophoresis shows the viral DNA substrate (21-mer oligonucleotide), 3′-processing product (19-mer oligonucleotide), and strand transfer products (>21-mer oligonucleotides).

Diketo acid integrase inhibitors are reported to be much more potent at inhibiting integrase-catalyzed strand transfer processes than 3′-processing reactions [35, 56]. Our MKA integrase inhibitor was also confirmed to be a selective strand transfer inhibitor (Figure 6.6). A 5′-end-labeled 21-mer double-stranded DNA oligonucleotide, corresponding to the last 21 bases of the U5 viral long terminal repeat (LTR), is converted via a 19-mer oligonucleotide (3′-processing product) to strand transfer products (>21-mer oligonucleotides) by HIV-1 integrase. MKA elvitegravir was more than 100-fold as potent at inhibiting integrase-catalyzed strand transfer processes as the 3′-processing reaction.

6.3 Conclusion

We discovered the coplanar MKA motif in the scaffold, 4-quinolone-3-carboxylic acid, to be an alternative to the diketo acid motif in integrase inhibitors. These novel quinolone integrase inhibitors were structurally optimized to provide the highly potent derivative elvitegravir, which was much more potent at inhibiting integrase-catalyzed strand transfer processes than 3′-processing reactions, as previously reported for compounds of the diketo acid class [35, 56]. The chelating ability of the diketo acids that can simultaneously coordinate with two divalent metal ions is considered to be important, but structurally modified MKAs cannot fully mimic the chelating function of the diketo acids. The MKAs have potentially weaker affinity for metal ions than the diketo acids, and cannot fully coordinate with the two metal ions (Figure 6.7). This would seem to be disadvantageous for achieving activity. In fact, the initial lead compound (MKA1) was less potent than the diketo acid. However, the highly potent elvitegravir was obtained by structural modification around the MKA moiety. Furthermore, this potentially weaker chelating ability of the MKAs might be advantageous in terms of selectivity and safety. Elvitegravir along with raltegravir was entered into clinical studies in 2004. Elvitegravir has indeed exhibited a very safe profile along with significant antiviral activities in the clinical studies. Twice-daily administrations of elvitegravir at a dose of 400 mg demonstrated mean reductions from baseline in HIV-1 RNA of 1.91 $\log_{10}$ copies/ml or greater [57, 58]. Elvitegravir is metabolized mainly via cytochrome P450 3A (CYP3A) along with minor pathways including primary or secondary glucuronidation. Co-administration of a CYP3A inhibitor, such as ritonavir or cobicistat, substantially increases elvitegravir plasma exposures and prolongs its elimination half-life from 3.5 to 9.5 h [59]. Therefore, it was possible to give elvitegravir once daily when administered with a pharmacokinetic booster. Standard treatment of HIV infection is a combination of at least three active agents from two or more classes of antiretroviral drugs because such antiretroviral therapy (ART) maintains a reduction in viral load to undetectable levels and a decrease in HIV-associated mortality [60]. However, complicated and inconvenient regimens often interfere with adherence to the ART. Thus, elvitegravir 150 mg has been co-formulated with cobicistat 150 mg (CYP3A4

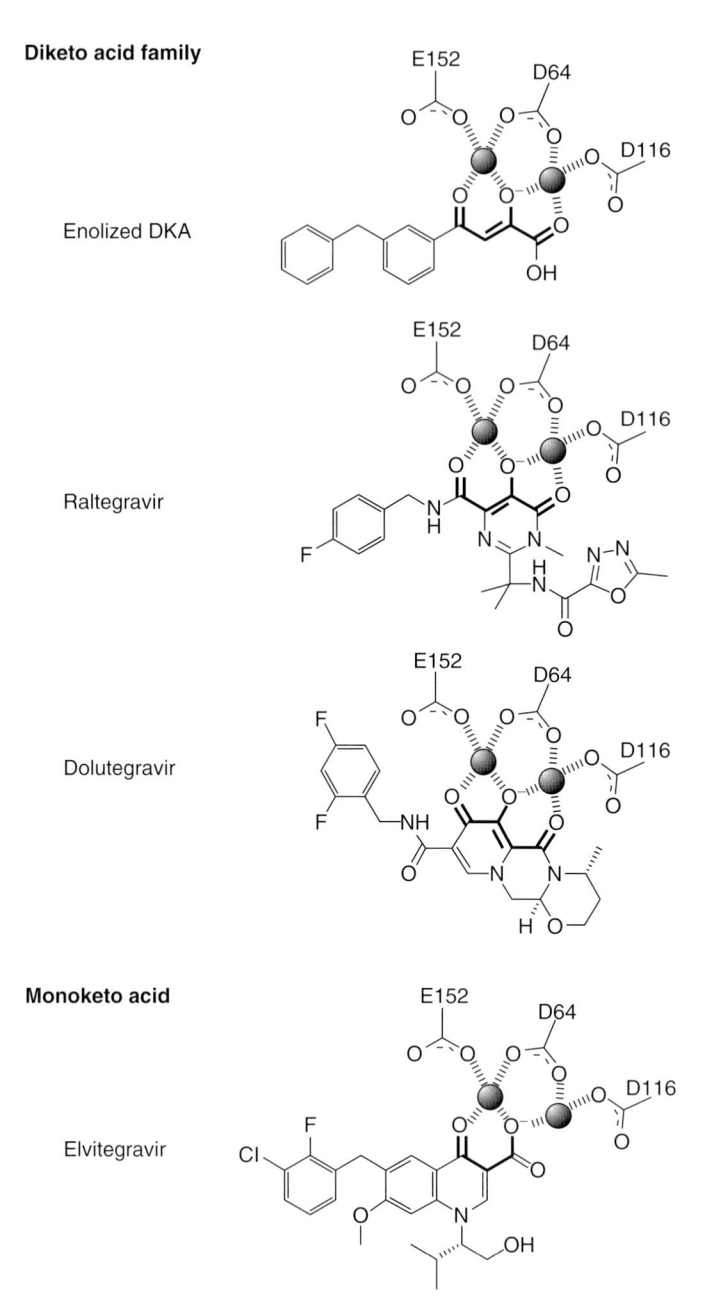

Figure 6.7 Metal chelating functions of diketo acid family and monoketo acid. Dolutegravir is the third integrase inhibitor approved in 2013.

inhibitor), emtricitabine 200 mg (nucleoside reverse transcriptase inhibitor), and tenofovir disoproxil fumarate 300 mg (nucleotide reverse transcriptase inhibitor) in a single tablet given once daily. The once-daily single tablet, stribild, showed potent and durable antiretroviral efficacy and favorable safety in clinical studies. Stribild demonstrated to produce a comparable level of virologic suppression with fewer adverse events compared with preferred regimens (co-formulated efavirenz 600 mg/emtricitabine 200 mg/tenofovir disoproxil fumarate 300 mg, and ritonavir-boosted atazanavir 300 mg plus co-formulated emtricitabine 200 mg/tenofovir disoproxil fumarate 300 mg) [61–63]. Elvitegravir was granted approval as part of stribild by the US FDA in 2012 and EMA in 2013. Elvitegravir 85 and 150 mg as stand-alone agents were also approved by EMA in 2013. Moreover, the health and human services panel on antiretroviral guidelines for the use of antiretroviral agents in HIV-1-infected adults and adolescents recommended stribild as a preferred regimen for ART-naïve patients.

List of Abbreviations

AIDS	acquired immunodeficiency syndrome
ART	antiretroviral therapy
AZT	zidovudine
EMA	European medicines agency
FDA	US Food and Drug Administration
HIV	human immunodeficiency virus

References

1. Barre-Sinoussi, F., Chermann, J.C., Rey, F., Nugeyre, M.T., Chamaret, S., Gruest, J., Dauguet, C., Axler-Blin, C., Vezinet-Brun, F., Rouzioux, C., Rozenbaum, W., and Montagnier, L. (1983) Isolation of a T-lymphotropic retrovirus from a patient at risk for acquired immune deficiency syndrome (AIDS). *Science*, **220**, 868–871.
2. Fischl, M.A., Richman, D.D., Grieco, M.H., Gottlieb, M.S., Volberding, P.A., Laskin, O.L., Leedom, J.M., Groopman, J.E., Mildvan, D., Schooley, R.T., Jackson, G.G., Durack, D.T., King, D., and The AZT Collaborative Working Group (1987) The efficacy of Azidothymidine (AZT) in the treatment of patients with AIDS and AIDS-related complex. *N. Engl. J. Med.*, **317**, 185–191.
3. Kitchen, C.M., Kitchen, S.G., Dubin, J.A., and Gottlieb, M.S. (2001) Initial virological and immunologic response to highly active antiretroviral therapy predicts long-term clinical outcome. *Clin. Infect. Dis.*, **33**, 466–472.
4. Valenti, W.M. (2001) HAART is cost-effective and improves outcomes. *AIDS Read.*, **11**, 260–262.
5. King, J.T. Jr., Justice, A.C., Roberts, M.S., Chang, C.C., Fusco, J.S., Collaboration in HIV Outcomes Research, and U.S. Program Team (2003) Long-term HIV/AIDS survival estimation in the highly active antiretroviral therapy era. *Med. Decis. Making*, **23**, 9–20.
6. Carr, A. (2003) Toxicity of antiretroviral therapy and implications for drug development. *Nat. Rev. Drug Discov.*, **2**, 624–634.
7. Martinez-Picado, J., De Pasquale, M.P., Kartsonis, N., Hanna, G.J., Wong, J., Finzi, D., Rosenberg, E., Gunthard, H.F., Sutton, L., Savara, A., Petropoulos, C.J., Hellmann, N., Walker, B.D.,

Richman, D.D., Siciliano, R., and D'Aquila, R.T. (2000) Antiretroviral resistance during successful therapy of HIV type 1 infection. *Proc. Natl. Acad. Sci. U.S.A.*, **97**, 10948–10953.

8. Little, S.J., Holte, S., Routy, J.P., Daar, E.S., Markowitz, M., Collier, A.C., Koup, R.A., Mellors, J.W., Connick, E., Conway, B., Kilby, M., Wang, L., Whitcomb, J.M., Hellmann, N.S., and Richman, D.D. (2002) Antiretroviral-drug resistance among patients recently infected with HIV. *N. Engl. J. Med.*, **347**, 385–394.

9. Craigie, R. (2001) HIV integrase, a brief overview from chemistry to therapeutics. *J. Biol. Chem.*, **276**, 23213–23215.

10. Pommier, Y., Johnson, A.A., and Marchand, C. (2005) Integrase inhibitors to treat HIV/AIDS. *Nat. Rev. Drug Discov.*, **4**, 236–248.

11. Esposito, D. and Craigie, R. (1999) HIV integrase structure and function. *Adv. Virus Res.*, **52**, 319–333.

12. Drelich, M., Wilhelm, R., and Mous, J. (1992) Identification of amino acid residues critical for endonuclease and integration activities of HIV-1 IN protein in Vitro. *Virology*, **188**, 459–468.

13. Bushman, F.D., Engelman, A., Palmer, I., Wingfield, P., and Craigie, R. (1993) Domains of the integrase protein of human immunodeficiency virus type 1 responsible for polynucleotidyl transfer and zinc binding. *Proc. Natl. Acad. Sci. U.S.A.*, **90**, 3428–3432.

14. Yao, Q.Y., Tierney, R.J., Croom-Carter, D., Cooper, G.M., Ellis, C.J., Rowe, M., and Rickinson, A.B. (1996) Isolation of intertypic recombinants of Epstein-Barr virus from T-cell- immunocompromised individuals. *J. Virol.*, **70**, 4585–4597.

15. Vink, C., Groeneger, A.M.O., and Plasterk, R.H. (1993) Identification of the catalytic and DNA-binding region of the human immunodeficiency virus type I integrase protein. *Nucleic Acids Res.*, **21**, 1419–1425.

16. Schauer, M. and Billich, A. (1992) The N-terminal region of HIV-1 integrase is required for integration activity, but not for DNA-binding. *Biochem. Biophys. Res. Commun.*, **185**, 874–880.

17. Kulkosky, J., Katz, R.A., Merkel, G., and Skalka, A.M. (1995) Activities and substrate specificity of the evolutionarily conserved central domain of retroviral integrase. *Virology*, **206**, 448–456.

18. Engelman, A., Bushman, F.D., and Craigie, R. (1993) Identification of discrete functional domains of HIV-1 integrase and their organization within an active multimeric complex. *EMBO J.*, **12**, 3269–3275.

19. Engelman, A. and Craigie, R. (1992) Identification of conserved amino acid residues critical for human immunodeficiency virus type 1 integrase function in vitro. *J. Virol.*, **66**, 6361–6369.

20. Van Gent, D.C., Vink, C., Groeneger, A.A., and Plasterk, R.H. (1993) Complementation between HIV integrase proteins mutated in different domains. *EMBO J.*, **12**, 3261–3267.

21. Mazumder, A., Engelman, A., Craigie, R., Fesen, M., and Pommier, Y. (1994) Intermolecular disintegration and intramolecular strand transfer activities of wild-type and mutant HIV-1 integrase. *Nucleic Acids Res.*, **22**, 1037–1043.

22. Shibagaki, Y., Holmes, M.L., Appa, R.S., and Chow, S.A. (1997) Characterization of feline immunodeficiency virus integrase and analysis of functional domains. *Virology*, **230**, 1–10.

23. Vink, C., van Gent, D.C., Elgersma, Y., and Plasterk, R.H. (1991) Human immunodeficiency virus integrase protein requires a subterminal position of its viral DNA recognition sequence for efficient cleavage. *J. Virol.*, **65**, 4636–4644.

24. Hazuda, D.J., Anthony, N.J., Gomez, R.P., Jolly, S.M., Wai, J.S., Zhuang, L., Fisher, T.E., Embrey, M., Guare, J.P. Jr., Egbertson, M.S., Vacca, J.P., Huff, J.R., Felock, P.J., Witmer, M.V., Stillmock, K.A., Danovich, R., Grobler, J., Miller, M.D., Espeseth, A.S., Jin, L., Chen, I.W., Lin, J.H., Kassahun, K., Ellis, J.D., Wong, B.K., Xu, W., Pearson, P.G., Schleif, W.A., Cortese, R., Emini, E., Summa, V., Holloway, M.K., and Young, S.D. (2004) A naphthyridine carboxamide provides evidence for discordant resistance between mechanistically identical

inhibitors of HIV-1 integrase. *Proc. Natl. Acad. Sci. U.S.A.*, **101**, 11233–11238.
25. Ellison, V. and Brown, P.O. (1994) A Stable complex between integrase and viral DNA ends mediates human immunodeficiency virus integration in vitro. *Proc. Natl. Acad. Sci. U.S.A.*, **91**, 7316–7320.
26. Thomas, A. and Steitz, T.A. (1998) Structural biology: a mechanism for all polymerases. *Nature*, **391**, 231–232.
27. Wlodawer, A. (1999) Crystal structures of catalytic core domains of retroviral integrases and role of divalent cations in enzymatic activity. *Adv. Virus Res.*, **52**, 335–350.
28. Thomas, A. and Steitz, T.A. (1999) DNA polymerases: structural diversity and common mechanisms. *J. Biol. Chem.*, **274**, 17395–17398.
29. Horton, N.C. and Perona, J.J. (2001) Making the most of metal ions. *Nat. Struct. Biol.*, **8**, 290–293.
30. Feng, M., Patel, D., Dervan, J.J., Ceska, T., Suck, D., Haq, I., and Sayers, J.R. (2004) Roles of divalent metal ions in flap endonuclease-substrate interactions. *Nat. Struct. Biol.*, **11**, 450–456.
31. Sechi, M., Bacchi, A., Carcelli, M., Compari, C., Duce, E., Fisicaro, E., Rogolino, D., Gates, P., Derudas, M., Al-Mawsawi, L.Q., and Neamati, N. (2006) From ligand to complexes: inhibition of human immunodeficiency virus type 1 integrase by β-diketo acid metal complexes. *J. Med. Chem.*, **49**, 4248–4260.
32. Marchand, C., Johnson, A.A., Semenova, E., and Pommier, Y. (2006) Mechanism and inhibition of HIV integration. *Drug Discov. Today*, **3**, 253–260.
33. Grobler, J.A., Stillmock, K., Binghua, H., Witmer, M., Felock, P., Espeseth, A.S., Wolfe, A., Egbertson, M., Bourgeois, M., Melamed, J., Wai, J.S., Young, S., Vacca, J., and Hazuda, D.J. (2002) Diketo acid inhibitor mechanism and HIV-1 integrase: implications for metal binding in the active site of phosphotransferase enzymes. *Proc. Natl. Acad. Sci. U.S.A.*, **99**, 6661–6666.
34. Sato, M., Motomura, T., Aramaki, H., Matsuda, T., Yamashita, M., Ito, Y., Kawakami, H., Matsuzaki, Y., Watanabe, W., Yamataka, K., Ikeda, S., Kodama, E., Matsuoka, M., and Shinkai, H. (2006) Novel HIV-1 integrase inhibitors derived from quinolone antibiotics. *J. Med. Chem.*, **49**, 1506–1508.
35. Hazuda, D.J., Felock, P., Witmer, M., Wolfe, A., Stillmock, K., Grobler, J.A., Espeseth, A., Gabryelski, L., Schleif, W., Blau, C., and Miller, M.D. (2000) Inhibitors of strand transfer that prevent integration and inhibit HIV-1 replication in cells. *Science*, **287**, 646–650.
36. Wai, J.S., Egbertson, M.S., Payne, L.S., Fisher, T.E., Embrey, M.W., Tran, L.O., Melamed, J.Y., Langford, H.M., Guare, J.P. Jr., Zhuang, L., Grey, V.E., Vacca, J.P., Holloway, M.K., Naylor-Olsen, A.M., Hazuda, D.J., Felock, P.J., Wolfe, A.L., Stillmock, K.A., Schleif, W.A., Gabryelski, L.J., and Young, S.D. (2000) 4-Aryl-2,4-dioxobutanoic acid inhibitors of HIV-1 integrase and viral replication in cells. *J. Med. Chem.*, **43**, 4923–4926.
37. Espeseth, A.S., Felock, P., Wolfe, A., Witmer, M., Grobler, J., Anthony, N., Egbertson, M., Melamed, J.Y., Young, S., Hamill, T., Cole, J.L., and Hazuda, D.J. (2000) HIV-1 integrase inhibitors that compete with the target DNA substrate define a unique strand transfer conformation for integrase. *Proc. Natl. Acad. Sci. U.S.A.*, **97**, 11244–11249.
38. Neamati, N., Lin, Z., Karki, R.G., Orr, A., Cowansage, K., Strumberg, D., Pais, G.C.G., Voigt, J.H., Nicklaus, M.C., Winslow, H.E., Zhao, H., Turpin, J.A., Yi, J., Skalka, A.M., Burke, T.R. Jr., and Pommier, Y. (2002) Metal-dependent inhibition of HIV-1 integrase. *J. Med. Chem.*, **45**, 5661–5670.
39. Marchand, C., Johnson, A.A., Karki, R.G., Pais, G.C.G., Zhang, X., Cowansage, K., Patel, T.A., Nicklaus, M.C., Burke, T.R. Jr., and Pommier, Y. (2003) Metal-Dependent Inhibition of HIV-1 integrase by β-diketo acids and resistance of the soluble double-mutant (F185K/C280S). *Mol. Pharmacol.*, **64**, 600–609.
40. Maurin, C., Bailly, F., Buisine, E., Vezin, H., Mbemba, G., Mouscadet, J.F., and Cotelle, P. (2004) Spectroscopic studies of diketoacids-metal interactions. A probing tool for the pharmacophoric intermetallic distance in the HIV-1

integrase active site. *J. Med. Chem.*, **47**, 5583–5586.

41. Barreca, M.L., Ferro, S., Rao, A., Luca, L.D., Zappala, M., Monforte, A.M., Debyser, Z., Witvrouw, M., and Chimirri, A. (2005) Pharmacophore-based design of HIV-1 integrase strand-transfer inhibitors. *J. Med. Chem.*, **48**, 7084–7088.

42. Goldgur, Y., Craigie, R., Cohen, G.H., Fujiwara, T., Yoshinage, T., Fujishita, T., Sugimoto, H., Endo, T., Murai, H., and Davies, D.R. (1999) Structure of the HIV-1 integrase catalytic domain complexed with an inhibitor: a platform for antiviral drug design. *Proc. Natl. Acad. Sci. U.S.A.*, **96**, 13040–13043.

43. Zhuang, L., Wai, J.S., Embrey, M.W., Fisher, T.E., Egbertson, M.S., Payne, L.S., Guare, J.P. Jr., Vacca, J.P., Hazuda, D.J., Felock, P.J., Wolfe, A.L., Stillmock, K.A., Witmer, M.V., Moyer, G., Schleif, W.A., Gabryelski, L.J., Leonard, Y.M., Lynch, J.J. Jr., Michelson, S.R., and Young, S.D. (2003) Design and synthesis of 8-hydroxy-[1,6]naphthyridines as novel inhibitors of HIV-1 integrase in vitro and in infected cells. *J. Med. Chem.*, **46**, 453–456.

44. Hazuda, D.J., Young, S.D., Guare, J.P., Anthony, N.J., Gomez, R.P., Wai, J.S., Vacca, J.P., Handt, L., Motzel, S.L., Klein, H.J., Dornadula, G., Danovich, R.M., Witmer, M.V., Wilson, K.A.A., Tussey, L., Schleif, W.A., Gabryelski, L.S., Jin, L., Miller, M.D., Casimiro, D.R., Emini, E.A., and Shiver, J.W. (2004) Integrase inhibitors and cellular immunity suppress retroviral replication in rhesus macaques. *Science*, **305**, 528–532.

45. Wang, Y., Serradell, N., Bolos, J., and Rosa, E. (2007) MK-0518, HIV integrase inhibitor. *Drugs Future*, **32**, 118–122.

46. Herr, J.R. (2002) 5-Substituted-1H-terazoles as carboxylic acid isosteres: medicinal chemistry and synthetic methods. *Bioorg. Med. Chem.*, **10**, 3379–3393.

47. Deeks, S.G., Kar, S., Gubernick, S.I., and Kirkpatrick, P. (2008) Raltegravir. *Nat. Rev. Drug Discovery*, **7**, 117–118.

48. Summa, V., Petrocchi, A., Bonelli, F., Crescenzi, B., Donghi, M., Ferrara, M., Fiore, F., Gardelli, C., Paz, O.G., Hazuda, D.J., Jones, P., Kinzel, O., Laufer, R., Monteagudo, E., Muraglia, E., Nizi, E., Orvieto, F., Pace, P., Pescatore, G., Scarpelli, R., Stillmock, K., Witmer, M.V., and Rowley, M. (2008) Discovery of raltegravir, a potent, selective orally bioavailable HIV-integrase inhibitor for the treatment of HIV-AIDS infection. *J. Med. Chem.*, **51**, 5843–5855.

49. Summa, V., Petrocchi, A., Matassa, V.G., Gardelli, C., Muraglia, E., Rowley, M., Paz, O.G., Laufer, R., Monteagudo, E., and Pace, P. (2006) 4,5-Dihydroxypyrimidine carboxamides and N-alkyl-5-hydroxypyrimidinone carboxamides are potent, selective HIV integrase inhibitors with good pharmacokinetic profiles in preclinical species. *J. Med. Chem.*, **49**, 6646–6649.

50. Eron, J.J. Jr., Rockstroh, J.K., Reynes, J., Andrade-Villanueva, J., Ramalho-Madruga, J.V., Bekker, L.G., Young, B., Katlama, C., Gatell-Artigas, J.M., Arribas, J.R., Nelson, M., Campbell, H., Zhao, J., Rodgers, A.J., Rizk, M.L., Wenning, L., Miller, M.D., Hazuda, D., DiNubile, M.J., Leavitt, R., Isaacs, R., Robertson, M.N., Sklar, P., and Nguyen, B.Y. (2011) Raltegravir once daily or twice daily in previously untreated patients with HIV-1: a randomised, active-controlled, phase III non-inferiority trial. *Lancet Infect. Dis.*, **11** (12), 907–915.

51. Dyda, F., Hickman, A.B., Jenkins, T.M., Engelman, A., Craigie, R., and Davies, D.R. (1994) Crystal structure of the catalytic domain of HIV-1 integrase: similarity to other polynucleotidyl transferases. *Science*, **266**, 1981–1986.

52. Rice, P.A. and Baker, T.A. (2001) Comparative architecture of transposase and integrase complexes. *Nat. Struct. Biol.*, **8**, 302–307.

53. Yang, W. and Steitz, T.A. (1995) Recombining the structures of HIV integrase, RuvC and RNase H. *Structure*, **3**, 131–134.

54. Kirschberg, T. and Parrish, J. (2007) Metal chelators as antiviral agents. *Curr. Opin. Drug Discovery Dev.*, **10**, 460–472.

55. Sato, M., Kawakami, H., Motomura, T., Aramaki, H., Matsuda, T., Yamashita, M., Ito, Y., Matsuzaki, Y., Yamataka, K., Ikeda, S., and Shinkai, H. (2009)

Quinolone carboxylic acids as a novel monoketo acid class of human immunodeficiency virus type 1 integrase inhibitors. *J. Med. Chem.*, **52**, 4869–4882.

56. Sechi, M., Derudas, M., Dallocchio, R., Dessi, A., Bacchi, A., Sannia, L., Carta, F., Palomba, M., Ragab, O., Chan, C., Shoemaker, R., Sei, S., Dayam, R., and Neamati, N. (2004) Design and synthesis of novel indole β-diketo acid derivatives as HIV-integrase inhibitors. *J. Med. Chem.*, **47**, 5298–5310.

57. DeJesus, E., Berger, D., Markowitz, M., Cohen, C., Hawkins, T., Ruane, P., Elion, R., Farthing, C., Zhong, L., Cheng, A., McColl, D., and Kearney, B.P. (2006) Antiviral activity, pharmacokinetics, and dose response of the HIV–1 integrase inhibitor GS–9137 (JTK–303) in treatment-naive and treatment-experienced patients. *J. Acquir. Immune. Defic. Syndr.*, **43**, 1–5.

58. Molina, J.M., LaMarca, A., Andrade-Villanueva, J., Clotet, B., Clumeck, N., Liu, Y.P., Zhong, L., Margot, N., Cheng, A.K., and Chuck, S.L. (2012) Efficacy and safety of once daily elvitegravir versus twice daily raltegravir in treatment-experienced patients with HIV–1 receiving a ritonavir-boosted protease inhibitor: randomised, double-blind, phase 3, non-inferiority study. *Lancet Infect. Dis.*, **12**, 27–35.

59. Wills, T. and Vega, V. (2012) Elvitegravir: a once-daily inhibitor of HIV-1 integrase. *Expert Opin. Investig. Drugs*, **21** (3), 395–401.

60. Palella, F.J. Jr., Delaney, K.M., Moorman, A.C., Loveless, M.O., Fuhrer, J., Satten, G.A., Aschman, D.J., and Holmberg, S.D. (1998) Declining morbidity and mortality among patients with advanced human immunodeficiency virus infection. *N. Engl. J. Med.*, **338**, 853–860.

61. Cohen, C., Elion, R., Ruane, P., Shamblaw, D., DeJesus, E., Rashbaum, B., Chuck, S.L., Yale, K., Liu, H.C., Warren, D.R., Ramanathan, S., and Kearney, B.P. (2011) Randomized, phase 2 evaluation of two single-tablet regimens elvitegravir/cobicistat/emtricitabine/tenofovir disoproxil fumarate versus efavirenz/emtricitabine/tenofovir disoproxil fumarate for the initial treatment of HIV infection. *AIDS*, **25**, F7–F12.

62. Sax, P.E., DeJesus, E., Mills, A., Zolopa, A., Cohen, C., Wohl, D., Gallant, J.E., Liu, H.C., Zhong, L., Yale, K., White, K., Kearney, B.P., Szwarcberg, J., Quirk, E., and Cheng, A.K. (2012) Co-formulated elvitegravir, cobicistat, emtricitabine, and tenofovir versus co-formulated efavirenz, emtricitabine, and tenofovir for initial treatment of HIV-1 infection: a randomised, double-blind, phase 3 trial, analysis of results after 48 weeks. *Lancet*, **379**, 2439–2448.

63. DeJesus, E., Rockstroh, J.K., Henry, K., Molina, J.M., Gathe, J., Ramanathan, S., Wei, X., Yale, K., Szwarcberg, J., White, K., Cheng, A.K., and Kearney, B.P. (2012) Co-formulated elvitegravir, cobicistat, emtricitabine, and tenofovir disoproxil fumarate versus ritonavir-boosted atazanavir plus co-formulated emtricitabine and tenofovir disoproxil fumarate for initial treatment of HIV-1 infection: a randomised, double-blind, phase 3, non-inferiority trial. *Lancet*, **379**, 2429–2438.

Hisashi Shinkai received a PhD from Tohoku University in Japan and joined Ajinomoto Co., Inc. as medicinal chemist in 1982 and discovered a new antidiabetic drug, nateglinide. Since 1992 he has been leading medicinal chemistry at JT Inc., Central Pharmaceutical Research Institute, He has authored over 55 published articles, patents, and book chapters covering various aspects of drug discovery with an emphasis on infectious diseases, diabetes, and cardiovascular diseases.

7
Discovery of Linagliptin for the Treatment of Type 2 Diabetes Mellitus

Matthias Eckhardt, Thomas Klein, Herbert Nar, and Sandra Thiemann

7.1
Introduction

First isolated in 1966, dipeptidyl peptidase-4 (DPP-4) is a widely distributed, membrane-bound, but also secreted, serine protease that is responsible for the rapid inactivation of glucagon-like peptide (GLP)-1 and glucose-dependent insulinotropic peptide (GIP) through cleavage of a dipeptide from the N-terminus of these oligopeptides [1, 2]. GLP-1 and GIP are released from the gut in response to food intake and exert a potent glucose-dependent insulinotropic action and thereby contribute to the maintenance of postprandial glycemic control [3]. GLP-1 is also involved with inhibition of postprandial glucagon secretion from pancreatic α-cells, retardation of gastric emptying, suppression of appetite leading to reduction in food intake, and direct beneficial effects on pancreatic β-cells [4, 5]. Inhibition of DPP-4 prolongs the half-life, and thus the activity of GLP-1 and GIP. Inhibition of GLP-1 and GIP metabolic breakdown was shown to improve glycemic control in patients with type 2 diabetes mellitus (T2DM), making it an excellent therapeutic target for T2DM [6]. The search for small molecules that can provide safe and effective inhibition of DPP-4 has resulted in the discovery and regulatory approval of eight DPP-4 inhibitor drugs to date [7, 8].

The DPP-4 inhibitor linagliptin is characterized by high selectivity for its biological target, in addition to its efficacious properties, and has achieved marketing approval for treatment of diabetes in numerous countries worldwide. Linagliptin, licensed as a once-daily, one dose strength medication for all adult patients without any need for dose adjustments, is available to physicians as stand-alone (Trajenta®, Tradjenta®, Trazenta®, Trayenta®) or combination treatment (with metformin: Jentadueto®) [9].

Successful Drug Discovery, First Edition. Edited by János Fischer and David P. Rotella.
© 2015 Wiley-VCH Verlag GmbH & Co. KGaA. Published 2015 by Wiley-VCH Verlag GmbH & Co. KGaA.

7.2
Discovery of Linagliptin – High Throughput Screening Hit Optimization

Our research toward the discovery of an oral DPP-4 inhibitor started by screening over 500 000 different molecules using high throughput screening (HTS), which identified several compounds belonging to different structural classes with DPP-4 inhibitory activity in the low micromolar range [10]. Of these candidates, one in particular, compound **II**, based on the xanthine scaffold, which is elaborated with four residues of different structural nature (Figure 7.1), demonstrated characteristics that were considered to represent a promising starting point for the search for an effective DPP-4 inhibitor.

It was first envisaged that the substituents attached to the xanthine core should be varied to study the structure–activity relationship (SAR) in the process of optimizing DPP-4 inhibition. A further inspection of the xanthine core itself was considered after elucidation of the contribution to the inhibitory profile of each substituent. Our research plan was to scrutinize the residues on xanthine successively starting with the residues on N-1, N-7, and C-8, as initial modifications to the structure had indicated their pronounced impact on DPP-4 activity; the residue on N-3 was then to be examined with a compound favorably decorated at all other sites.

The general synthesis route depicted in Scheme 7.1 was established as a means of delivering an efficient and selective pathway for the proposed structural modifications. Compound **III** was identified as a candidate molecule that would permit the selective introduction of a residue attached to N-7 and N-1, respectively, via a carbon atom by reacting with an alkyl electrophile, and of a residue attached

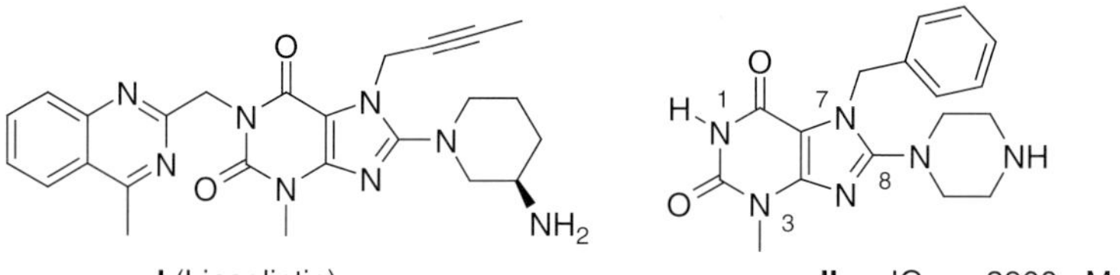

Figure 7.1 Structures of linagliptin and the HTS hit selected for in-depth evaluation.

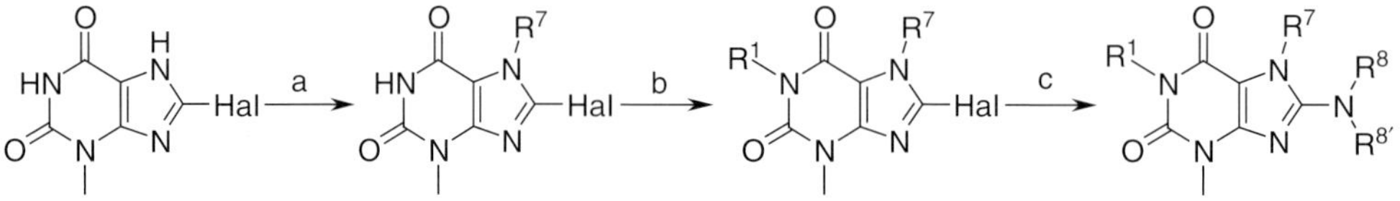

Scheme 7.1 Reagents and conditions: (a) an alkyl group – Halogen' (R^7-Hal'), N,N-diisopropyl-ethylamine or triethylamine, dimethylformamide, room temperature; (b) R^1-Hal', potassium carbonate (K$_2$CO$_3$), dimethylformamide, room temperature; and (c) H-NR^{8}R$^{8'}$, K$_2$CO$_3$, dimethylformamide, or dimethylsulfoxide, 75 °C; Halogen (Hal) = Cl, Br; Halogen' (Hal') = Cl, Br, I.

to C-8 via a nitrogen atom by displacing the halogen atom with an amine. Reaction of compound **III** with an alkyl halide in the presence of a tertiary amine base furnished the monoalkylated product derivatized selectively at N-7; neither the competing N-9 nor the N-1 alkylated product was detected. Alkylation of N-1 was then achieved using a carbonate base to deliver compound **V**; O-alkylation at C-6 instead of N-alkylation was not observed. Replacement of the halogen atom at C-8 with an amino group was accomplished using a carbonate base in dimethylsulfoxide, or *N,N*-dimethylformamide at elevated temperature. The order of introducing the substituents on N-1 and C-8 could be reversed, allowing specific variations at either site at a later stage of the reaction sequence. In addition, it was feasible to carry out the last two steps sequentially in one reaction vessel without the need for work-up in between. Accordingly, compound **IV** was first reacted with the alkyl halide in the presence of potassium carbonate in *N,N*-dimethylformamide at ambient temperature. After complete consumption of the starting material, the amine was added and the temperature increased so as to deliver yields that were comparable to those obtained by the step-by-step procedure.

The ability of compounds to inhibit DPP-4 activity was determined using a preparation of human DPP-4 derived from Caco-2 cells. Data on DPP-4 inhibition obtained for close analogs of compound **II**, also tested in the HTS process, indicated the important contribution of a basic amino group incorporated in the residue attached to C-8: replacing piperazine with morpholine or piperidine in compound **II** resulted in derivatives with little or no inhibitory activity on DPP-4. In addition, the first modifications to compound **II** (Figure 7.2) suggested that a favorable arrangement with 3,3-dimethyl-prop-2-enyl or but-2-ynyl at N-7 provided good potency on DPP-4 inhibition. Consequently, the residue on C-8 underwent a comprehensive series of variations based on the core structure of compound **2** (Table 7.1). These investigations suggested 2-amino-ethyl-amines to be the most suitable embodiment at this site. In particular, installation of 3-aminopiperidine (**2–11**) resulted in compounds with markedly increased potency. In fact, compounds with only minor deviations from this putatively optimal constellation, such as replacing the amino group with a less basic or tertiary amino group or with an altered alignment of the amino group with respect to the scaffold, showed considerably decreased DPP-4 inhibition.

Compound **3a**, featuring beneficial residues on N-7 and C-8, was chosen as the platform from which to study the group on N-1 (Table 7.2). Attachment of benzyl or phenethyl to N-1 yielded compounds of inferior DPP-4 activity compared with

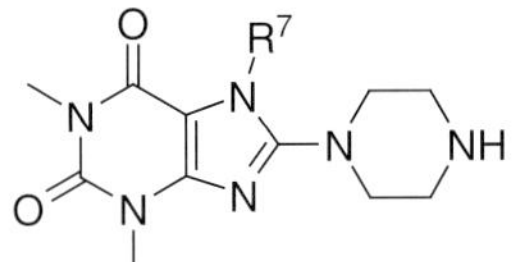

a: $R^7 = C_6H_5\text{-}CH_2$: $IC_{50} = 2800$ nM
b: $R^7 = C_6H_5\text{-}CH_2CH_2$: $IC_{50} > 10\,000$ nM
c: $R^7 = C_6H_{11}\text{-}CH_2$: $IC_{50} > 10\,000$ nM
d: $R^7 = (o\text{-}NC\text{-}C_6H_4)CH_2$: $IC_{50} = 6006$ nM
e: $R^7 = (H_3C)_2CCHCH_2$: $IC_{50} = 580$ nM
f: $R^7 = H_3CCCCH_2$: $IC_{50} = 200$ nM

Figure 7.2 DPP-4 inhibitory activity of compounds varied at N-7.

Table 7.1 DPP-4 inhibitory activity of xanthines varied at C-8.

Compound	R⁸	DPP-4 IC$_{50}$ (nM)
2-1	piperazine	580
2-2	homopiperazine	287
2-3	N-methyl-N-(2-methylaminoethyl)amine	12 130
2-4	N-(2-methylaminoethyl)amine	2600
2-5	N-(2-aminoethyl)amine	4600
2-6	N-(3-aminopropyl)amine	>10 000
2-7	N-methyl-N-(2-aminoethyl)amine	95
2-8	N-methyl-N-(2-hydroxyethyl)amine	>100 000
2-9	N-methyl-N-(2-amino-2-methylpropyl)amine	381

(*continued overleaf*)

Table 7.1 (Continued)

Compound	R⁸	DPP-4 IC$_{50}$ (nM)
2-10	3-aminophenyl	>100 000
2-11	3-aminopiperidin-1-yl	35
(R)-2-11	(R)-3-aminopiperidin-1-yl	52
(S)-2-11	(S)-3-aminopiperidin-1-yl	22
2-12	3-amino-3-methylpiperidin-1-yl	361
2-13	3-aminoazepan-1-yl	~5000
2-14	3-aminoazepan-1-yl (isomer)	580
2-15	3-aminopyrrolidin-1-yl	~5000
2-16	2-aminocyclohexyl (d.r. ~1:1)	100
2-17	3-(methylamino)piperidin-1-yl	4620
2-18	3-(dimethylamino)piperidin-1-yl	>10 000

Table 7.2 DPP-4 inhibitory activity of xanthines varied at N-1.

a: R^7 = $CH_2CH=C(CH_3)_2$
b: R^7 = CH_2CCCH_3

Compound[a]	R^1	R^7	DPP-4, IC_{50} (nM)	hERG[b] (%)	M1, IC_{50} (nM)
3a-1	Phenyl-CH_2	a	284	n.d.[c]	n.d.[c]
3a-2	Phenyl-CH_2CH_2	a	56	n.d.[c]	n.d.[c]
3a-3	(naphthyl-CH_2)	a	22	16	n.d.[c]
3b-3	(naphthyl-CH_2)	b	15	n.d.[c]	451
(R)-3a-4	(isoquinolinyl-CH_2)	a	4	51	5
(S)-3a-4	(isoquinolinyl-CH_2)	a	2	51	26
(R)-3a-5	(phenacyl)	a	6	31	25
(S)-3a-5	(phenacyl)	a	3	23	50
(R)-3b-5	(phenacyl)	b	4	88[d]	6161
(S)-3b-5	(phenacyl)	b	9	88[d]	1112
(R)-3b-6	(2-(CONHCH$_3$-methoxy)phenacyl)	b	1	88	1174
(S)-3b-6	(2-(CONHCH$_3$-methoxy)phenacyl)	b	2	93[d]	240
(R)-3b-7	(methylisoquinolinyl-CH_2)	b	3	78	430
(S)-3b-7	(methylisoquinolinyl-CH_2)	b	3	93[d]	283
(R)-3b-8 (I)	(methylquinazolinyl-CH_2)	b	1	97	295
(S)-3b-8	(methylquinazolinyl-CH_2)	b	7	n.d.[c]	546

a) Compound numbers prefixed with (R) or (S) denote the enantiomerically pure isomer of the compound.
b) Current remaining at a compound concentration of 1 µM.
c) Denotes not determined.
d) Value obtained with racemic compound.

the methyl derivative **2–11**. Conversely, attaching phenacyl or fused bicyclic aryl- or heteroarylmethyl residues resulted in compounds of high potency in the one-digit nanomolar range. Unfortunately, further characterization of the enantiomerically pure compounds revealed unacceptably high interaction with the muscarinic receptor 1 (M1) and the human *ether-à-go-go*-related gene (hERG) channel that were expressed by all of these variations on N-1. Because it was not possible to reduce either of the unwanted characteristics sufficiently while retaining the desired high inhibitory activity based on the xanthine bearing dimethylallyl on N-7, it was postulated whether switching to an alternative N-7 derivatized scaffold would address this issue. Indeed, attaching but-2-ynyl to N-7 proved to be an effective means of abolishing interaction with the hERG channel and also decreased interaction with the M1 receptor to acceptable levels. Various phenacyl and fused bicyclic aryl- and heteroarylmethyl residues attached to N-1 of this scaffold exhibited high DPP-4 inhibition with no perturbing interactions with the hERG channel or M1 receptor (series 3b in Table 7.2). Intriguingly, when the respective enantiomeric pairs were compared, the *R*-configured compounds of the 7-butynyl series were more active in almost all cases tested. This finding contrasted with the results obtained from the 7-dimethylallyl series, where the *S*-configured compounds showed higher potency.

It was concluded that but-2-ynyl is the most suitable substituent at N-7 due to the associated favorable characteristics it conveys to the compounds, particularly their decreased interaction with the M1 receptor and hERG channel. In addition, a series of suitable substituents for N-1 was identified that, when combined with the knowledge of the superior group at C-8, provided an understanding of SAR sufficient to permit the rational development of new xanthine-based compounds with promising DPP-4 inhibitory activity. Nevertheless, to obtain a more complete understanding of the SAR and to generate potential molecular alternatives, we proceeded according to our research plan, addressing next the role of the substituent at N-3. The task required a new synthesis route to access compounds varying at N-3, in a fashion that was equal in speed and efficacy to the previous means of assembling derivatives. Compound **VII** was considered to be a suitable starting position, allowing the synthesis of a derivative in two steps: attachment of a group via its alkylene carbon atom to N-3, accomplished using the respective alkyl halide and a carbonate base, followed by removal of the protecting group (Scheme 7.2).

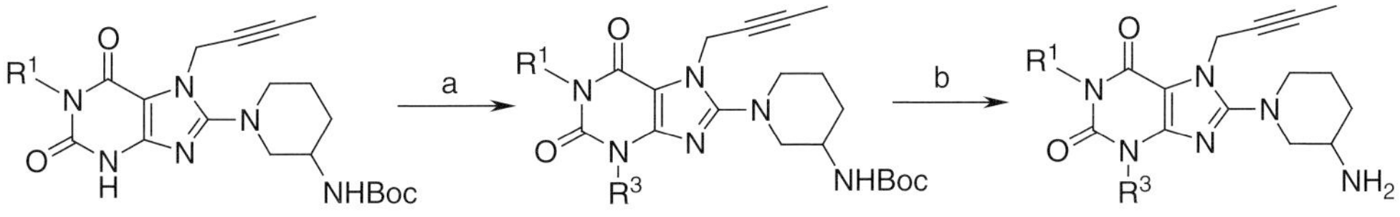

Scheme 7.2 Reagents and conditions: (a) R^3-Hal, K$_2$CO$_3$, dimethylformamide, 60 °C and (b) F$_3$CCO$_2$H, CH$_2$Cl$_2$, room temperature; Hal = Cl, Br, I; Boc = (H$_3$C)$_3$COC(=O).

Table 7.3 DPP-4 inhibitory activity of xanthines varied at N-3.

4a: R^1 = naphth-1-ylmethyl
4b: R^1 = 4-methylquinazolin-2-ylmethyl

Compound	R^3	DPP-4 IC$_{50}$ (nM)
4a-1	H$_3$C	15
4a-2	Phenyl-CH$_2$CH$_2$	170
4a-3	Phenyl-CH$_2$	141
4a-4	Phenyl-C(O)CH$_2$	87
4a-5	H$_2$C=CHCH$_2$	23
4a-6	HC≡CCH$_2$	12
4a-7	NC-CH$_2$	11
4a-8	H$_3$CO$_2$CCH$_2$	10
(R)-4b-1[a]	H$_3$C	1
(R)-4b-2[a]	H$_3$CO$_2$CCH$_2$	3
(R)-4b-3[a],[b]	Phenyl	5
(R)-4b-4[a],[b]	Cyclopropyl	3

a) Compound was pure R enantiomer.
b) Compound was synthesized as described in Scheme 7.1 starting with the respective substituent already installed at N-3.

Several candidate derivatives were prepared based on the scaffold bearing the favorable residues at N-7 and C-8 and naphth-1-ylmethyl or 4-methyl-quinazolin-2-ylmethyl at N-1 (Table 7.3). The series provided only a few compounds with the potency of the respective comparators **4a-1** and (R)-**4b-1**. Furthermore, *in vitro* and *in vivo* characterization of these compounds revealed their inferiority to the respective methyl derivatives, in particular in terms of the duration of sufficient DPP-4 inhibition in rats.

Study of the impact of alterations within the xanthine scaffold was expected to complement the data set and provide a comprehensive SAR picture in terms of DPP-4 inhibition [11]. A series of compounds with different scaffolds was designed and synthesized. These compounds satisfied the basic requirements that were deduced from the available SAR data, and X-ray crystal structures of selected xanthine derivatives complexed with human DPP-4: spatial alignment of the residues at N-1, N-7, and C-8, comparable with the one achieved with xanthine, including the C-6 carbonyl group of xanthine. The residue ensemble selected for the scaffold study comprised 3-aminopiperidinyl on C-8, but-2-ynyl on N-7, and naphth-1-yl methyl or 3-methyl isoquinolin-1-yl methyl on N-1 (Table 7.4).

7.2 Discovery of Linagliptin – High Throughput Screening Hit Optimization

Table 7.4 DPP-4 inhibitory activity of xanthine-related scaffolds.

a: R^1 = naphth-1-ylmethyl
b: R^1 = 3-methyl-isoquinolin-1-ylmethyl

Compound	Scaffold	DPP-4 IC$_{50}$ (nM) a	DPP-4 IC$_{50}$ (nM) b
5-1		15	2
5-2		13	5
5-3		—	7
5-4		52	19
5-5		89	21
5-6		107	99
5-7		3059	—

(*continued overleaf*)

Table 7.4 (Continued)

Compound	Scaffold	DPP-4 IC$_{50}$ (nM) a	b
5-8	(imidazole scaffold with Br)	559	—
5-9	(imidazole scaffold with NC)	582	—
5-10	(triazole scaffold)	—	24

Initial structural alterations varied the uracil substructure of xanthine but retained the integrity of the cyclic nature and the attachment site of the N-1 residue. Replacement of the uracil substructure with a pyridazone with an additional methyl group (**5-3**) or without (**5-2**), had little effect on the inhibitory activity, whereas installing a methoxy-substituted pyridazone (**5-4**) or a pyridone (**5-5**) resulted in compounds of significantly lower potency. More substantial alterations of the uracil moiety included disbanding the ring structure, resulting in monocyclic imidazoles with even lower activities. It became clear from the data that a co-planar relationship of the imidazole ring and the amide group, with the carbonyl oxygen positioned next to the butynyl group, is a prerequisite for optimal heteroarylmethyl and carbonyl group alignment and, in turn, high DPP-4 inhibitory activity.

Based on the global SAR thus obtained, the profile of those compounds that had the best fit with the SAR conclusions were compared with the inhibitors identified during the SAR evaluation process as having the most promising molecular profile. Four compounds emerged from the selection process as complying with all criteria applied; compounds (*R*)-**3b-6**, (*R*)-**3b-7**, (*R*)-**3b-9**, and linagliptin (**I**) (Figure 7.3).

Although all compounds effected high DPP-4 inhibition and exhibited no unwanted side effects, such as interaction with the M1 receptor or hERG channel, they showed different characteristics during the in-depth profiling process. Compound **I** (linagliptin) prevailed, showing high efficacy in all animal models assayed, superior selectivity toward other enzymes and receptors, and a favorable pharmacokinetic profile (*vide infra*).

I (Linagliptin)

(*R*)-**3b-6**

(*R*)-**3b-7**

(*R*)-**3b-9**

Figure 7.3 The four compounds identified from the selection process complying with all criteria applied.

7.3
Rationalization of DPP-4 Inhibition Potency by Crystal Structure Analysis and Studies of Binding Kinetics

The DPP-4 ectodomain is a functional homodimer whose subunits exhibit an N-terminal β-propeller domain and a catalytic protease domain of the α,β-hydrolase fold (Figure 7.4). The active site of the protease is contained within an interior cavity that is accessible via the central pore of the β-propeller domain or a larger lateral opening between the two subdomains.

The X-ray crystal structure of compound **I** in complex with human DPP-4, makes it possible to highlight the main interactions of the inhibitor within the

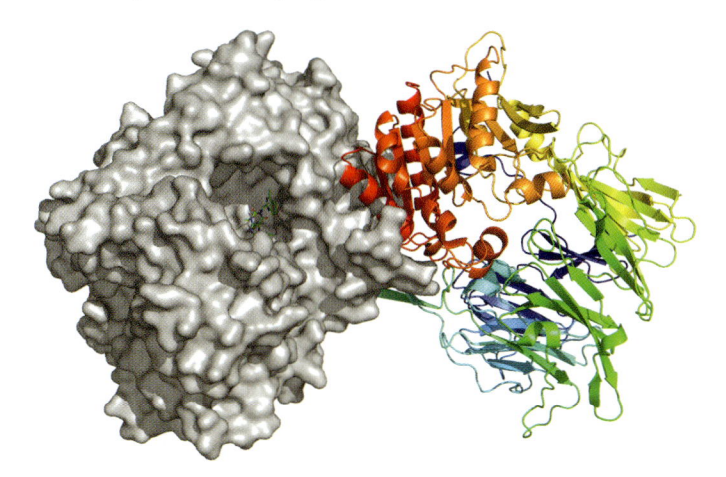

Figure 7.4 DPP-4 homodimer structure as determined by X-ray crystallography (surface representation with bound linagliptin – left; ribbon representation with rainbow coloring – right).

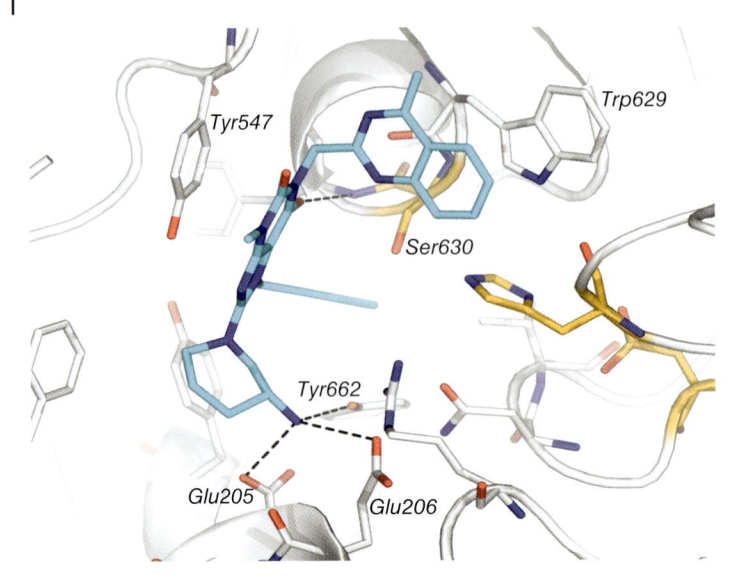

Figure 7.5 Compound I (light-blue carbon atoms) bound to DPP-4. Active site residues Ser630, His740, and Asp708 are shown in orange. Three hydrogen bonds (shown by black dashes) are formed by the amino function on the piperidine ring with acceptor groups on the protein Glu205, Glu206, and Tyr662. A fourth hydrogen bond is formed between the C-6 carbonyl of the xanthine scaffold and the backbone amide of residue Tyr631. Aromatic stacking interactions are formed between the xanthine ring system and Tyr547 as well as between the quinazoline ring and Trp629.

active site of the enzyme and to rationalize the observed SAR (Figure 7.5). The aminopiperidine substituent at C-8 of the xanthine scaffold occupies the S2 subsite. The primary amine forms a network of charge-reinforced hydrogen bonds to the Glu205, Glu206, and Tyr662 amino acid residues that constitute the recognition site for the amino terminus of peptide substrates of DPP-4. The butynyl substituent at N-7 occupies the hydrophobic S1 pocket of the enzyme. The xanthine moiety is positioned such that its uracil moiety lies on top of Tyr547, forming aromatic π-stacking interactions with the phenol of Tyr547. Thereby, the side chain of Tyr547 is pushed from its "relaxed" position in the uncomplexed enzyme and the substrate bound form. A similar conformational change has been reported for related xanthine-based inhibitors and for inhibitors from other structural classes. The C-6 carbonyl function of the xanthine scaffold forms a hydrogen bond with the backbone NH of Tyr631. Finally, the quinazoline substituent at N-1 is placed on a hydrophobic surface patch of the protein, and interacts with Trp629 by π-stacking its phenyl ring with the pyrrole ring of the amino acid side chain.

The crystal structure explains much of the observed SARs described in Section 7.2.

The substituent at the C-8 of the xanthine scaffold is a crucial factor for improving affinity (Table 7.1). The molecule originally identified by HTS, piperazine derivative **II**, as well as a close analog **2-1**, can form only two hydrogen bonds with the protein, hence it needs to adopt an unfavorable conformation to facilitate binding. Open chain derivatives with a primary amine, capable of forming three hydrogen bonds with the protein, showed much weaker affinity (e.g., **2-5**), which suggests that a rigid scaffold is beneficial to keep the entropic penalty of binding low. Methylation at the exocyclic nitrogen of the open chain derivatives alleviates this loss of activity, probably due to stabilization of receptor bound conformations around the exocyclic bond at C-8. The "boost" in affinity that follows introduction of the aminopiperidine group at C-8 appears to result from both intimate interactions of the positively charged terminal primary ammonium that can form three strong hydrogen bonds as well as conformational restriction. The observed bound chair conformation of the piperidine ring is a low-energy conformation.

No clear stereoselectivity is observed in the xanthine derivatives (Tables 7.1 and 7.2). Both stereoisomers are able to bind almost equally well to the protein. The three-dimensional structures of DPP-4 complexes with linagliptin and its enantiomer show that adoption of distinct piperidine conformations allows for both enantiomers positioning of the crucial primary amine substituent at the hot spot position between the Glu205, Glu206, and Tyr662 amino acid residues.

An optimization of the geometry of π-stacking interactions with Trp629 explains the SAR for R1-substituents (Table 7.2). The increasing activities of the phenacyl moiety (**3a/b-5**) or bicyclic ring systems (**3b-6/7/8**) are due to the presence of planar arrangements of the substituents that place an aromatic ring above the Trp629 indole ring.

Finally, relatively flat SAR is observed for small R3-substituent variations (Table 7.3). The structures show that exocyclic bond vector at N-3 points away from the protein surface into bulk solvent such that it is not expected that specific ligand–protein interactions would be formed that would significantly impact activity of the inhibitors.

Linagliptin dissociates slowly from human recombinant DPP-4 as measured by enzyme kinetics [12] ($k_{off} = 3.0 \times 10^{-5}$ s^{-1}). The extremely slow off-rate is the main factor for the high affinity and results in a prolonged drug-target residence time of 9 h (see Sections 7.4 and 7.5).

7.4
Basic Physicochemical, Pharmacological, and Kinetic Characteristics

Linagliptin (**I**) has a molecular weight of 472.5 Da, a log D value of 0.4 at pH 7.4, and pK_a values corresponding to the protonation of the quinazoline and the primary amino group of 1.9 and 8.6, respectively. A favorable crystalline modification of the free base of linagliptin (**I**) is characterized by a high melting point (202 °C) and high aqueous solubility at physiological pH value (pH 7.4, >5 g l^{-1}).

Linagliptin is further characterized by high stability in human liver cytosolic and microsomal compartments ($t_{1/2} > 90$ min), low interaction with cytochrome P450 enzymes (3A4/2D6/2C9/2C19/1A2: $IC_{50} > 50\,\mu M$) and human muscarinic receptors (M1/M2/M3 receptor, IC_{50}: 295/1038/836 nM) as well as its low affinity for the hERG channel ($IC_{50} > 30\,\mu M$). Linagliptin has a mean residence time ($MRT_{tot,oral}$) of 14 h in rats and 17 h in cynomolgus monkeys, a volume of distribution at steady state (V_{ss}) of $>5\,l\,kg^{-1}$ in the former and about $10\,l\,kg^{-1}$ in the latter species, and an oral bioavailability of around 50% in both species at a dose of $5\,mg\,kg^{-1}$. Once absorbed, linagliptin is distributed into tissues with high DPP-4 expression such as the kidney and liver. In contrast to other known DPP-4 inhibitors, which possess linear pharmacokinetics, linagliptin is unique in having non-linear pharmacokinetics in the therapeutic dose range (see Section 7.6.1). This was demonstrated by saturable affinity binding of linagliptin to soluble DPP-4 in the plasma of mice, rats, and humans. DPP-4 is a membrane-bound protein and is expressed in almost all tissues (strongly expressed in tissues such as kidney, liver, lung, and intestine). It also exists in a soluble form in the plasma. The binding characteristics of linagliptin were shown to be absent in DPP-4 deficient animals [13]. Linagliptin is a P-gp (P-glycoprotein) substrate and does not cross the blood brain barrier. The limited absorption and the reduced bioavailability of linagliptin could be explained by P-gp. Despite extended tissue retention of linagliptin, accumulation appears to be limited [14]. High protein/DPP-4 binding explains the primarily hepatic route of elimination that occurs in the absence of significant metabolism and also the low dissociation rate (k_{off}) from DPP-4 accounts for the long half-life: 36 h in rats, 39 h in monkeys, and more than 100 h in humans (see Section 7.6.1).

The IC_{50} half maximal inhibitory concentration of linagliptin on its primary target, DPP-4, is 1 nM, which is at least 1 order of magnitude lower than that of other DPP-4 inhibitors. Linagliptin possesses high selectivity for DPP-4, with IC_{50} values of $40\,\mu M$ for DPP-8 and $>10\,\mu M$ for DPP-9. Inhibition of DPP-8/9 is reported to be associated with undesirable effects such as alopecia, thrombocytopenia, and skin lesions in one single study but was never reported in humans [15]. In addition, selectivity versus other proteases such as aminopeptidase N/P, prolyloligopeptidase, trypsin, plasmin, and thrombin was more than 10 000-fold. Selectivity versus fibroblast activating protein (FAP), a matrix-degrading enzyme with 52% amino acid sequence homology to DPP-4, was less pronounced ($IC_{50} = 89$ nM) than for other proteases. In contrast to DPP-4, which is ubiquitously expressed, FAP is selectively expressed in activated tumor stoma fibroblasts found in carcinomas (e.g., breast, colorectal) and in regeneration tissue [16]. It is also involved in tissue remodeling such as wound healing and fibrosis. Binding characteristics of linagliptin for DPP-4 ($k_{off} = 3.0 \times 10^{-5}\,s^{-1}$) compared with FAP ($k_{off} = 0.06\,s^{-1}$) differ markedly, making significant and relevant inhibition of FAP at the therapeutic oral linagliptin dose of 5 mg (plasma concentration of around 20 nM) unlikely.

7.5
Preclinical Studies

7.5.1
Glucose Regulation by Linagliptin

Secretion of the incretin hormones GLP-1 and GIP by enteroendocrine cells in the gut in response to postprandial hyperglycemia and concomitant insulin secretion is impaired in patients with diabetes. By prolonging the half-life of GLP-1 and GIP, DPP-4 inhibitors appear to compensate for this diabetic condition thus benefiting glucose homeostasis. This was first demonstrated for linagliptin in the Zucker (fa/fa) rat model of T2DM [17]. Linagliptin (3 mg kg^{-1}) reduced total glucose excursion by approximately 30% after Zucker rats were challenged with an oral glucose tolerance test. Active GLP-1 levels also increased two to threefold with a concomitant increase in insulin secretion. The long-lasting effect of linagliptin on glucose tolerance compared with other DPP-4 inhibitors was further shown in mice. Equal doses of linagliptin, sitagliptin, or vildagliptin were dosed 45 min before the glucose challenge. Relative efficacies were similar, with a 50% reduction in glucose excursion. However, when the time between drug administration and glucose challenge was increased to 16 h, only linagliptin maintained similar levels of efficacy. This also reflects the longer half-life of linagliptin compared with other DPP-4 inhibitors.

Repeat doses of linagliptin were administered for 4 weeks to streptozotocin-treated mice on a high fat diet (HFD) [12]. Linagliptin reduced DPP-4 activity in a dose-dependent manner, increased active GLP-1 levels threefold and markedly reduced glucose excursions. After 4 weeks, the change in glycated hemoglobin (HbA1C) was reduced by 1% compared with vehicle controls. Linagliptin was also tested in the more severe diabetic model, the male Zucker diabetic fatty rat (ZDF rat). This model is characterized by severe insulin resistance and progressive β-cell failure and also affects male animals more severely than females. Five weeks of administration of linagliptin 3 mg kg^{-1} reduced glucose excursion by 50%, and GLP-1 levels were increased 2.5-fold. As with other DPP-4 inhibitors, linagliptin did not influence body weight, in contrast with the effects of stable GLP-1 analogs (e.g., exenatide, liraglutide).

Interestingly, in a study where Wistar rats fed on a HFD were switched from exenatide to linagliptin, time to regain any weight lost using exenatide therapy was significantly increased compared with placebo [18].

Two studies have investigated the impact of hepatic steatosis on the efficacy of linagliptin in animal models [19, 20]. Accumulation of fat in hepatocytes (fatty liver, non-alcoholic fatty liver disease) often occurs in patients with diabetes and may progress further to hepatosteatosis, fibrosis, and/or liver failure. Experimental models of non-alcoholic fatty liver disease can be derived by long-term feeding with high fat food [21]. In one of these studies, the adoption of different feeding and treatment regimens showed that therapeutic intervention with linagliptin (3–4 weeks), in animals fed on a HFD (2–4 months' feeding),

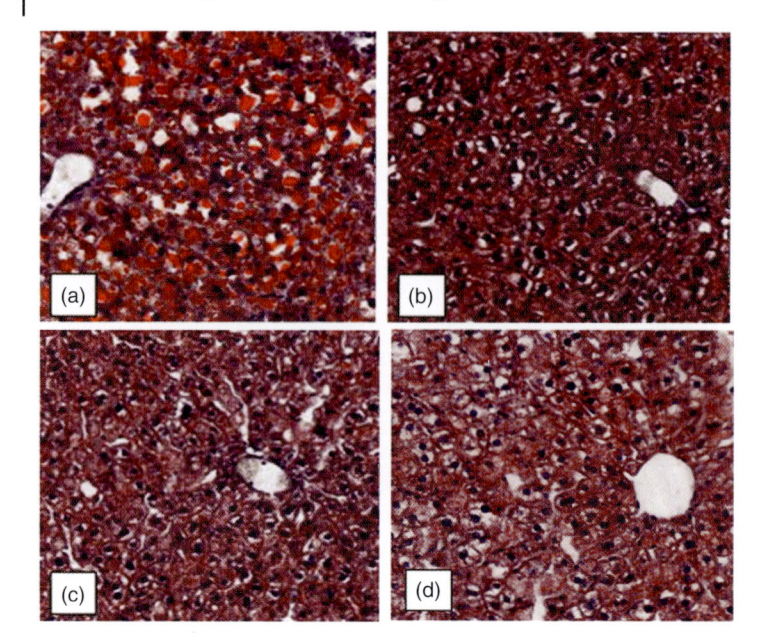

Figure 7.6 Oil Red O staining (intracellular fat positively stains red by this method, see (a)) of liver specimens from mice with diet-induced obesity (4 months) and those treated for 4 weeks with (a) vehicle, (b) linagliptin 3 mg kg^{-1} day^{-1}, (c) linagliptin 30 mg kg^{-1} day^{-1}, and (d) chow-fed mice. No fat accumulation (red staining) was seen in (b–d).

significantly reduced fat accumulation in hepatocytes by up to 30% (detected via histology, Figure 7.6 and magnetic resonance spectroscopy) [19]. These changes correlated with significant dose-dependent improvements in glucose disposal rates during the steady state of the euglycemic-hyperinsulinemic clamps at the doses tested (3 and 30 mg kg^{-1}). The improved insulin-sensitivity associated with linagliptin was also accompanied by a marked 62–83% suppression of hepatic glucose production compared with control. Active GLP-1 plasma levels increased up to 18-fold in the linagliptin groups, compared with control animals.

A study investigating liver steatosis (by means of magnetic resonance spectroscopy) was performed in Wistar rats that were fed on a HFD for 3 months [20]. Animals received vehicle, linagliptin (10 mg kg^{-1}), or the appetite suppressant sibutramine (5 mg kg^{-1}) for 6 weeks, instead of continuing with the HFD. Exposure to linagliptin was associated with a profound reduction in hepatic fat (−59.0% from baseline) compared with vehicle, with an effect that was comparable with that of sibutramine (−54.3% from baseline). However, significant reductions in body weight were only seen in the sibutramine group (−15.7%). Thus, the data suggest that linagliptin may reduce fat accumulation in the liver, thereby improving insulin sensitivity without impacting body weight and providing evidence for a further potential use of the drug for liver diseases.

7.5.2
Effects of Linagliptin on the Kidney

One of the major late stage complications in patients with type 1 diabetes mellitus (T1DM) or T2DM is the development of diabetic nephropathy, most frequently occurring after 6–20 years of disease. A dramatic increase of urinary albumin excretion (>100-fold) and marked decrease (50%) in glomerular filtration rate (GFR) are observed in patients with advanced diabetic nephropathy. Due to a decrease in renal function (filtration) and the subsequent increase in plasma concentrations of renally cleared drugs, dose adjustment is often required in nephrotic patients, whereas this is not the case with linagliptin, which is eliminated by predominantly nonrenal routes.

Despite its nonrenal elimination linagliptin binds and inhibits DPP-4 in the rodent kidney, where DPP-4 is mainly expressed in glomeruli and proximal tubulus compartments.

Animal models of diabetic nephropathy are somewhat limited and it is often necessary to derive understanding from investigation of several different ones. One model which appropriately mimics the loss of renal function is the rat 5/6 nephrectomy model (5/6N) in which a whole kidney is removed along with two-thirds of the functionality of the remaining kidney. In this model the remaining kidney compensates for the missing kidney for several weeks until nephrons begin to fail [22]. In rats administered with linagliptin 0.23 or 2.3 mg kg^{-1} for 4 days, there were no differences in plasma levels of linagliptin in rats losing renal function compared with sham-operated animals. In contrast, renally cleared drugs such as alogliptin and sitagliptin showed an approximate twofold increase in plasma levels, correlating well with the reduction (∼50%) in renal filtration in this model. In addition, there was no change in markers of glomerular or tubular stress following exposure to linagliptin, but a significant increase in active GLP-1 area under the plasma concentration–time curve (AUC) over the observed time period of 96 h compared with the sham-operated animals.

The 5/6 nephropathy model has been previously used in a chronic regimen. Animals were treated for 4 months with linagliptin (added to food), or the angiotensin receptor blocker (ARB) telmisartan (5 mg kg^{-1}, added in drinking water). At the end of treatment, the urinary albumin/creatinine ratio (ACR) had increased 14-fold in 5/6N rats compared with sham-operated rats. In contrast, ACR significantly decreased (66%) in the linagliptin group and by 92% in telmisartan-treated animals compared with 5/6N rats administered placebo. One important observation is that these effects were achieved in normoglycemic animals and therefore it may be concluded that these are independent of the effect on glucose control. Another set of experiments was conducted in endothelial nitric oxide synthase knockout mice treated with streptozocin [23]. This model matches more closely what happens in T1DM, where blood glucose levels reach values above 600 mg dl^{-1} due to β-cell toxicity of streptozocin. Neither treatment with linagliptin (3 mg kg^{-1}) or telmisartan (1 mg kg^{-1}) alone nor in combination improved the blood glucose plasma, but combined treatment significantly

(42%) reduced albuminuria, an early, sensitive, and prognostic marker of renal dysfunction, compared with diabetic controls.

Most patients suffering from diabetic nephropathy receive anti-hypertensive co-therapy, such as ARBs, but a large proportion of patients do not react adequately with improved renal function. Thus, the authors emphasized the clinical relevance and the treatment on top of ARB with the DPP-4 inhibitor linagliptin as a potential new therapeutic approach for patients with diabetic nephropathy. Little is yet known about the mechanism behind the apparent renal-protective effects of linagliptin on renal function. A recent *in vitro* study proposed an interaction of nanomolar concentrations of linagliptin with profibrotic pathways in the signaling pathways of transforming growth factor β (TGFβ) [24]. An interaction of linagliptin with TGFβ pathways was also confirmed by an independent group from *in vivo* experiments. Fibrosis and TGF-β induced endothelial-to-mesenchymal transition were reduced after 4 weeks of administration of therapeutic doses of linagliptin (5 mg kg^{-1}) in diabetic CD-1 mice [25].

In summary, data from several preclinical studies describe linagliptin as a potent anti-diabetic drug, also positively affecting liver steatosis and liver inflammation. In addition, therapeutic effects in addition to glycemic control have been shown in different animal models. These correlate with benefits for the diabetic kidney, most likely via an increase in GLP-1. In this respect, linagliptin's non-renal elimination pathway makes it a first-line candidate for use in diabetic patients with associated renal complications: not only in terms of safety and convenience of a fixed dose but also due to probably direct and glucose-independent protective effects on the kidney.

7.6
Clinical Studies

Linagliptin was first approved in the US, Europe, Japan, and other territories in 2011 to improve glycemic control in adults with T2DM, as the first DPP-4 inhibitor with a single-strength once-daily dose (5 mg).

7.6.1
Clinical Pharmacokinetics

The development of linagliptin included a clinical pharmacology program encompassing several single- and multiple-dose randomized studies of the absorption and disposition of linagliptin in healthy subjects and patients with T2DM.

In general, the pharmacokinetic characteristics of linagliptin in humans are similar to those observed previously in animals.

Linagliptin exhibits non-linear pharmacokinetics after both oral and intravenous administration of single or multiple oral doses of 1–10 mg with less than dose-proportional increases in maximum plasma concentration (Cmax) and

AUC [13, 26]. The nonlinear pharmacokinetics of linagliptin is best described by a two-compartmental model. While linagliptin plasma concentrations are low, linagliptin is predominately bound to its target DPP-4, and the free eliminatable linagliptin fraction is low. As linagliptin plasma concentrations increase DPP-4 becomes saturated, linagliptin's free fraction increases and therefore its volume of distribution and clearance increase, that is, those pharmacokinetic parameters that remain constant independent of dose for drugs with linear pharmacokinetics (e.g., clearance, volume of distribution, and fraction excreted renally) increase in the case of linagliptin when the linagliptin dose is increased. The nonlinearity of linagliptin in the therapeutic dose range is a characteristic that is not observed with other currently licensed DPP-4 inhibitors (sitagliptin [27]; saxagliptin [28]; vildagliptin [29]; alogliptin [30]), which exhibit linear pharmacokinetics. This difference is most likely due to the fact that linagliptin's binding affinity to DPP-4 is higher and its dissociation rate from DPP-4 is slower than other DPP-4 inhibitors' [17], resulting in target-mediated drug disposition.

As a result of its nonlinear pharmacokinetics, elimination of linagliptin from plasma after oral dosing occurs in a biphasic manner, with an initial rapid decline followed by a long terminal elimination phase [26]. At a dose of 5 mg, linagliptin displays a long plasma terminal half-life (>100 h) [26, 31, 32]; however, its accumulation half-life is much shorter (approximately 10 h) [26, 32], accounting for a rapid attainment of steady state. These observations are consistent with an initial rapid distribution of linagliptin into a large peripheral compartment and initial rapid elimination of non–DPP-4-bound drug (short accumulation half-life), followed by a slow dissociation of the linagliptin/DPP-4 complex (long plasma terminal half-life, likely due to the slow dissociation rate). Linagliptin is eliminated primarily in feces, with only around 5% of the oral therapeutic dose excreted in the urine at steady state [33].

Linagliptin is absorbed rapidly after oral administration, with C_{max} occurring after approximately 90 min [26, 31, 32, 34]. Steady-state concentrations are achieved at the therapeutic dose of 5 mg within 4 days [26]. With the therapeutic dose, steady-state C_{max} (11–12 nmol l^{-1}) and AUC (~150 nmol h l^{-1}) are approximately 1.3-fold greater than after a single dose, indicating little drug accumulation with repeat dosing [26].

In a randomized study in 36 healthy males, the absolute bioavailability of linagliptin was estimated to be approximately 30% [13]. The apparent volume of distribution at steady state following an intravenous infusion of linagliptin 0.5–10 mg was 380–1540 l [13], indicating extensive distribution into tissues. The dose dependency is likely to be a consequence of target-mediated drug disposition.

Metabolism plays a minor role in the overall pharmacokinetics of linagliptin in humans. Following single oral dosing of [^{14}C]-linagliptin in healthy volunteers, approximately 90% of the recovered radioactivity was identified as unchanged parent compound [33]. The main metabolite detected, CD1790, an S-3-hydroxypiperidinyl derivative of linagliptin, accounted for approximately

17% of the total drug-related compounds in plasma. CD1790 is considered to be pharmacologically inactive [33].

7.6.2 Clinical Pharmacodynamics

7.6.2.1 Inhibition of DPP-4

Linagliptin is a potent inhibitor of DPP-4 in clinical studies. In male patients with T2DM, inhibition of ≥80% of plasma DPP-4 activity for at least 24 h post dose was achieved with therapeutic dosing of linagliptin at steady state. The 10 mg once-daily dosage led only to a modest increase in inhibition of DPP-4 compared with the 5 mg dosage [26, 35]. These studies provided a pharmacodynamic rationale for the selection of 5 mg once daily as the therapeutic dose.

7.6.2.2 Effects on Glucagon-Like Peptide-1 and Hyperglycemia

The potent DPP-4 inhibition observed following once-daily dosing with linagliptin 5 mg is associated with clinically meaningful increases in plasma GLP-1 and consistent reductions in elevated plasma glucose and HbA1c in various patient populations and across various background therapies.

In a randomized, double-blind, placebo-controlled Phase II study, a meal tolerance test was conducted after 28 days of treatment (clinicaltrials.gov, NCT00328172). Linagliptin treatment doubled plasma GLP-1 concentrations, lowered 24-h weighted mean glucose, and reduced fasting plasma glucose.

In Phase III studies in patients with T2DM and inadequate glycemic control, linagliptin 5 mg significantly reduced HbA1c and fasting plasma glucose when given as either a monotherapy [36], in addition to metformin [37], in triple combination with metformin and a sulfonylurea [38], or in combination with basal insulin [39]. After 24 weeks, placebo-adjusted changes in HbA1c ranged between −0.62% and −0.69% [36, 38]. The clinical efficacy and tolerability profile of linagliptin is consistent in elderly patients [40], over different races, and independent of body mass index, body weight, age, and diabetes duration.

7.6.3 Clinical Use in Special Patient Populations

7.6.3.1 Patients with Renal Impairment

As linagliptin is mainly eliminated nonrenally, its pharmacokinetics are largely unaltered in patients with declining renal function [41]. As a consequence, dosage adjustment of linagliptin is not required in patients with declining renal function [31, 42, 43], in contrast to sitagliptin [27], saxagliptin [28], vildagliptin [29], and alogliptin [30].

The unaltered pharmacokinetics were observed in a Phase I study, which included individuals with different degrees of renal impairment (normal renal function, or mild, moderate, or severe renal impairment, or end-stage renal disease) [41]. Although mean exposure was slightly greater (20–60%) in patients with renal impairment than in those with normal renal function, the steady-state AUC and C_{max} values showed a large overlap and only a weak correlation with the degree of renal impairment. In patients with end-stage renal disease, steady-state conditions were predicted to be comparable with those observed in patients with severe renal impairment. These Phase I data are supported by a post hoc pooled analysis of linagliptin pharmacokinetics in 969 T2DM patients with renal impairment, which showed that renal impairment has a minor effect on linagliptin exposure [44].

Renal impairment is a frequent and serious T2DM complication that restricts options for managing hyperglycemia, and has an associated increased cardiovascular (CV) risk. Therefore, a lack of clinically relevant change in pharmacokinetics in patients with renal impairment is advantageous for an anti-hyperglycemic agent to facilitate its convenient use in a broad population of patients with T2DM.

The safety and efficacy of linagliptin in patients with T2DM and renal impairment was prospectively evaluated in two randomized, double-blind, placebo-controlled Phase III trials [45, 46]. In patients with severe renal impairment, linagliptin elicited a mean HbA1c reduction of −0.8% from a baseline value of 8.2%, which was maintained through at least 52 weeks (placebo-adjusted mean change in HbA1c at Week 52 −0.72%), a similar incidence of adverse events as with placebo, and stable renal function [45, 47]. These findings were confirmed in patients with moderate to severe renal impairment [46]. Patients received linagliptin for 12 weeks, then placebo patients were switched to glimepiride 1–4 mg once daily, an agent that is frequently used in renal impairment, and treatment continued to week 52. Linagliptin treatment resulted in a mean placebo-corrected HbA1c reduction of −0.42% from baseline 8.1% with a tolerability profile similar to placebo and no changes in renal function. In the 40-week extension, HbA1c was lower following linagliptin than glimepiride with fewer drug-related adverse events, less hypoglycemia and relative weight loss versus glimepiride.

These data are further supported by several post hoc pooled analyses of the Phase III study program showing clinical efficacy and tolerability of linagliptin independent of renal function ($n = 2141$ patients with T2DM were grouped by renal function, as assessed by eGFR, as normal, mild renal impairment, or moderate renal impairment) [48], in elderly patients with renal impairment on various background therapies [49], as well as in patients with renal impairment on a background of insulin [50].

The albuminuria-lowering effect of linagliptin observed in preclinical studies was investigated in a pooled analysis of four Phase III clinical trials. In patients with T2DM and diabetic nephropathy receiving a background of recommended standard treatment for renin-angiotensin-aldosterone system (RAAS) inhibition (ACE inhibitor or ARB), addition of linagliptin 5 mg once daily significantly lowered adjusted urine albumin-to-creatinine ratio (UACR), by 33% (95% CI: 22–42%; $p < 0.05$) with a between-group difference versus placebo of −29% (−3% to −48%; $p < 0.05$). The overall effect on UACR occurred as early as 12 weeks and continued over 24 weeks. This effect was independent of glucose control or blood pressure reduction.

Renal safety and outcomes of linagliptin were further evaluated in a meta-analysis including all randomized, double blind, placebo-controlled Phase III trials of more than 12 weeks' duration and 5466 patients with T2DM [51]. Linagliptin treatment led to a statistically significant 16% reduction [0.84, CI: 0.72–0.97] for the composite renal endpoint (new onset of (i) micro- or (ii) macroalbuminuria, (iii) chronic kidney disease (CKD), (iv) worsening of CKD – loss in eGFR > 50% (estimated glomerular filtration rate) versus baseline) versus placebo. The main driver of this reduction was the new onset of either micro- or macroalbuminuria.

To substantiate linagliptin's potential effect on renal function, prospective, randomized, controlled clinical trials are now needed to assess the renal effects of incretin-based therapies in patients with T2DM. The MARLINA-T2D™ study (Efficacy, Safety & Modification of Albuminuria in Type 2 Diabetes Subjects with Renal Disease with LINAgliptin; NCT01792518) will address the albuminuria-lowering potential of linagliptin. Whether treatment with linagliptin leads to long-term renal benefit will be evaluated in the long term CV and renal outcome trial CARMELINA™ (CArdiovascular safety and Renal Microvascular outcomE with LINAgliptin in patients with T2DM at high vascular risk; (NCT01897532)).

7.6.3.2 Patients with Hepatic Impairment

As linagliptin is eliminated primarily by non-renal pathways into the feces, hepatic impairment may have clinically important effects on its pharmacokinetics. However, an open label Phase I study found that linagliptin exposure was not increased in patients with mild, moderate, or severe hepatic impairment compared with those with normal hepatic function [52]. As a consequence, dosage adjustment of linagliptin is not required in patients with hepatic impairment [31, 42, 43].

7.6.4
Cardiovascular Safety

Linagliptin showed a favorable CV safety profile during clinical development. In healthy volunteers, administration of up to 20-fold the clinical dose of linagliptin did not prolong the frequency corrected QT-time interval [53].

A CV meta-analysis of clinical trials including 9459 patients (5847 received linagliptin and 3612 pooled comparators) supports the theory that linagliptin is not associated with an increased CV risk [54]. However, any meta-analysis of this nature has limitations as the number of events seen in the clinical studies was low and duration of exposure was limited. Therefore the CV safety of linagliptin is to be prospectively evaluated in the aforementioned long-term cardiovascular outcome trial (CVOT), CARMELINA™.

Interestingly, linagliptin was associated with fewer CV events (12 versus 26 patients; relative risk 0.46, 95% CI: 0.23–0.91; $p = 0.0213$) than glimepiride in a randomized controlled head-to-head trial over 2 years [55]. To establish consecutively whether linagliptin reduces CV risk compared with a sulfonylurea, a large trial (CAROLINA®, CARdiovascular Outcome study of LINAgliptin versus glimepiride in patients with T2DM; NCT01243424), designed specifically to evaluate the effect of linagliptin versus glimepiride on CV outcomes, is under way. CAROLINA® is the first head-to-head, long-term CVOT comparing a DPP-4 inhibitor (linagliptin) with a single active comparator (NCT01243424).

In summary, compared with other available DPP-4 inhibitors, linagliptin has a unique pharmacokinetic/pharmacodynamic profile that is characterized by target-mediated nonlinear pharmacokinetics, concentration-dependent protein binding, minimal renal clearance, and no requirements for dose adjustment for any intrinsic or extrinsic factors. In the clinical setting, linagliptin shows consistent glucose-lowering effects in a broad range of patients over various background therapies, and is generally well tolerated. Whether linagliptin shows differentiated clinical benefits, as a result of its chemical structure and unique pharmacological profile, will be further explored in the MARLINA-T2D™, CARMELINA™, and CAROLINA® trials.

7.7 Conclusion

Linagliptin is a highly potent, selective, and long-acting DPP-4 inhibitor of a novel chemotype that was attained upon structural optimization of a modest DPP-4 inhibitor discovered through HTS. Linagliptin's structure is based on the xanthine scaffold, which is elaborated with four different residues essential for its high potency and specificity. Its excellent inhibitory potency on DPP-4 translates into high *in vivo* potency demonstrated by substantial reduction of glucose excursion and, in turn, incidences of diabetic complications (e.g., stroke, myocardial infarction, diabetic nephropathy, and diabetic retinopathy) in various diabetic animal models. Likewise, once daily treatment of diabetic patients with different background therapies with the recommended linagliptin dose of 5 mg has been shown to result in clinically meaningful and statistically significant reductions in HbA1c, fasting plasma glucose, and postprandial glucose. Linagliptin is further characterized by a beneficial non-linear pharmacokinetic and nonrenal elimination profile, unique within the class of approved DPP-4 inhibitors, making dose adjustment for renally impaired patients unnecessary and treatment regimens convenient.

List of Abbreviations

5/6 N	5/6 nephrectomy model
ACE	angiotensin-converting enzyme
ACR	albumin/creatinine ratio
ARB	angiotensin receptor blocker
AUC	area under the plasma concentration-time curve
C_{max}	maximum plasma concentration
CV	cardiovascular
CKD	chronic kidney disease
DPP	dipeptidyl peptidase

eGFR   estimated glomerular filtration rate
FAP   fibroblast activating protein
GIP   glucose-dependent insulinotropic peptide
GLP   glucagon-like peptide
GFR   glomerular filtration rate
HbA1C   glycated hemoglobin
hERG   *ether-à-go-go*-related gene
HFD   high-fat diet
HTS   high throughput screening
IC_{50} half maximal inhibitory concentration
M1 receptor muscarinic receptor 1
MRT   mean residence time
P-gp   P-glycoprotein
RAAS   Renin-angiotensin-aldosterone system
SAR   structure–activity relationship
T1DM   type 1 diabetes mellitus
T2DM   type 2 diabetes mellitus
TGF   transforming growth factor
UACR   urine albumin-to-creatinine ratio
V_{ss} volume of distribution at steady state
ZDF rat   Zucker diabetic fatty rat

References

1. Kieffer, T.J., McIntosh, C.H.S., and Pederson, R.A. (1995) Degradation of glucose-dependent insulinotropic polypeptide and truncated glucagon-like peptide 1 in vitro and in vivo by dipeptidyl peptidase IV. *Endocrinology*, **136** (8), 3585–3597.
2. Peters, J.U. and Mattei, P. (2010) in *Analogue-Based Drug Discovery II* (eds J. Fischer and C.R. Ganellin), Wiley-VCH Verlag, GmbH & Co. KGaA, Weinheim, pp. 109–134.
3. Meier, J.J., Nauck, M.A., Schmidt, W.E. *et al.* (2002) Gastric inhibitory polypeptide: the neglected incretin revisited. *Regul. Pept.*, **107** (1–3), 1–13.
4. Egan, J.M., Bulotta, A., Hui, H. *et al.* (2003) GLP-1 receptor agonists are growth and differentiation factors for pancreatic islet beta cells. *Diabetes Metab. Res. Rev.*, **19** (2), 115–123.
5. Murphy, K.G., Dhillo, W.S., and Bloom, S.R. (2006) Gut peptides in the regulation of food intake and energy homeostasis. *Endocr. Rev.*, **27** (7), 719–727.
6. McIntosh, C.H.S. (2008) Dipeptidyl peptidase IV inhibitors and diabetes therapy. *Front. Biosci.*, **13** (1), 1753–1773.
7. Augustyns, K., Van der Veken, P., Haemers, A. *et al.* (2005) Inhibitors of proline-specific dipeptidyl peptidases: DPP IV inhibitors as a novel approach for the treatment of Type 2 diabetes. *Expert Opin. Ther. Pat.*, **15** (10), 1387–1407.
8. Deacon, C.F. and Holst, J.J. (2013) Dipeptidyl peptidase-4 inhibitors for the treatment of type 2 diabetes: comparison, efficacy and safety. *Expert Opin. Pharmacother.*, **14** (15), 2047–2058.
9. Deeks, E.D. (2012) Linagliptin: a review of its use in the management of type 2 diabetes mellitus. *Drugs*, **72** (13), 1793–1824.
10. Eckhardt, M., Langkopf, E., Mark, M. *et al.* (2007) 8-(3-(*R*)-Aminopiperidin-1-yl)-7-but-2-ynyl-3-methyl-1-(4-methyl-quinazolin-2-ylmethyl)-3,7-dihydropurine-2,6-dione (BI 1356), a highly potent, selective, long-acting, and orally bioavailable DPP-4 inhibitor for

the treatment of type 2 diabetes. *J. Med. Chem.*, **50** (26), 6450–6453.
11. Eckhardt, M., Hauel, N., Himmelsbach, F. et al. (2008) 3,5-Dihydro-imidazo[4,5-d] pyridazin-4-ones: a class of potent DPP-4 inhibitors. *Bioorg. Med. Chem. Lett.*, **18** (11), 3158–3162.
12. Thomas, L., Tadayyon, M., and Mark, M. (2009) Chronic treatment with the dipeptidyl peptidase-4 inhibitor BI 1356 [(R)-8-(3-amino-piperidin-1-yl)-7-but-2-ynyl-3-methyl-1-(4-methyl-quinolin-2-ylmethyl)-3,7-dihydro-purine-2,6-dione] increases basal glucagon-like peptide-1 and improves glycemic control in diabetic rodent models. *J. Pharmacol. Exp. Ther.*, **328** (2), 556–563.
13. Retlich, S., Withopf, B., Greischel, A. et al. (2009) Binding to dipeptidyl peptidase-4 determines the disposition of linagliptin (BI 1356) – investigations in DPP-4 deficient and wildtype rats. *Biopharm. Drug Dispos.*, **30** (8), 422–436.
14. Fuchs, H., Binder, R., and Greischel, A. (2009) Tissue distribution of the novel DPP-4 inhibitor BI 1356 is dominated by saturable binding to its target in rats. *Biopharm. Drug Dispos.*, **30** (5), 229–240.
15. Lankas, G.R., Leiting, B., Roy, R.S. et al. (2005) Dipeptidyl peptidase IV inhibition for the treatment of type 2 diabetes: potential importance of selectivity over dipeptidyl peptidases 8 and 9. *Diabetes*, **54** (10), 2988–2994.
16. Jacob, M., Chang, L., and Puré, E. (2012) Fibroblast activation protein in remodeling tissues. *Curr. Mol. Med.*, **12** (10), 1220–1243.
17. Thomas, L., Eckhardt, M., Langkopf, E. et al. (2008) (R)-8-(3-amino-piperidin-1-yl)-7-but-2-ynyl-3-methyl-1-(4-methyl-quinolin-2-ylmethyl)-3,7-dihydro-purine-2,6-dione (BI 1356), a novel xanthine-based dipeptidyl peptidase 4 inhibitor, has a superior potency and longer duration of action compared with other dipeptidyl peptidase-4 inhibitors. *J. Pharmacol. Exp. Ther.*, **325** (1), 175–182.
18. Vickers, S.P., Cheetham, S.C., Birmingham, G.D. et al. (2012) Effects of the DPP-4 inhibitor, linagliptin, in diet-induced obese rats: a comparison in naive and exenatide-treated animals. *Clin. Lab.*, **58** (7–8), 787–799.
19. Kern, M., Klöting, N., Niessen, H.G. et al. (2012) Linagliptin improves insulin sensitivity and hepatic steatosis in diet-induced obesity. *PLoS One*, **7** (6). doi: 10.1371/journal.pone.0.038744.
20. Klein, T., Niessen, H.G., Ittrich, C. et al. (2012) Evaluation of body fat composition after linagliptin treatment in a rat model of diet-induced obesity: a magnetic resonance spectroscopy study in comparison with sibutramine. *Diabetes Obes. Metab.*, **14** (11), 1050–1053.
21. Buettner, R., Schölmerich, J., and Bollheimer, L.C. (2007) High-fat Diets: modelling the metabolic disorders of human obesity in rodents. *Obesity*, **15** (4), 798–808. doi: 10.1038/oby.2007.608
22. Chaykovska, L., von Websky, K., Rahnenführer, J. et al. (2011) Effects of DPP-4 inhibitors on the heart in a rat model of uremic cardiomyopathy. *PLoS One*, **6** (11). doi: 10.1371/journal.pone.0027861
23. Alter, M.L., Ott, I.M., von Websky, K. et al. (2012) DPP-4 inhibition on top of angiotensin receptor blockade offers a new therapeutic approach for diabetic nephropathy. *Kidney Blood Press. Res.*, **36** (1), 119–130.
24. Panchapakesan, U., Gross, S., Komala, M.G. et al. (2013) DPP4 inhibition in human kidney proximal tubular cells – renoprotection in diabetic nephropathy? *J. Diabetes Metab.*, **S9** (007). doi: 10.4172/2155-6156.S9-007.
25. Kanasaki, K., Shi, S., Kanasaki, M. et al. (2014) Linagliptin-mediated DPP-4 inhibition ameliorates kidney fibrosis in streptozotocin-induced diabetic mice by inhibiting endothelial-to-mesenchymal transition in a therapeutic regimen. *Diabetes*, **63** (6), 2120–2131.
26. Heise, T., Graefe-Mody, E.U., Hüttner, S. et al. (2009) Pharmacokinetics, pharmacodynamics and tolerability of multiple oral doses of linagliptin, a dipeptidyl peptidase-4 inhibitor in male type 2 diabetes patients. *Diabetes Obes. Metab.*, **11** (8), 786–794.
27. Januvia (sitagliptin) (2014) Prescribing Information, Merck and Co. Inc., Whitehouse Station, NJ,

http://www.merck.com/product/ usa/pi_circulars/j/januvia/januvia_pi.pdf (accessed 21 October 2014).

28. Onglyza (saxagliptin) (2013) Prescribing Information, Bristol-Myers Squibb Company, Princeton, NJ, http://www.astrazeneca-us.com/medicines/astrazeneca-medications (accessed 21 October 2014).

29. EMC (2014) Galvus (Vildagliptin) Prescribing Information Camberley, Novartis Pharmaceuticals UK Ltd, Surrey, http://www.novartis.com.au/PI_PDF/gal.pdf (accessed 21 October 2014).

30. Christopher, R., Covington, P., Davenport, M. et al. (2008) Pharmacokinetics, pharmacodynamics, and tolerability of single increasing doses of the dipeptidyl peptidase-4 inhibitor alogliptin, in healthy male subjects. *Clin. Ther.*, **30** (3), 513–527.

31. Tradjenta (linagliptin) (2014) Prescribing Information, Boehringer Ingelheim Pharmaceuticals Inc., Ridgefield, CT.

32. Forst, T., Uhlig-Laske, B., Ring, A. et al. (2011) The oral DPP-4 inhibitor linagliptin significantly lowers HbA1c after 4 weeks of treatment in patients with type 2 diabetes mellitus. *Diabetes Obes. Metab.*, **13** (6), 542–550.

33. Blech, S., Ludwig-Schwellinger, E., Grafe-Mody, E.U. et al. (2010) The metabolism and disposition of the oral dipeptidyl peptidase-4 inhibitor, linagliptin, in humans. *Drug Metab. Dispos.*, **38** (4), 667–678.

34. Hüttner, S., Graefe-Mody, E.U., Withopf, B. et al. (2008) Safety, tolerability, pharmacokinetics, and pharmacodynamics of single oral doses of BI 1356, an inhibitor of dipeptidyl peptidase 4, in healthy male volunteers. *J. Clin. Pharmacol.*, **48** (10), 1171–1178.

35. Sarashina, A., Sesoko, S., Nakashima, M. et al. (2010) Linagliptin, a dipeptidyl peptidase-4 inhibitor in development for the treatment of type 2 diabetes mellitus: a phase I, randomized, double-blind, placebo-controlled trial of single and multiple escalating doses in healthy adult male Japanese subjects. *Clin. Ther.*, **32** (6), 1188–1204.

36. Del Prato, S., Barnett, A.H., Huisman, H. et al. (2011) Effect of linagliptin monotherapy on glycaemic control and markers of beta-cell function in patients with inadequately controlled type 2 diabetes: a randomized controlled trial. *Diabetes Obes. Metab.*, **13** (3), 258–267.

37. Taskinen, M.R., Rosenstock, J., Tamminen, I. et al. (2010) Safety and efficacy of linagliptin as add-on therapy to metformin in patients with type 2 diabetes: a randomized, double-blind, placebo-controlled study. *Diabetes Obes. Metab.*, **13** (1), 65–74.

38. Owens, D.R., Swallow, R., Dugi, K.A. et al. (2011) Efficacy and safety of linagliptin in persons with type 2 diabetes inadequately controlled by a combination of metformin and sulphonylurea: a 24-week randomized study. *Diabet. Med.*, **28** (11), 1352–1361.

39. Yki-Järvinen, H. and Kotronen, A. (2013) Is there evidence to support use of premixed or prandial insulin regimens in Insulin-naïve or previously insulin-treated type 2 diabetic patients? *Diabetes Care*, **36** (Suppl. 2), s205–211.

40. Barnett, A.H., Huisman, H., Jones, R. et al. (2013) Linagliptin for patients aged 70 years or older with type 2 diabetes inadequately controlled with common antidiabetes treatments: a randomised, double-blind, placebo-controlled trial. *Lancet*, **382** (9902), 1413–1423.

41. Graefe-Mody, U., Friedrich, C., Port, A. et al. (2011) Effect of renal impairment on the pharmacokinetics of the dipeptidyl peptidase-4 inhibitor linagliptin. *Diabetes Obes. Metab.*, **13** (10), 939–946.

42. Trajenta (linagliptin) (2014) *Summary of Product Characteristics*, Boehringer Ingelheim International GmbH, Ingelheim, http://www.ema.europa.eu/ema/index.jsp?curl=pages/medicines/human/medicines/002110/human_med_001482.jsp&mid=WC0b01ac058001d124 (accessed 21 October 2014).

43. Trazenta (linagliptin) (2013) *Prescribing Information*, Nippon Boehringer Ingelheim Co., Ltd, Tokyo, (accessed 8 August 2014).

44. Friedrich, C., Jungnik, A., Retlich, S., et al. (2013) Bioequivalence of Linagliptin 5 mg Once Daily and 2.5 mg Twice Daily: Pharmacokinetics and Pharmacodynamics in an Open-label Crossover Trial.
45. Sloan, L., Newman, J., Sauce, C. et al. (2011) Safety and efficacy of linagliptin in type 2 diabetes patients with severe renal impairment [poster]. *Diabetes*, **60** (Suppl. 1), A114.
46. Laakso, M., Rosenstock, J., Groop, P.H. et al. (2013) Linagliptin vs placebo followed by glimepiride in type 2 diabetes patients with moderate to severe renal impairment. *Diabetes*, **62** (Suppl. 1), A281–282.
47. McGill, J., Sloan, L., Newman, J. et al. (2012) Long-term efficacy and safety of linagliptin in patients with type 2 diabetes and severe renal impairment: a 1-year, randomized, double-blind, placebo-controlled study. *Diabetes Care*, **36** (2), 237–44.
48. Cooper, M., von Eynatten, M., Emser, A., et al. (2011) Efficacy and safety of linagliptin in patients with type 2 diabetes with or without renal impairment: results from a global phase 3 program. 71th Scientific Sessions of the American Diabetes Association, San Diego, CA.
49. Patel, S., Schernthaner, G., Barnett, A.H. et al. (2013) Renal safety of linagliptin in elderly patients with type 2 diabetes: analysis of pooled patient data from seven Phase 3 clinical trials. *Diabetologia*, **56** (Suppl. 1), S370.
50. McGill, J.B., Sloan, L., Newman, J. et al. (2012) Long term efficacy and safety of linagliptin in patietns with type 2 diabetes and severe renal impairment. *Diabetes Care*. doi: 10.2337/dc12-0706.
51. Von Eynatten, M., Hehnke, U., Cooper, M. et al. (2013) Renal safety and outcomes with linagliptin: meta-analysis of individual data for 5466 patients with type 2 diabetes. *Diabetologia*, **56** (Suppl. 1), S364.
52. Graefe-Mody, E.U., Rose, P., Ring, A. et al. (2012) Pharmacokinetics of linagliptin in subjects with hepatic impairment. *Br. J. Clin. Pharmacol.*, **74** (1), 75–85. doi: 10.1111/j.1365-2125.2012.04173.x
53. Ring, A., Port, A., Graefe-Mody, E.U. et al. (2011) The DPP-4 inhibitor linagliptin does not prolong the QT interval at therapeutic and supratherapeutic doses. *Br. J. Clin. Pharmacol.*, **2** (1), 39–50.
54. Johansen, O.E. et al. (2013) Cardiovascular (CV) safety of linagliptin in patients with Type 2 Diabetes (T2D): a pooled comprehensive analysis of prospectively adjudicated CV events in phase 3 studies. Presented at the American Diabetes Association 73rd Scientific Sessions, Chicago, IL, June 21–25, 2013, 376-OR, and the 49th Annual Meeting of the European Association for Study of Diabetes (EASD), Barcelona, Spain, September 23–27.
55. Gallwitz, B., Rosenstock, J., Rauch, T. et al. (2012) Two year efficacy and safety of linagliptin compared with glimepiride in patients with type 2 diabetes inadequately controlled on metformin: a randomised, double-blind, non-inferiority trial. *Lancet*, **380** (9840), 475–483. doi: 10.1016/S0140-6736(12)60691-6

Matthias Eckhardt studied Chemistry at the Universities of Marburg and Göttingen and received his PhD from the latter university in 1996. After postdoctoral studies in New York, he started working in the chemical industry. He is presently a Principal Scientist at the medicinal chemistry department of Boehringer Ingelheim at Biberach, Germany. He has been working in the field of cardiometabolic diseases since 2001 and has been involved in the discovery of several antidiabetic drugs currently in clinical studies or on the market.

Thomas Klein received his PhD in 1995 in Biochemistry and Pharmacology from the University of Konstanz, Germany. He has worked in the pharmaceutical industry (Altana, Nycomed, currently Boehringer-Ingelheim) since 1997 in the research areas of inflammatory, gastrointestinal, and metabolic diseases.

Herbert Nar, PhD, graduated from the Technical University Munich in Chemistry and obtained his PhD from the same institute with structural studies on blue copper electron transfer proteins. In his postdoctoral work with Robert Huber at the Max-Planck-Institute für Biochemie, he determined the 3D structures of proteins involved in pterin biosynthesis and studied their enzymatic mechanisms. He joined Boehringer Ingelheim in 1995 to establish a protein crystallography laboratory. In 2000, he took over responsibility for the Structural Research Group that comprises units for protein expression and purification, biophysics of ligand binding, and NMR and protein crystallography.

Sandra Thiemann, PhD, is Global Medical Director in the Corporate Division Medicine in Boehringer Ingelheim Pharma GmbH & Co. KG. She is the Team Leader Linagliptin in Global Medical Affairs Metabolism at Boehringer Ingelheim. She joined Boehringer Ingelheim as Global Medical Advisor in 2011. In her current role, Dr. Thiemann is responsible for the linagliptin product maintenance and optimization strategy, including phase IV trials, scientific activities, and scientific communication. She was active in research for 5 years focusing on pharmacokinetics and −dynamics of motile protein complexes, before she joined the pharmaceutical industry in the diabetes field. She worked as medical advisor for Sanofi in diabetes for 4 years before joining Boehringer Ingelheim.

8
The Discovery of Alimta (Pemetrexed)
Edward C. Taylor

I started my college education at Hamilton, a small liberal arts college in upstate New York, in order to major in English Literature and in writing. I had no interest whatsoever in science, for I had attended a classical high school that majored in Latin, Greek, literature, and (most fortunately) mathematics but offered only one poorly taught science course (physics), and no chemistry. Most fortunately, Hamilton had a distribution requirement that mandated some courses in science. Only two were available at the time, chemistry and biology, and because I had no interest in either one, I flipped a coin and it came down "chemistry."

I was fascinated, charmed, and excited from the very first lecture, and from that point on I took every chemistry course Hamilton offered, one after the other, until there were no more left. At that point I transferred to Cornell University, graduated in 1946, and immediately entered graduate school at Cornell to major in organic chemistry.

In searching for a research topic, I ran across an article in *Science* from Lederle (a small pharmaceutical company on America's east coast) describing a compound that they had isolated from human liver that possessed the unexpected property of being essential for the growth of several types of microorganisms. Lederle proceeded to determine the structure of this strange compound and then to synthesize it [1]. Structural studies, synthesis, and identification of this material as folic acid have been summarized in a fascinating review of this and earlier work [2]. What struck me as utterly bizarre was the identity of the structure of the left-hand bicyclic heterocyclic ring found in this liver factor with the structure of the yellow pigment in the wings of the brimstone butterfly [1–7]. Only one substituent differentiated the wing pigment from the "liver *L. casei* factor" from human liver.

Completely fascinated by my first encounter with heterocycles and with the mysteries of natural products, I decided to dedicate my graduate studies to this field. Because the butterfly wing pigment had been characterized as a "pteridine" derivative [1], my research focus started with the chemistry and synthesis of these most unusual (to me) bicyclic heterocycles. And thus I was introduced to the strange history of pteridines. This area of natural product chemistry has been heavily referenced and reviewed. Several reviews also reference earlier ones [5–7]. Our own efforts and results in the field of exploring inhibitors of folate-dependent

Successful Drug Discovery, First Edition. Edited by János Fischer and David P. Rotella.
© 2015 Wiley-VCH Verlag GmbH & Co. KGaA. Published 2015 by Wiley-VCH Verlag GmbH & Co. KGaA.

enzymes have been previously summarized [8, 9]. The most comprehensive extant treatment of the chemistry of pteridines is the volume by D. J. Brown published in 1988 (730 pp., 1750 references) [10].

Interest in this extremely esoteric field of natural products started in 1889 with the efforts of Frederick Gowland Hopkins (he would go on to win a Nobel Prize for the discovery of vitamins) who, at the age of 29 as a medical student in London, presented a paper at a meeting of the Royal Society on his efforts to isolate and characterize the pigment in the wings of the white cabbage butterfly. He incorrectly concluded that it was a derivative of uric acid [11–15].

Interest in butterfly wing pigments then lapsed until Heinrich Wieland (who would later win a Nobel Prize for his work on the bile acids) in the late 1920s, together with his colleague Clemens Schöpf, took up the challenge. They ultimately isolated and purified the white pigment from the cabbage butterfly that they named leucopterin [16] and the yellow pigment from the brimstone butterfly that they named xanthopterin [17]. Frustrated with structure determinations because of the physical characteristics of these pigments (insolubility, resistance to combustion), they too abandoned the project. It was not until the early 1940s that Robert Purrmann, a young organic chemist in the Ludwig-Maximilians-Universität in Munich, finally determined the structures of leucopterin and xanthopterin (see Scheme 8.1), and devised a method of synthesis for each. This work appeared in a series of papers in Liebig's Annalen der Chemie [3, 18, 19]. A fascinating essay on the history of these historic explorations of butterfly wing pigments was provided by Schöpf [20]. It was Lederle's recognition of the

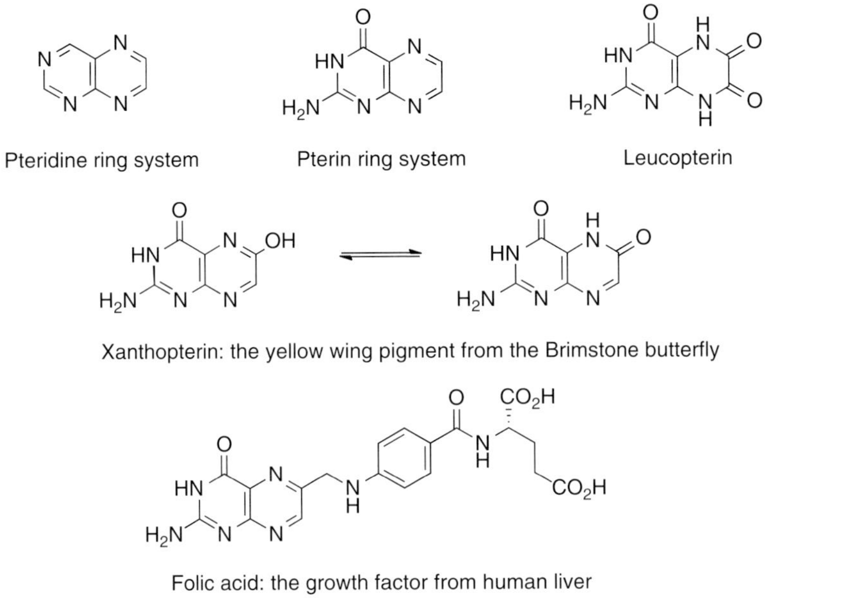

Pteridine ring system   Pterin ring system   Leucopterin

Xanthopterin: the yellow wing pigment from the Brimstone butterfly

Folic acid: the growth factor from human liver

Scheme 8.1

remarkable structural relationship of these butterfly wing pigments with that of the "liver growth factor" (now referred to as folic acid) that had so fascinated me.

I quickly realized why it had taken more than 50 years, despite the efforts of the above two Nobel Prize winners, for the final elucidation of the structures of these pigments. These compounds were not easy to work with. Their solubility, for example, rivaled that of Vermont granite, so that spectroscopic examinations were extremely difficult and limited. The pigments were so difficult to burn that the usual techniques of combustion analysis were essentially ineffective. Normal techniques for purification were worthless, so I devised new methods such as, *inter alia*, high vacuum sublimation. I did not realize it at the time, but my 3-year experience with pteridines and related heterocycles proved to be an invaluable training for my lifelong fascination with heterocyclic chemistry.

Present-day cancer chemotherapy can be said to have had its beginnings with the discovery that aminopterin, synthesized by Lederle as a potential antimetabolite of folic acid, brought about remissions of acute lymphoblastic leukemia in children [21]. This compound, as well as methotrexate (MTX), was designed on the basis of the "antimetabolite theory" that suggested that administration of an almost-exact "duplicate" of a natural metabolite that had a specific biological function or activity, might result in uptake of the "duplicate," thus leading either to inactivity, or to the opposite biological activity either by displacement of the natural metabolite from the active site, or by competition with the natural metabolite for the active site. The "duplicate" might also be incorporated in place of the natural metabolite as a building block (as in the incorporation of sulfonamide in place of *p*-aminobenzoic acid in the intracellular synthesis of folic acid) [22]. In Lederle's case, their synthesis of folic acid, coupled with their discovery that this material, found in human liver, was, most surprisingly, necessary for the growth of bacteria, placed them in a position to put the above theory to test; that is, a deliberately falsified structural analog of folic acid might prove to be an antibacterial drug.

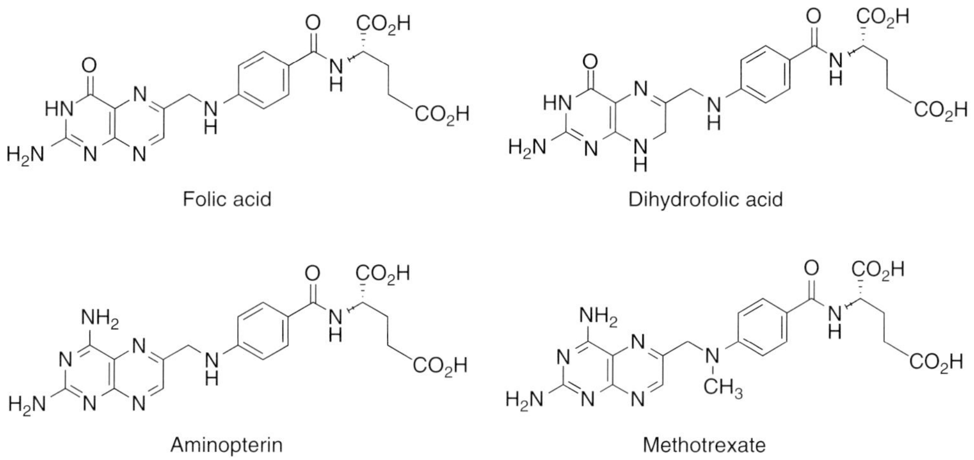

Scheme 8.2

Lederle synthesized aminopterin and MTX and both compounds, as anticipated, were extremely active as antibacterials (albeit also very toxic). It was known at the time that a deficiency of folic acid had a beneficial effect in human leukemia [23], so Farber [21] was able to show that aminopterin (and later MTX as well) brought about remissions of acute lymphocytic leukemia in children. The world sat up and took notice of this remarkable situation where an antibacterial drug demonstrated anticancer activity. MTX was approved by the FDA early in the 1950s and is still prescribed for some types of cancer and for rheumatoid arthritis (see Scheme 8.2).

The biological basis of the cytotoxicity of these two compounds was found to be their potent inhibitory activity against dihydrofolate reductase (DHFR) [24, 25]. This enzyme reduces dihydrofolic acid (as well as folic acid itself) to tetrahydrofolic acid (THF), thereby maintaining a constant cellular supply of the latter. Tetrahydrofolate coenzymes play critical roles in cellular one-carbon transfer reactions in intermediary metabolism, including critical steps in the de novo pathways for thymidylate and purine biosynthesis, the regeneration of methionine from homocysteine, the interconversion of serine and glycine, and the catabolism of certain amino acids [26]. The toxicity exhibited by MTX, aminopterin, and other DHFR inhibitors undoubtedly is due in part to their lack of specificity for targeting cancer cells instead of normal cells. As a consequence, there was increasing interest in the development of less toxic drugs with greater selective transport properties that might lead to their effectiveness against a broader range of human cancers. There was evidence that an increase in lipophilicity in molecules designed as inhibitors of thymidylate synthase (TS) and DHFR results in an increase in enzyme binding, [27]; in line with this observation, an increase in polar character at N-10 of quinazoline analogs normally decreases binding, (Terry R. Jones, personal communication). Various deaza analogs of aminopterin have been reported to be effective antitumor agents against L210 leukemia and sarcoma 180 tumors. For example, 10-deazaaminopterin is a more effective antitumor agent against L1210 leukemia and sarcoma 180 tumors than either MTX or aminopterin, probably because of its transport properties [28–30]. 8,10-Dideazaaminopterin appears to be comparable to MTX as an inhibitor of L1210 leukemia in mice [31] while the N-10-propargyl derivative of 5,8-dideazafolic acid is a potent inhibitor of TS [32]. There are several thorough reviews on biological evaluations of the deaza analogs of MTX [33, 34].

By the time we had decided to enter this field of potential antifolate chemotherapeutic agents, much had already been discovered about the reason folic acid was essential for life. Enzymatic reduction of folic acid gives THF that then serves as the "mother" of a series of so-called folate-dependent enzyme cofactors, all of them derived from THF by addition of a single carbon atom in the oxidation states of methanol, or formaldehyde, or formic acid (see Scheme 8.3) [24, 26]. These cofactors, coupled with the appropriate enzyme, then transfer one-carbon units in a broad variety of essential intracellular biosynthetic processes, as enumerated earlier.

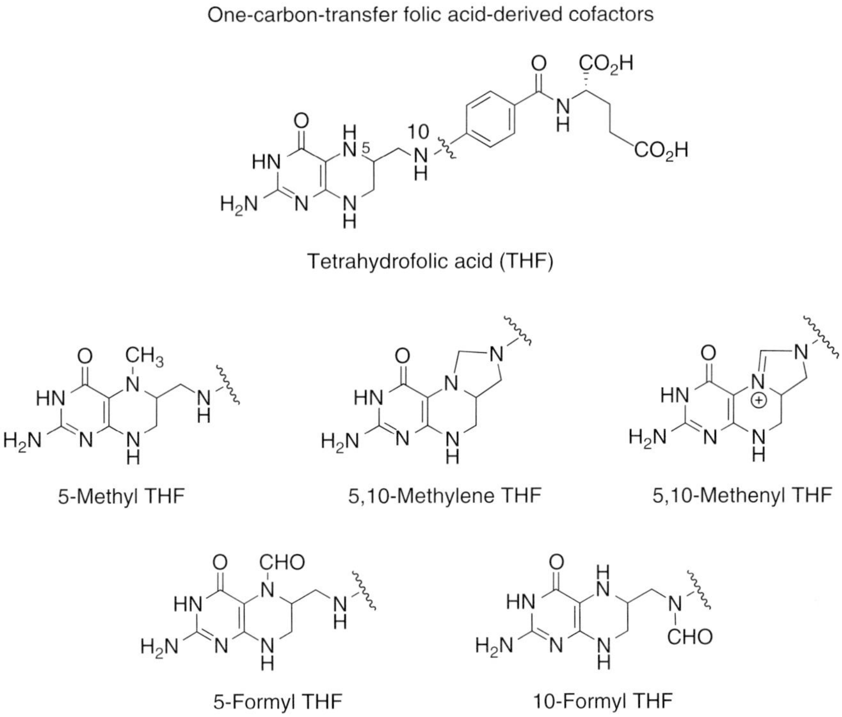

8 The Discovery of Alimta (Pemetrexed) | 161

One-carbon-transfer folic acid-derived cofactors

Tetrahydrofolic acid (THF)

5-Methyl THF   5,10-Methylene THF   5,10-Methenyl THF

5-Formyl THF   10-Formyl THF

Scheme 8.3

It is striking that only two (N-5 and N-10) of the five nitrogen atoms in the skeleton of THF are involved in the formation of these cofactors. This focused our attention on the synthesis of deaza analogs of folic acid and aminopterin, and their corresponding tetrahydro derivatives, that lacked one or both of these specific nitrogen atoms. Because these objectives in heterocyclic chemistry were not at the time a part of our major programs, their synthesis literally took us years.

We were particularly intrigued by the tetrahydro derivative of 5,10-tetrahydrofolic acid (Scheme 8.4), for this compound lacks *both* of the involved nitrogen atoms of THF, the "mother" of all of the folate-dependent cofactors

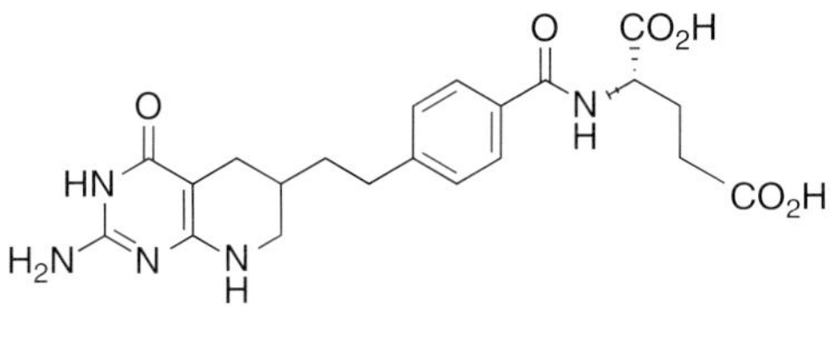

5,10-Dideazatetrahydrofolic acid (DDATHF)

Scheme 8.4

(see Scheme 8.4). We coined the acronym DDATHF (for 5,10-dideaza-5,6,7,8-tetrahydrofolic acid) for this intriguing compound and envisioned that the absence of these nitrogen atoms would preclude its utilization in the cell as a THF surrogate for intracellular synthesis of one-carbon transfer reactions in cellular metabolism. In addition, an increase in lipophilicity, coupled with increased solubility and decreased basicity, might result in better transport properties and possibly greater selectivity. The absence of the 4-amino group meant that it would have been extremely unlikely to be an inhibitor of DHFR, at least as a primary target, and as a tetrahydropyridine rather than a tetrahydropyrazine derivative, it should be much more stable than the oxidation-sensitive tetrahydropyrazine ring of THF. The presence of a normal p-benzoylglutamate moiety would allow conversion by folylpolyglutamate synthase (FPGS) to polyglutamates [35].

Our synthesis of this compound commenced with condensation of α-cyanothioacetamide with 2-methyl-3-ethoxyacrolein to give 3-cyano-5-methyl-2(1H)-pyridinethione that was arylated on sulfur with 4-nitrofluorobenzene. The resulting sulfide was brominated under free radical conditions to give a mixture

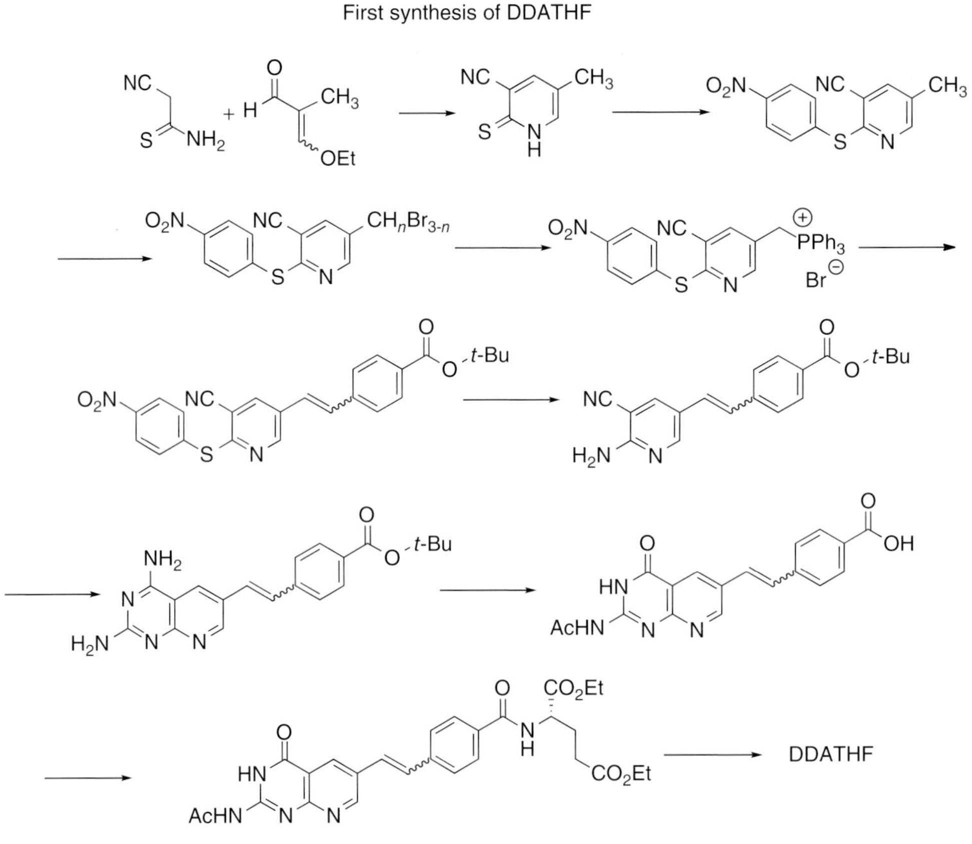

Scheme 8.5

of mono-, di-, and tribromo derivatives, and the mixture was treated with triphenylphosphine. This resulted in the direct crystallization from the reaction mixture of the triphenylphosphonium salt that was condensed with *t*-butyl 4-formylbenzoate under normal Wittig conditions. The resulting olefin was converted into the *o*-aminonitrile shown in Scheme 8.5 by reaction with liquid ammonia and cupric bromide at room temperature for 14 days. Subsequent annulation of the pyrimidine ring was accomplished by reaction with guanidine free base in refluxing *t*-butanol, and hydrolysis to the free acid was accomplished with dry hydrochloric acid in nitromethane. The 4-amino group was then hydrolyzed with base to give the desired 2-amino-4(3*H*)-pyrimidinone unit. Acetylation, coupling with diethyl L-glutamate using phenyl *N*-phenylphosphoramidochloridate as the coupling agent in *N*-methylpyrrolidone as solvent, catalytic reduction to the tetrahydropyridine stage and careful saponification finally gave DDATHF [36].

In preliminary *in vitro* tests, as DDATHF proved to be extremely active as a cell growth inhibitor, we sent a sample to Eli Lilly and Co. for a more complete evaluation. I had already been for some 20 years a consultant to the discovery and development groups at Lilly; they knew me well, and I knew them, and their programs well. This long-term relationship greatly facilitated our proposed collaboration. Lilly's results on the initial testing of DDATHF were sensational, for this compound was the most active antitumor agent Lilly had ever seen. The Table in Scheme 8.6 summarizes some of the initial results obtained by Dr. Gerald B. Grindey of Lilly against murine and human tumor xenograft models. Reversal studies of toxicity showed that addition of hypoxanthine to the medium completely protected L1210 cells from the cytotoxicity of DDATHF, whereas addition

Murine solid tumor models	DDATHF	MTX
X-5563 Myeloma	+++	++
CA 755 Adenocarcinoma	+	−
MS Ovarian Carcinoma	−	−
6C3HED Lymphosarcoma	+++	NT
Colon 26 Carcinoma	+++	−
B-16 Melanoma	+++	−
Lewis Lung Carcinoma	++	−
Madison Lung Carcinoma	++	−
C3H Mammary Carcinoma	+++	−
Human tumor xenograft models		
LX-1 Lung Carcinoma	++	−
CX-1 Colon Carcinoma	+	−
MX-1 Mammary Carcinoma	+	−
HC1 Colon Carcinoma	+++	−
GC3 Colon Carcinoma	+++	−
VRC5 Colon Carcinoma	+++	−

Key:
+++ = 95–100% inhibition − = <60% inhibition
++ = 80–95% inhibition   NT = not tested
+ = 60–80% inhibition

Scheme 8.6

of thymidine had no effect. DDATHF produced a nearly 10-fold decline in adenosine triphosphate (ATP) and guanosine triphosphate (GTP) with a concentration dependence very similar to that seen for cytotoxicity. In addition, DDATHF was completely active against tumors that had developed resistance to MTX. These and further studies pointed strongly to glycinamide ribonucleotideformyltransferase (GARFT) as the primary site of DDATHF action in L1210 cells, and that the mechanism of tumor inhibition was blockage of the first formylation step in de novo purine biosynthesis (see Scheme 8.7).

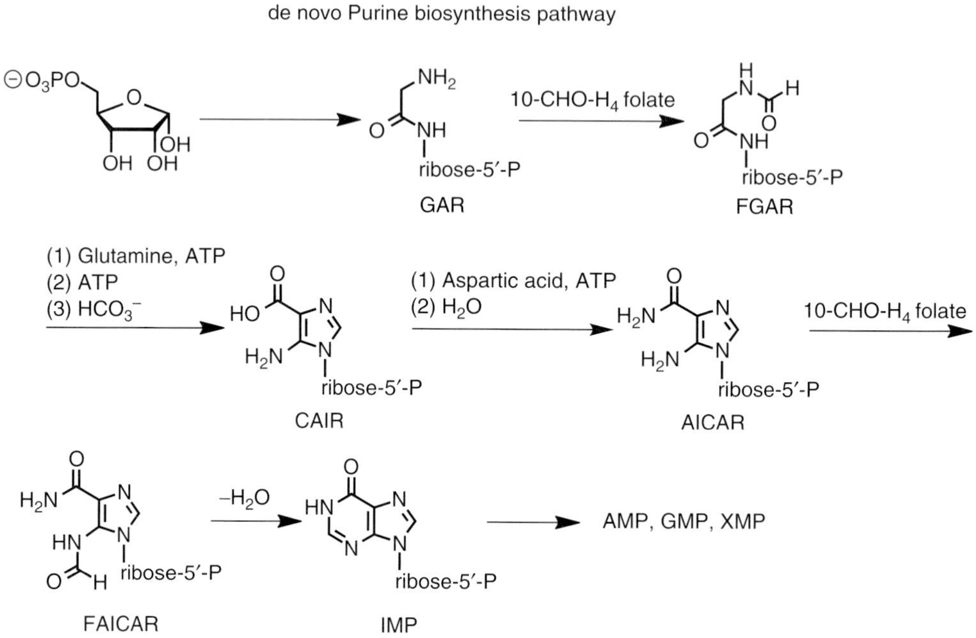

Scheme 8.7

DDATHF therefore was functioning as an antitumor agent through inhibition of a target different from DHFR, the primary target of traditional antifolates [37]. These results appeared to open up a very promising new area for cancer chemotherapy obviously worthy of a full and intensive investigation.

Neither Princeton nor Lilly was prepared to tackle this challenge alone. Lilly had no prior experience or familiarity with this area of synthetic heterocyclic chemistry that was Princeton's specialty. Princeton has no medical school, and no capabilities involving the many disciplines and expertise involved in pursuing preclinical and clinical studies, while these were Lilly's constant activities and responsibilities.

The two organizations therefore proposed a formal collaborative scientific partnership, with the design and synthetic aspects of this program put primarily into Princeton's hands, and evaluation and hoped-for drug development studies into Lilly's hands. Patents would be assigned to Princeton, with Lilly the exclusive licensee. All information from all of these studies would be fully shared and discussed. In the event, the resulting academic/industrial partnership proved to be extraordinarily successful, with three compounds in as many years reaching clinical trial, and one of them an FDA-approved anticancer drug of major importance.

One reason for the length of time required, at the beginnings of our synthetic work, to carry out each individual synthesis was the extraordinary insolubility of a great many of the intermediates involved in each of these syntheses (I am reminded of the book by Ernest Thompson Seton [38] dealing with his trip in the Arctic Prairies. Chapter IX started as follows: "Reference to my Smith Landing Journal for June 17 shows the following: The Spring is now on in full flood, the grass is high, the trees are fully leaved, flowers are blooming, birds are nesting, *and the mosquitoes are a terror to man and beast.* If I were to repeat all the entries in that last key, it would make dreary and painful reading; I shall rather say the worst right now, and henceforth avoid the subject."). And so it is with the insolubility of pteridines and most related heterocycles. Their generic insolubility was the equivalent, for us, of the mosquito problem for Seton. As an example, we wrote in one paper that "The extraordinary insolubility of the pteroic acid analogs 1 and 2 has thus far frustrated all attempts to effect coupling with diethyl (L)-glutamate. In fact, 2 proved to be so insoluble that the only way found to compare the sample of 2 prepared from 1 with that prepared from 15 was to convert each into the somewhat more soluble 2-acetyl derivative 17, which permitted an FT NMR spectrum to be obtained. We have not even been able to acetylate 1 because of its insolubility, and the acetyl derivative 17 was too insoluble to be coupled with diethyl L-glutamate. Some method for solubilizing these extremely insoluble fused 2-amino-4(3*H*)-pyrimidinones and 2,4-diaminopyrimidines must be found" (see Scheme 8.8) [39].

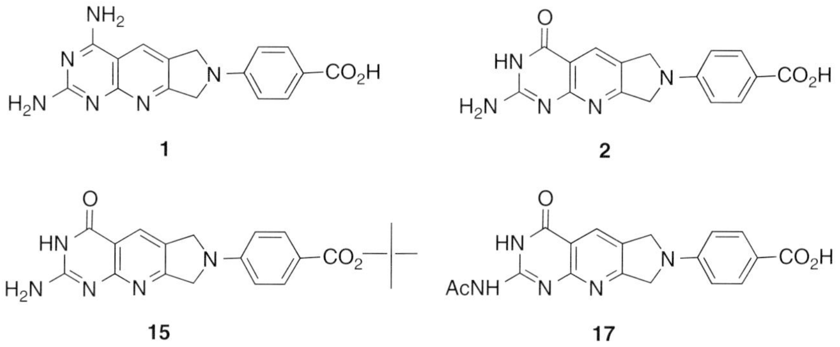

Scheme 8.8

A surprisingly effective solution eventually turned out to be 2-aminopivaloylation, effected by digesting the heterocycle with a mixture of pyridine and pivaloyl chloride or pivalic anhydride until eventual dissolution occurred. Workup gave the

2-pivavoylamino derivative, usually readily (and gratefully) soluble in the usual organic solvents [40–43].

It was immediately obvious that our very exciting biological results, unveiled almost daily, demanded a thorough structure–activity relationships (SARs) examination of DDATHF, and this could be accomplished only with a much improved synthesis. Our original synthesis was multistep, linear, and totally unsuitable for the synthesis of analogs. A much improved, convergent synthesis [44] (see Scheme 8.9) was then devised that provided a high percentage of the many hundreds of analogs prepared by Princeton and Lilly over the next several years.

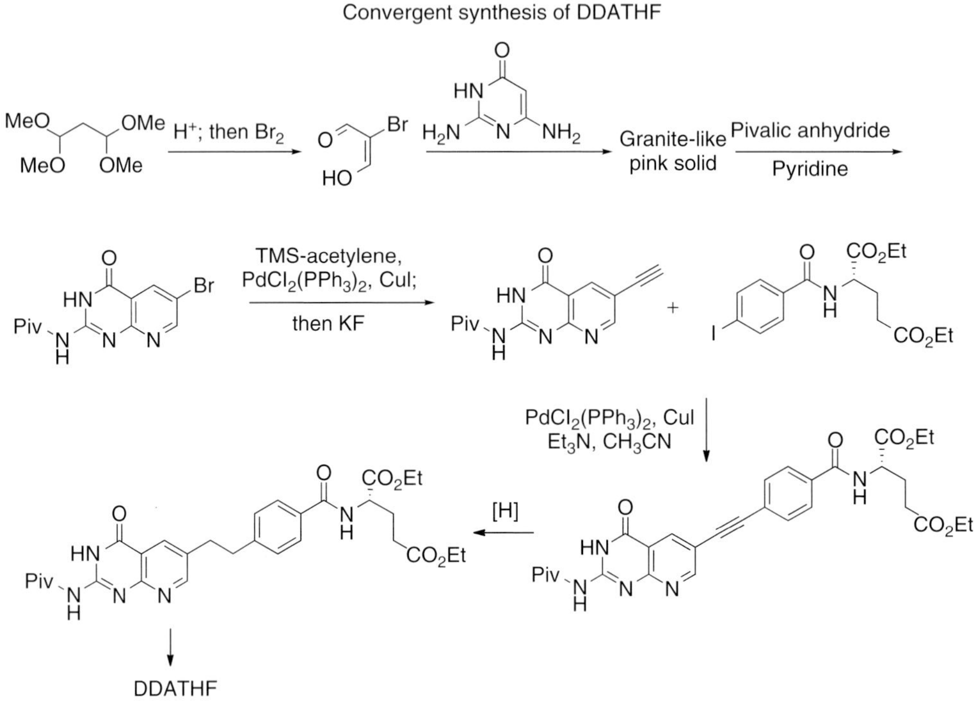

Scheme 8.9

A key intermediate in this new synthesis was 2-pivaloyl-6-bromo-5-deazapterin, efficiently prepared by condensation of 2,4-diamino-6(1H)-pyrimidinone with bromomalondialdehyde, followed by pivaloylation. Palladium-catalyzed carbon–carbon coupling with trimethylsilylacetylene followed by desilylation gave the 6-ethynyl derivative that was then subjected to a second palladium-catalyzed carbon–carbon coupling reaction, this time with diethyl 4-iodobenzoylglutamate, to give an intermediate that required only reduction of the triple bond and the pyridine ring, and then careful alkaline deprotection of the 2-amino grouping and the glutamate esters, to give DDATHF. A versatile alternative route involved palladium-catalyzed coupling of diethyl 4-ethynylglutamate (itself prepared by coupling of diethyl 4-iodobenzoylglutamate with trimethylsilylacetylene followed by desilylation) with 2-pivaloyl-6-bromo-5-deazapterin.

During the following several years, we successfully developed several further independent syntheses of DDATHF via: (i) an intermolecular inverse electron-demand strategy [45, 46]; (ii) an intramolecular inverse-demand strategy [8, 45]; (iii) a novel synthesis involving an activated carbonyl approach [47]; (iv) a synthesis using isoxazoles [8]; and (v) a four-step high-yield synthesis from 4-vinylbenzoic acid [48].

It seemed most unlikely that the first compound prepared (DDATHF) would prove to be the optimum, so our first project as a team was to determine the structural requirements for maximum activity for inhibition of GARFT. This required a step-by-step examination of the consequences of structural changes in every part of the structure of DDATHF. Ultimately the team made and examined over 800 new compounds, including 10-hydroxymethyl-DDATHF, deazatetrahydrofolic acid (5-DATHF), and homo-5-DATHF [46]. Several summaries of structural features of DDATHF that determine inhibition of mammalian GARFT have been published [49–51].

Astonishingly, the most active of all of the compounds prepared proved to be DDATHF itself, the very first compound synthesized at Princeton, together with an analog in which thiophene replaced phenyl in the linker group of DDATHF (giving LY309887) (see Scheme 8.10). These two candidates were selected for clinical trial.

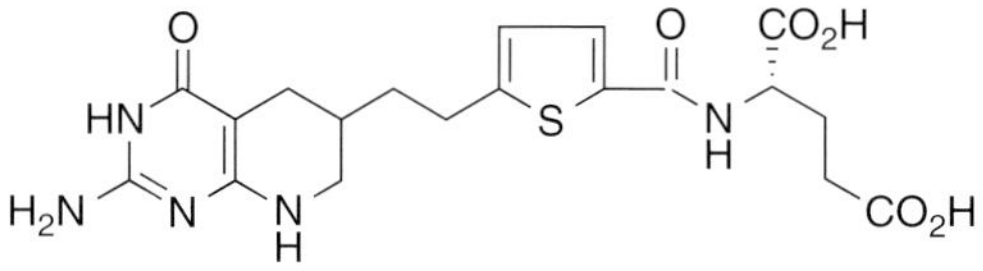

LY309887

Scheme 8.10

Both DDATHF and LY309887 are 50 : 50 mixtures of diastereomers (as were all the exploratory targets possessing an sp^3 attachment of the side chain at an equivalent C-6 position). Our decision to take these two mixtures of diastereomers into clinical trial therefore meant that, eventually, either resolution of the 50 : 50 diastereomeric mixtures or an asymmetric synthesis of each diastereomer might become mandatory, for we had to know whether the observed activity was due to one or the other of the diastereomers, or perhaps to both. A parallel objective was the design of analogs that lacked the stereogenic center at C-6, thus bypassing the necessity of resolution or the total synthesis of each component of each diastereomeric pair. The Princeton/Lilly team decided to put the resolution of DDATHF and of the thiophene analog LY309887 into Lilly's hands, and the synthetic work into Princeton's hands. The resolution of DDATHF (and ultimately of

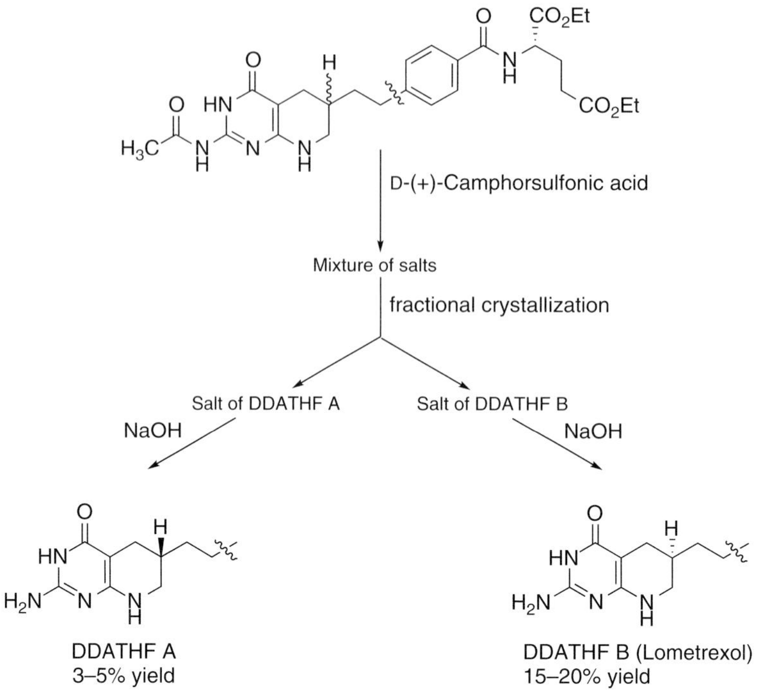

Scheme 8.11

LY309887, that was synthesized several years later) was carried out at Lilly by Dr. Joe Shih (See Scheme 8.11) [52].

Diastereomer A could be obtained in only 2–5% yield, but diastereomer B (later shown to possess the same absolute configuration at C-6 as THF) was obtained in 15–20% yield. Both diastereomers were found to be potent inhibitors of leukemia cell growth due to effects on de novo purine synthesis. Both diastereomers were efficient substrates for mouse liver folypolyglutamate synthetase. The two diastereomers are remarkably similar and equiactive antimetabolites inhibitory to de novo purine synthesis, and the biochemical processes involved in their cytotoxicity display little stereochemical specificity. Diastereomer B was given the designation lometrexol and chosen for clinical trials. It was clear to all, however, that resolution of diastereomers would be an exceedingly expensive process for large scale synthesis.

In the meantime, all attempts at Princeton to carry out an asymmetric reduction of a pyridopyrimidine precursor to a chiral tetrahydropyridopyrimidine failed. Chiral catalysts, whether soluble or insoluble, were ineffective, as were all explorations with chiral auxiliaries. We therefore turned our efforts to the design and synthesis of possible alternatives to DDATHF, but lacking the stereogenic C-6 carbon. What appeared to be a simple approach was the introduction of a double bond in the tetrahydropyridine ring of DDATHF that would convert C-6 from sp^3 to sp^2. Compounds C and D (see Scheme 8.12) were thus prepared, but both

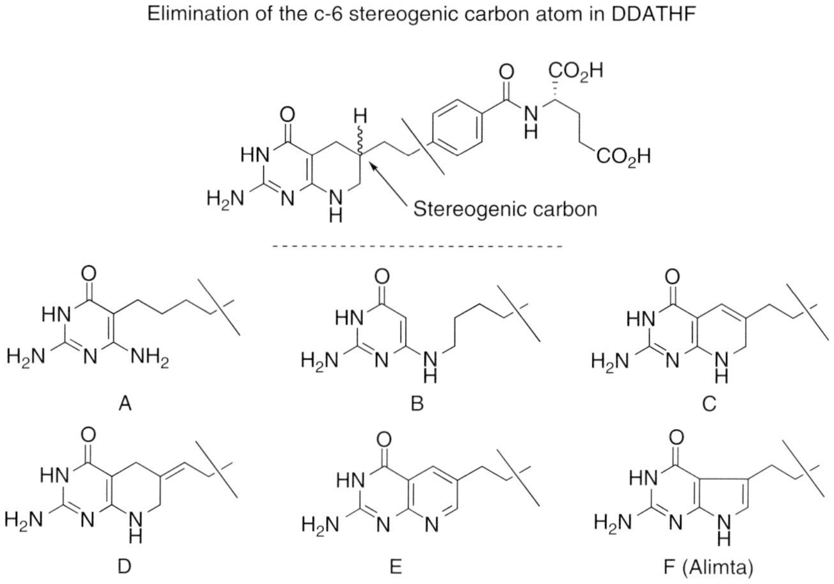

Elimination of the c-6 stereogenic carbon atom in DDATHF

Scheme 8.12

were extremely unstable, and underwent very rapid air oxidation to give the pyridine precursor E that was inactive as a cell growth inhibitor [36]. Another apparently simple approach was deletion of the C-7 methylene group in DDATHF to give what we termed an "open-chain" or des-methylene analog (7-DM-DDATHF, structure A) [53–56]. An isomeric "open-chain" analog would result from deletion of the C-5 methylene group to give 5-DM-DDATHF (structure B) [57]. Many compounds in both series were prepared and evaluated; the latter 5-DM-DDATHF compounds were essentially inactive as cytotoxic agents, but there was considerable activity in the isomeric 7-DM series. Cell culture cytotoxicity studies against CCRF-CEM human lymphoblastic leukemic cells gave encouraging IC_{50}s, while inhibition reversal studies against isolated enzymes suggested that these compounds were inhibitors of de novo purine biosynthesis. In contrast to DDATHF, hypoxanthine and aminoimidazolecarboamide were each able to reverse the cytotoxicity of 7-DM-DDTHF at all dose levels examined. Surprisingly, studies against GARFT isolated from L1210 cells showed that this flexible monocyclic pyrimidine analog was surprisingly only threefold less inhibitory than was DDATHF itself. Thus, removal of the methylene group at C-7 from DDATHF affected only slightly the binding efficacy of the resulting acyclic analog toward GARFT. Other studies with 7-DM-DDATHF confirmed previous results with DDATHF analogs that the nitrogen atom at position 8 was critical both for enzyme inhibitory activity as well as for cellular cytotoxicity. Our conclusion was that hydrogen bond donation (i.e., from the -NH group at position 8 in bicyclic DDATHF and analogs, and from the 4-amino group of the pyrimidine ring in 7-DM-DDATHF) might be providing critical hydrogen bonding interactions with active site residues of GARFT. In summary, these acyclic analogs were less efficient substrates for FPGS than their bicyclic counterparts, and this presumably is responsible for the only moderate antitumor activity exhibited by 7-DM-DDATHF. We were thus back to ground zero in our search for an effective DDATHF analog lacking the C-6 stereogenic center.

Our SAR results thus far had led to the following conclusions:

1) Essential for effective GARFT inhibition is replacement of N-5 with carbon. 5-Deazatetrahydrofolate was as good an inhibitor of GARFT as DDATHF, both in cell culture and against murine solid tumors in indicating that removal of the N-10 nitrogen was actually not critical either to the overall biochemical properties or to the antitumor activity of DDATHF [43].
2) The 2-amino-4(3H)-pyrimidinone ring is essential.
3) There must be a minimum of a 2-atom bridge between the tetrahydropyridine ring and the phenyl group.
4) The side-chain phenyl group can be replaced by methylene groups, substituted phenyl groups, and various five- or six-membered heterocyclic rings [50, 51].
5) A rigid, bicyclic framework must be present. Flexible "open-chain" analogs in which C-7 of DDATHF is removed were effective inhibitors of GARFT but lacked good *in vivo* activity as tumor growth inhibitors. The C-5 desmethylene isomers were general inactive.
6) Ring B cannot be aromatic [36].
7) The pyrimidine C-6 amino group, or its appearance as an -NH unit in any second ring fused to the pyrimidine ring, is mandatory.
8) Polyglutamation plays a major role in the potent cytotoxicity of DDATHF, so a successful candidate must be an extremely good substrate for FPGS.
9) The side-chain must be connected to ring B through a non-stereogenic (sp^2) carbon.

It was immediately clear that the incompatibilities of some of the above "requirements" meant that we had to ignore at least a few of them. Observation #6 came from the inactivity of a pyridine ring as ring B, as well as the inactivity of 5,10-dideazafolic acid. Perhaps "aromaticity," when defined as a pyridine ring, is actually a reflection of the electron-deficiency of pyridine. Pyridine itself, with a dipole moment of 2.2 D, has the negative end on nitrogen, and electrophiles react with pyridine on nitrogen. Pyrrole, on the other hand, has a dipole moment with its negative end on carbon, and electrophiles react with pyrrole and fused pyrroles at carbon. Thus perhaps observation #6 can therefore be ignored when considering a pyrrolopyrimidine such as structure F in Scheme 8.12. Attachment of the DDATHF sidechain at position 3 of the above pyrrolopyrimidine would satisfy all of the remaining guidelines except for #6.

Dr. Dietmar Kuhnt arrived at this time from Germany as a postdoc to work with me on our anticancer project. I already had on hand a bottle containing

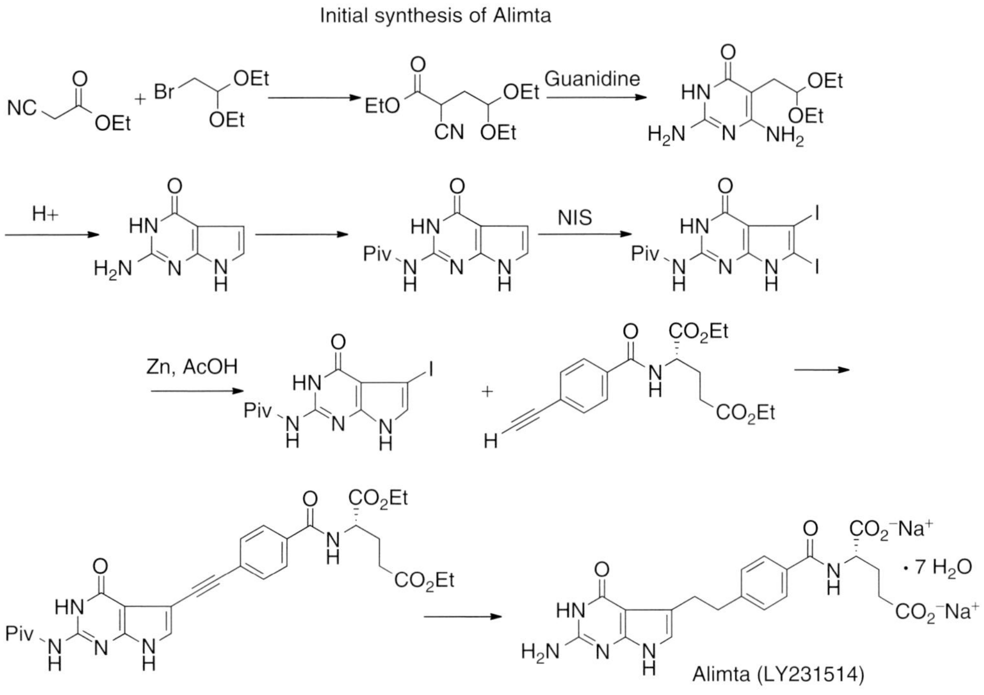

Scheme 8.13

Initial synthesis of Alimta

the DDATHF sidechain (diethyl 4-ethynylbenzoylglutamate), left over from the DDATHF analog project, so we embarked on the synthesis outlined in Scheme 8.13 even though the resulting compound would violate observation #6. The synthesis was completed in the fall of 1987, and the resulting compound was sent to Lilly for evaluation early in January, 1988.

This compound, now a single enantiomer and given the designation LY231514, was quickly shown to be an extremely potent antitumor agent with a locus of action distinctly different from that of DDATHF or lometrexol. The primary enzymatic target of this new compound was TS [58] with secondary effects shown later to include inhibition of DHFR, GARFT, aminoimidazole carboxamide-ribonucleotide-formyltransferase (AICARFT), 5,10-methylenetetrahydrofolate dehydrogenase, and 10-formyltetrahydrofolate synthetase [59–62]. It was a superb substrate for FPGS, and studies confirmed that LY231514 requires polyglutamation and transport via the reduced folate carrier for cytotoxic potency. This new antitumor agent was thus a classical antifolate, but unique in its activity toward many different targets that control both transport and intracellular folate levels as a result of effects on cellular pyrimidine and purine biosynthesis. It was thus a true multitargeted antifolate and was referred to in many initial publications as MTA (Scheme 8.14) [63–71].

Mechanism of Alimta™ as antitumor agent

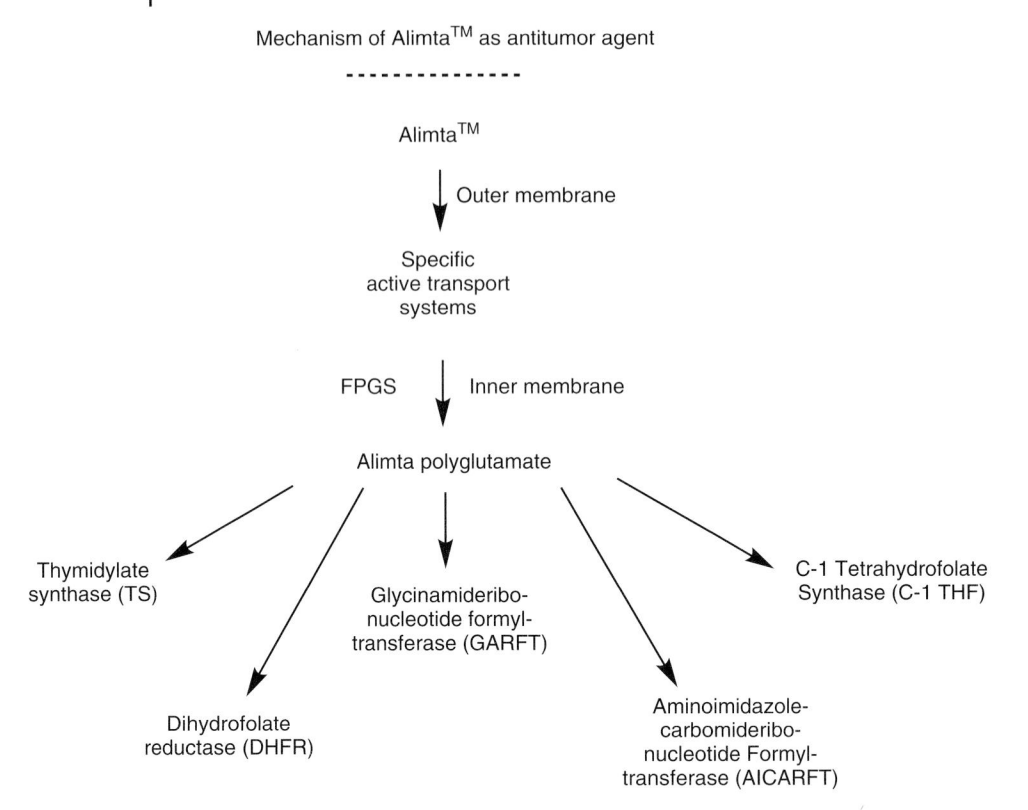

Alimta™
↓ Outer membrane

Specific active transport systems

FPGS ↓ Inner membrane

Alimta polyglutamate

→ Thymidylate synthase (TS)
→ Glycinamideribo-nucleotide formyl-transferase (GARFT)
→ C-1 Tetrahydrofolate Synthase (C-1 THF)
→ Dihydrofolate reductase (DHFR)
→ Aminoimidazole-carbomideribo-nucleotide Formyl-transferase (AICARFT)

Scheme 8.14

Once again, we were faced with the discovery of a remarkable antitumor agent that was, however, the first compound of its kind to be evaluated. Could its properties possibly be surpassed with another compound derived from a SAR study of LY231514? Some 50 variations involving other six to five ring systems with the same or different or additional ring substituents were synthesized and studied, but summation of the entire effort convinced us that the remarkable properties of LY231514 were uniquely attributable to its specific structure, and (in our hands, at least) could not be improved upon.

Initial Phase I studies indicated positive responses against bladder, breast, cervical, esophageal, gastric, genitourinary, head and neck, non-small cell lung, small-cell lung, malignant pleural mesothelioma, pancreatic, and renal cancers [66]. In early Phase II clinical trials, in confirmation of the earlier work, LY231514 exhibited antitumor activity in malignant pleural mesothelioma, non-small cell lung cancer (NSCLC), bladder, breast, cervical, colon, pancreas, and head and neck carcinomas [72]. The discovery that patients suffering from malignant pleural mesothelioma responded to LY231514 was extraordinarily exciting, because a treatment for this devastating disease had never been found. However, during the final stages of a large scale mesothelioma clinical trial (Phase II with 90 patients), several patients inexplicably died, and Lilly immediately informed the FDA that it was stopping the trial and withdrawing the drug. Both Dr. Paolo Paoletti, the Lilly physician running the trial, and Dr. Clet Niyikiza, a statistician working on the LY231514 team at Lilly, pleaded with Lilly, and with the FDA to hold off for a 2-week period while Dr. Niyikiza undertook an exhaustive stochastic examination of available biological data for all the patients who had ever been treated with MTA, including the ones who had died. It turned out that those patients who had exhibited toxicity had high homocysteine plasma levels, a clearcut indication of a deficiency of folic acid, as well as a marked increase in urine methylmalonic acid, indicating a deficiency of Vitamin B_{12} [73]. The FDA agreed to a suspension of the trial until all patients involved demonstrated normal blood levels of these two vitamins, and then allowed the trials to be resumed under a regimen that called for administration of MTA as a 10-min infusion once every 3 weeks preceded by vitamin therapy of each of the above vitamins ((oral FA) and Vitamin B_{12} (as an intramuscular injection)), for a normal course of 18 weeks. So the trial continued, with minimal reports of toxicity and no deaths. These results justified the FDA's permission for Lilly, before formal approval, to provide the drug free of charge to more than 1000 medically eligible patients through a compassionate use program. LY231514 was given the generic name pemetrexed disodium, and the proprietary name Alimta, and was approved by the FDA in early February 2004 for treatment of first line malignant pleural mesothelioma together with cisplatin. This was followed by approval in August 2004 for second line NSCLC (later changed by the FDA to nonsquamous NSCLC), then in September 2008 for first line nonsquamous NSCLC plus cisplatin, and then in July 2009 for nonsquamous NSCLC continuation maintenance (non-Alimta-containing platinum doublet followed by Alimta). In October 2012, more data were added to the label that now states:"first line nonsquamous NSCLC continuation maintenance with Alimta/cisplatin followed by Alimta alone." In all cases, before the administration of Alimta, the FDA mandates supplementation with oral folic acid and intramuscular vitamin B_{12}, that is continued throughout treatment. Corticosteroids are given the day before, the day of, and the day after Alimta administration.

The following is a brief summary of the characteristics of Alimta that emphasize its unique features:

- Alimta not only provides a much broader range of effectiveness against resistant cells, but also slows or prevents their development. A 1998 study showed that resistance to Alimta was multifactorial, and that the drug appears to have the unique ability to shift target enzymes through resistance development [74].
- Alimta, because it inhibits at least five targets simultaneously, is in effect acting by itself as a combination therapy with all of the conventional advantages of a combination of effective drugs (multiple targets) with none of its serious disadvantages (a summation of the toxicities of each individual drug in the combination).

Prediction of toxicity can be assessed by measuring homocysteine plasma levels and urine methylmalonic acid concentrations, and indications of folate deficiency can be handled by pretreatment with folic acid and vitamin B_{12} [75–77].

The above vitamin administration effectively reduces toxicity without reducing antitumor efficacy.

Treatment consists only of a 10-min infusion once every 3 weeks.

Based on sales, Lilly states that Alimta achieved the most successful launch of a new cancer drug in history. Despite more than 50 years of intensive world-wide investigations in this field of cancer chemotherapy, Alimta is the first antifolate launched in the United States since MTX in the 1950s.

> Alimta is the first FDA-approved drug for treatment of malignant pleural mesothelioma.
> Alimta has received subsequent FDA approvals for non-squamous first line, second line, and maintenance therapy in NSCLC.
> Alimta is the first NSCLC drug to demonstrate clinically relevant efficacy in specific histology groups.

The synthetic work involved in this project was carried out primarily at Princeton University, up to and including the discovery of the pyrrolopyrimidine N-{4-[2-(2-amino-3,4-dihydro-4-oxo-7H-pyrrolo[2,3-d]pyrimidin-5-yl)ethyl]benzoyl}-L-glutamic acid (LY231514). This compound, in Lilly's hands, became the cancer drug Alimta (pemetrexed disodium). My fascination was sparked at the beginning of my graduate studies when I realized that the bicyclic heterocyclic nucleus of xanthopterin, the yellow pigment in the wings of the brimstone butterfly, differed in only one substituent from the structure of folic acid. I was entranced by heterocyclic chemistry – with the role played by heterocycles as natural products, as latent functionalities for organic synthesis, as protecting groups, as leaving groups, or as highly activated intermediates. Control of reactivity through the choice and placement of heteroatoms became a versatile and extremely productive research focus, and it was inevitable that medicinal chemistry applications would become a central theme in our research group.

I would like to pay a sincere and grateful tribute to my many dedicated, talented, and enthusiastic coworkers – undergraduate, graduate, and postdoctoral students, to my colleagues in other institutions, and to a very large group of superb scientists at Lilly. The background chemistry, the synthetic methodologies discovered and developed, the diverse disciplines required and woven into our research efforts, all constituted a rich and versatile research background that ultimately made possible the discoveries outlined in this essay.

Special thanks are due to the following members of my research team:

At Princeton University:

Zareen Ahmed, Beena Bhatia, Zen-yu Chang, Rejendra Chaudhuri, James Dowling, Donald Dumas, Inci Durucasu, Steven Fitzjohn, Stephen Fletcher, Thomas George, Paul Gillespie, James Hamby, Peter Harrington, Philip Harrington, Baihua Hu, Evelyn Jackson, Lee Jennings, Jong-Gab Jun, Robert Kempton, Dietmar Kuhnt, Koo Lee, Xiaobing Li, Bin Liu, Hshiou-Tiong Liu, Zenmin Mao,

Keith McDaniel, Shashank Otiv, David Palmer, Maria Papadopoulou, Andrew Papoulis, Hemant Patel, Mona Patel, Partha Ray, Gowravaram Sabitha, Thomas Schrader, Jerauld Skotnicki, Carsten Spanka, Zdzislaw Szulc, Chi-ping Tseng, Markandu Vigneswaran, Loren Walensky, Yao Wang, Cary Ward, John Warner, Samuel Watson, George Wong, Cheol-min Yoon, Wendy Young, Sanbao Zhong, and Ping Zhou.

In other institutions:

Prof. G. Peter Beardsley (Yale Medical School), Prof. Richard G. Moran (Goodwin Cancer Research Facility, VCU, Dept. of Pharmacology and Toxicology).

At Lilly:

Charles Barnett, Gerald Grindey, Laura Mendelsohn, Clet Niyikiza, Paolo Paoletti, Homer Pearce, and Chuan (Joe) Shih.

In retrospect, Alimta was discovered in the nick of time. I seriously doubt whether this project could be carried out today for the discouraging reason that research funding now is available primarily, if not only, for projects with clearly defined objectives, utilizing methods that have solid experimental precedent. Curiosity seems to be neglected as a winning argument for financial support.

The ultimate success of this research effort critically depended, in its latter stages, on a remarkably successful and harmonious collaboration between Princeton and Lilly that deserves to be considered a model for the potential of purely academic basic research motivated entirely by curiosity, coupled with the effectiveness of a real scientific collaboration between academia and industry.

List of Abbreviations

5-DATHF	5-deazatetrahydrofolic acid
AMP	adenosine monophosphate
ATP	adenosine triphosphate
DDATHF	5,10-dideaza-5,6,7,8-tetrahydrofolic acid
DHFR	dihydrofolate reductase
FPGS	folylpolyglutamate synthase
GARFT	glycinamideribonucleotide formyltransferase
GMP	guanosine monophosphate
GTP	guanosine triphosphate
Homo-5-DATHF	see Ref. [46] for structure
Lometrexol	isomer B of DDATHF; see Ref. [52]
MTA	multitargeted antifolate
MTX	methotrexate
NSCLC	non-small-cell lung cancer
SAR	structure–activity relationships
THF	tetrahydrofolic acid
TS	thymidylate synthase
XMP	xanthosine monophosphate

References

1. Angier, R.B., Boothe, J.H., Hutchings, B.L., Mowat, J.H., Semb, J., Stokstad, E.L.R., SubbaRow, Y., Waller, C.W., Cosulich, D.B., Fahrenbach, M.J., Hultquist, M.E., Kuh, E., Northey, E.H., Seeger, D.R., Sickels, J.P., and Smith, J.M. Jr., (1946) The Structure and Synthesis of the Liver L. *casei* Factor. *Science*, **103**, 667–669.
2. Welch, A.D. (1983) *Folic Acid: Discovery and the Exciting First Decade*, The University of Chicago, pp. 64–75.
3. Purrmann, R. (1940) Die Synthese des Xanthopterins. *Liebigs. Ann. Chem.*, **546**, 98–102.
4. Schöpf, C. and Reichert, R. (1941) Zur Kenntnis des Leukopterins. *Liebigs. Ann. Chem.*, **548**, 82–94.
5. Nixon, J.C. (1985) in *Folates and Pterins*, Chemistry and Biochemistry of Pterins, vol. 2 (eds R.L. Blakley and S.J. Benkovic), John Wiley & Sons, Inc., New York, pp. 1–42.
6. Pfleiderer, W. (1985) Chemistry of naturally occurring pterins, in *Folates and Pterins*, Chemistry and Biochemistry of Pterins, vol. 2, Chapter 2, (eds R.L. Blakley and S.J. Benkovic), John Wiley & Sons, New York, pp. 43–114.
7. Albert, A. (1986) in *Chemistry and Biology of Pteridines* (eds B.A. Cooper and V.M. Whitehead), Walter de Gruyter & Co., Berlin, pp. 1–12.
8. Taylor, E.C. (1990) New pathways from pteridines. Design and synthesis of a new class of potent and selective antitumor agents. *J. Heterocycl. Chem.*, **27**, 1–12.
9. Taylor, E.C. (1993) in *Chemistry and Biology of Pteridines and Folates*, Advances in Experimental Medicine and Biology, vol. 338 (eds J.E. Ayling, M.G. Nair, and C.M. Baugh), Plenum Press, New York, pp. 387–408.
10. Brown, D.J. (1988) in *Fused Pyrimidines, Part Three, Pteridines*, Part 3 of The Chemistry of Heterocyclic Compounds, vol. 24 (ed. E.C. Taylor), John Wiley & Sons, Inc, New York.
11. Hopkins, F.G. (1889) Note on a yellow pigment in butterflies. *Abstr. Proc. Chem. Soc.*, **71**, 117–118.
12. Hopkins, F.G. (1889) Note on a yellow pigment in butterflies. *Nature*, **40**, 335.
13. Hopkins, F.G. (1891) Pigment in yellow butterflies. *Nature*, **45**, 197–198.
14. Hopkins, F.G. (1895) The pigments of the pieridae. A contribution to the study of excretory substances which function in ornament. *Proc. R. Soc. London*, **57**, 5–6.
15. Hopkins, F.G. (1895) The pigments of the pieridae: a contribution to the study of excretory substances which function in ornament. *Philos. Trans. Soc. Lond., Ser. B.*, **186**, 661–682.
16. Schöpf, C. and Wieland, H. (1926) Über das Leukopterin, das weisse Flügelpigment der Kohlweisslinge (Pieris brassicae und P.Napi). *Ber. Dtsch. Chem. Ges.*, **59**, 2067–2072.
17. Wieland, H. and Purrmann, R. (1939) Über die Flügelpigmente der Schmetterlinge. IV. Die Beziehungen zwischen Xanthopterin und Leukopterin. *Liebigs. Ann. Chem.*, **539**, 179–187.
18. Purrmann, R. (1940) Über die Flügelpigmente der Schmetterlinge. VII. Synthese des Leukopterins und Natur des Guanopterins. *Liebigs. Ann. Chem.*, **544**, 182–190.
19. Purrmann, R. (1941) Konstitution und Synthese des sogenannten Anhydroleukopterins. Über die Flügelpigmente der Schmetterlinge XII. *Liebigs. Ann. Chem.*, **548**, 284–292.
20. Schöpf, C. (1964) in *Pteridine Chemistry, Proceedings of the Third International Symposium held at the Institut für Organische Chemie der Technischen Hochschule Stuttgart, September 1962* (eds W. Pfleiderer and E.C. Taylor), Pergamon Press, pp. 3–14.
21. Farber, S., Diamond, L.K., Mercer, R.D., Sylvester, R.F. Jr., and Wolff, J.A. (1948) Temporary remissions in acute leukemia in children produced by folic acid antagonist, 4-aminopteroyl-glutamic acid (Aminopterin). *N. Engl. J. Med.*, **23** (238), 787–793.
22. Silverman, R.B. (1992) *The Organic Chemistry of Drug Design and Drug Action*, Academic Press, San Diego, CA.

23. Heine, R.W. and Welch, A.D. (1948) *J. Clin. Invest.*, **27**, 5639.
24. Huennekens, F.M. (1963) The role of dihydrofolic reductase in the metabolism of one-carbon units. *Biochemistry*, **2** (1), 151–159.
25. For an extensive review of DHFR inhibitors, see Palmer, D.C., Skotnicki, J.S., and Taylor, E.C. (1988) in *Progress in Medicinal Chemistry*, vol. 25 (eds G.P. Ellis and G.B. West), Elsevier, Amsterdam, pp. 85–231.
26. Kisliuk, R.L. (1999) in *Anticancer Drug Development Guide: Antifolate Drugs in Cancer Therapy* (ed. A.L. Jackman), Humana Press, Totowa, NJ, pp. 13–36.
27. Yan, S.J., Weinstock, L.T., and Cheng, C.C. (1979) Folic acid analogs. III. N-(2-[2-(2,4-diamino-6-quinazolinyl)ethyl]benzoyl)-L-glutamic acid. *J. Heterocycl. Chem.*, **16**, 541–544.
28. Sirotnak, F.M., DeGraw, J.I., and Chello, P.L. (1978) Pharmacokinetics and diagnostic procedures. Proceedings of the 10th International Congress of Chemotherapy and Diagnostic Procedures. Currrent Chemotherapy, Zürich, Switzerland, September 18–23, 1977, pp. 1128–1130.
29. Sirotnak, F.M., DeGraw, J.I., Chello, P.L., Moccio, D.M., and Dorick, D.M. (1982) Biochemical and pharmacologic properties of a new folate analog, 10-Deaza-aminopterin, in mice. *Cancer Treat. Rep.*, **66**, 351–358.
30. Kisliuk, R.L. (2003) Deaza analogs of folic acid as antitumor agents. *Curr. Pharm. Des.*, **9**, 2615–2625.
31. DeGraw, J.I., Kelly, L.F., Kisliuk, R.L., Gaumont, Y., and Sirotnak, F.M. (1982) Synthesis and antifolate activity of 8,10-dideazaminopterin. *J. Heterocycl. Chem.*, **19**, 1587–1588.
32. Jones, T.R., Calvert, A.H., Jackman, A.L., Brown, S.J., Jones, M., and Harrap, K.R. (1981) A potent antitumour quinazoline inhibitor of thymidylate synthetase: synthesis, biological properties and therapeutic results in mice. *Eur. J. Cancer*, **17**, 11–19.
33. Montgomery, J.A. and Piper, J.R. (1984) in *Folate Antagonists as Therapeutic Agents*, vol. 1 (eds F.M. Sirotnak, J.A. Burchall, and W.B. Ensininger), Academic Press, New York, pp. 219–260.
34. Rosowsky, A. (1989) in *Progress in Medicinal Chemistry*, vol. 26 (eds G.P. Ellis and G.B. West), Elsevier Science, pp. 1–252.
35. Beardsley, G.P., Taylor, E.C., Grindey, G.B., and Moran, R.G. (1986) *Chemistry and Biology of Pteridines*, Walter de Gruyter & Co, Berlin, pp. 953–957.
36. Taylor, E.C., Harrington, P.J., Fletcher, S.R., Beardsley, G.P., and Moran, R.G. (1985) Synthesis of the antileukemic agents 5,10-dideazaaminopterin and 5,10-dideaza-5,6,7,8-tetrahydroaminopterin. *J. Med. Chem.*, **28**, 914–921.
37. Moran, R.G., Baldwin, S.W., Shih, C., and Taylor, E.C. (1990) *Chemistry and Biology of Pteridines 1989*, Walter de Gruyter & Co., Berlin, pp. 1080–1085.
38. Seton, E.T. (1911) *The Arctic Prairies: A Canoe-Journey*, International University Press, New York, pp. 61–69.
39. Taylor, E.C., Fletcher, S.R., and Fitzjohn, S. (1985) Synthesis of 4-Amino-4-deoxy-7,10-methano-5-deazapteroic Acid and 7,10-Methano-5-deazapteroic Acid. *J. Org. Chem.*, **50**, 1010–1014.
40. Taylor, E.C. and Ray, P.S. (1987) Pteridines. 51. A new and unequivocal route to c-6 carbon-substituted pterins and pteridines. *J. Org. Chem.*, **52**, 3997–4000.
41. Taylor, E.C. and Yoon, C.-M. (1988) A convenient synthesis of 6-formyl-5-deazapterin. *Synth. Commun.*, **18**, 1187–1191.
42. Taylor, E.C. and Ray, P.S. (1987) Pteridines. 52, A convenient synthesis of 6-formylpterin. *Synth. Commun.*, **17**, 1865–1868.
43. Taylor, E.C., Hamby, J.M., Shih, C., Grindey, G.B., Rinzel, S.M., Beardsley, G.P., and Moran, R.G. (1989) Synthesis and antitumor activity of 5-deaza-5,6,7,8-tetrahydrofolic acid and its N10-substituted analogues. *J. Med. Chem.*, **32**, 1517–1522.
44. Taylor, E.C. and Wong, G.S.K. (1989) Convergent and efficient palladium-effected synthesis of 5,10-dideaza-5,6,7,8-tetrahydrofolic acid (DDATHF). *J. Org. Chem.*, **54**, 3618–3624.

45. Taylor, E.C., Harrington, P.M., and Warner, J.C. (1988) Diels-Alder reactions of 6-azapterins. An alternative strategy for the synthesis of 5,10-dideaza-5,6,7,8-tetrahydrofolic acid (DDATHF). *Heterocycles*, **27**, 1925–1928.
46. Taylor, E.C., Chang, Z.Y., Harrington, P.M., Hamby, J.M., Papadapoulou, M., Warner, J.C., Wong, G.S.K., and Yoon, C.-M. (1989) New synthetic studies on deazafolates, in *Chemistry and Biology of Pteridines 1989*, Walter de Gruyter & Co., Berlin, pp. 987–994.
47. Taylor, E.C. and Harrington, P.M. (1990) A convergent synthesis of 5,10-dideaza-5,6,7,8-tetrahydrofolic acid and 5,10-dideaza-5,6,7,8-tetrahydrohomofolic acid. An effective principle for carbonyl group activation. *J. Org. Chem.*, **55**, 3222–3227.
48. Taylor, E.C., Chaudhari, R., and Lee, K. (1996) A simplified and efficient synthesis of 5,10-dideaza-5,6,7,8-tetrahydrofolic acid. *Invest. New Drugs*, **14**, 281–285.
49. Shih, C., Grindey, G.B., Gossett, L.S., Moran, R.G., Taylor, E.C., and Harrington, P.M. (1989) Synthesis and structure-activity relationship studies of 5,10-dideazatetrahydrofolic acid (DDATHF), in *Chemistry and Biology of Pteridines 1989*, Walter de Gruyter & Co., Berlin, pp. 1035–1038.
50. Baldwin, S.W., Tse, A., Gossett, L.S., Taylor, E.C., Rosowsky, A., Shih, C., and Moran, R.G. (1991) Structural features of 5,10-dideaza-5,6,7,8-tetrahydrofolate that determine inhibition of mammalian glycinamide ribonucleotide formyltransferase. *Biochemistry*, **30**, 1997–2006.
51. Habeck, L.L., Mendelsohn, L.E., Shih, C., Taylor, E.C., Colman, P.D., Gossett, L.S., Leitner, T.A., Schultz, R.M., Andis, S.L., and Moran, R.G. (1995) Substrate specificity of mammalian folylpolyglutamate synthetase for 5,10-dideazatetrahydrofolate analogs. *Mol. Pharmacol.*, **48**, 326–333.
52. Moran, R.G., Baldwin, S.W., Taylor, E.C., and Shih, C. (1989) The 6S- and 6R-diastereomers of 5,10-dideaza-5,6,7,8-tetrahydrofolate are equiactive inhibitors of *de novo* purine synthesis. *J. Biol. Chem.*, **264** (35), 21047–21051.
53. Shih, C., Gossett, L.S., Worzalla, J.F., Rinzel, S.M., Grindey, G.B., Harrington, P.M., and Taylor, E.C. (1992) Synthesis and biological activity of acyclic analogues of 5,10-dideaza-5,6,7,8-tetrahydrofolic acid. *J. Med. Chem.*, **35**, 1109–1116.
54. Taylor, E.C., Harrington, P.M., and Shih, C. (1989) A facile route to "open chain" analogues of DDATHF. *Heterocycles*, **28**, 1169–1178.
55. Taylor, E.C. and Gillespie, P. (1992) Further acyclic analogues of 5,10-dideaza-5,6,7,8-tetrahydrofolic acid. *J. Org. Chem.*, **57**, 5757–5761.
56. Taylor, E.C., Schrader, T.H., and Walensky, L.D. (1992) Synthesis of 10-substituted "open-chain" analogues of 5,10-dideaza-5,6,7,8-tetrahydrofolic acid (DDATHF, Lometrexol). *Tetrahedron*, **48**, 19–32.
57. Taylor, E.C., Gillespie, P., and Patel, M. (1992) Novel 5-desmethylene analogues of 5,10-dideaza-5,6,7,8-tetrahydrofolic acid as potential anticancer agents. *J. Org. Chem.*, **57**, 3218–3225.
58. Jackman, A.L. (1997) Thymidylate synthase as a target for antifolate cancer chemotherapy, in *Chemistry and Biology of Pteridines and Folates 1997* (eds W. Pfleiderer and H. Rokos), Blackwell Science, pp. 93–98.
59. Taylor, E.C., Kuhnt, D., Shih, C., Rinzel, S.M., Grindey, G.B., Barredo, J., Jannatipour, M., and Moran, R.G. (1992) A dideazatetrahydrofolate analogue lacking a chiral center at C-6, N-[4-[2-(2-Amino-3,4-dihydro-4-oxo-7H-pyrrolo[2,3-d]pyrimidin-5-yl)ethyl]benzoyl]-L-glutamic acid, is an inhibitor of thymidylate synthase. *J. Med. Chem.*, **35**, 4450–4454.
60. Taylor, E.C., Kuhnt, D., and Shih, C. (1997) Synthesis of N-{4-[2-(2-amino, 3,4-dihydro-4-oxo-7H-pyrrolo[2,3-d]pyrimidin-5-yl)ethyl]benzoyl}-L-glutamic acid (LY231514) and analogues, in *Chemistry and Biology of Pteridines and Folates 1997* (eds W. Pfleiderer and H. Rokos), Blackwell Science, pp. 83–91.
61. Shih, C., Beardsley, G.P., Chen, V.J., Ehlhardt, W.J., MacKellar, W.C., Mendelsohn, L.G., Ratnam, M., Schultz, R.M., Taylor, E.C., and Worzalla, J.F.

(1997) Biochemical pharmacology studies of a multitargeted antifolate: LY231514, in *Chemistry and Biology of Pteridines and Folates 1997* (eds W. Pfleiderer and H. Rokos), Blackwell Science, pp. 181–187.

62. Shih, C., Chen, V.J., Gossett, L.S., Gates, S.B., MacKellar, W.C., Habeck, L.L., Shackelford, K.A., Mendelsohn, L.G., Soose, D.J., Patel, V.F., Andis, S.L., Bewley, J.R., Rayl, E.A., Moroson, B.A., Beardsley, G.P., Kohler, W., Ratnam, M., and Schultz, R.M. (1997) LY231514, a pyrrolo[2,3-d]pyrimidine-based antifolate that inhibits multiple folate-requiring enzymes. *Cancer Res.*, **57**, 1116–1123.

63. Calvert, H. (1999) MTA: summary and conclusions. *Semin. Oncol.*, **26** (2), 105–108.

64. Taylor, E.C. and Liu, B. (1999) A simple and concise synthesis of LY231514 (MTA). *Tetrahedron Lett.*, **40**, 4023–4026.

65. Taylor, E.C. and Liu, B. (2003) A new and efficient synthesis of pyrrolo[2,3-d]pyrimidine anticancer agents: alimta (LY231514), MTA), homo-alimta, TNP-351, and some aryl 5-substituted pyrrolo[2,3-d)pyrimidines. *J. Org. Chem.*, **68** (26), 9938–9947.

66. Taylor, E.C. and Liu, B. (1999) A novel synthetic route to 7-substituted derivatives of antitumor agent LY231514 (MTA). *Tetrahedron Lett.*, **40**, 5291–5294.

67. Taylor, E.C. and Liu, B. (2001) A new route to 7-substituted derivatives of N-{4-[2-(2-amino-3,4-dihydro-4-oxo-7H-pyrrolo[2,3-d]pyrimidin-5-yl)-ethyl]benzoyl}-L-glutamic acid [alimta (LY231514, MTA)]. *J. Org. Chem.*, **66**, 2726–3738.

68. Cripps, M.C., Burnell, M., Jolivet, J., Lofters, W., Fisher, B., Panasci, L., Iglesias, J., and Eisenhauer, E. (1997) Phase II study of a multi-targeted antifolate (LY231514) (MTA) as first line therapy in patients with locally advanced or metastatic colorectal cancer (MCC). *Eur. J. Cancer*, **S172**, 768.

69. Rusthoven, J., Eiserhauer, E., Butts, C., Gregg, R., Dancey, J., Fisher, B., and Iglesias, J. (1997) Phase II trial of the multi-targeted antifolate LY231514 (MTA) as first-line therapy for patients with advanced non-small lung cancer (NSCLC). *Eur. J. Cancer*, **S231**, 1045.

70. Thoedtmann, R., Kemmerich, M., Depenbrock, H., Blatter, J., Ohnmacht, U., Rastetter, J., and Hanauske, A.-R. (1977) A phase I study of MTA (multi-targeted antifolate, LY231514) plus cisplatin (CIS) in patients with advanced solid tumours. *Eur. J. Cancer*, **S247**, 1116.

71. Calvert, A.H. and Walling, J.M. (1998) Clinical studies with MTA. *Br. J. Cancer*, **78** (Suppl. 3), 35–40.

72. Hanauske, A.-T., Chen, V., Paoletti, P., and Niyikiza, C. (2001) Pemetrexed disodium: a novel antifolate clinically active against multiple solid tumors. *Oncologist*, **6**, 363–373.

73. Niyikiza, C., Hanauske, A.-R., Rusthoven, J.J., Calvert, A.H., Allen, R., Paoletti, P., and Bunn, P.A. Jr., (2002) Pemetrexed safety and dosing strategy. *Semin. Oncol.*, **29** (6, Suppl. 18), 24–29.

74. Schultz, R.M., Chen, V.J., Bewley, J.R., Roberts, E.F., Shih, C., and Dempsey, J.A. (1998) Resistance to the multi-targeted antifolate (MTA, LY231514): Multifactorial in human leukemia, breast, and colon carcinoma cells. *Ann. Oncol.*, **9** (Suppl. 4), 645P.

75. Niyikiza, C., Baker, S.D., Seitz, D.E., Walling, J.M., Nelson, K., Rusthoven, J.J., Stabler, S.P., Paoletti, P., Calvert, A.H., and Allen, R.H. (2002) Homocysteine and methylmalonic acid: markers to predict and avoid toxicity from pemetrexed therapy. *Mol. Cancer Ther.*, **1**, 545–552.

76. Niyikiza, C., Baker, S., Johnson, R., Walling, J., Seitz, D., and Allen, R. (1998) MTA (LY231514): relationship of vitamin metabolite profile, drug exposure, and other patient characteristics to toxicity. *Ann. Oncol.*, **9** (Suppl. 4), 609P.

77. Niyikiza, C. (2007) Antifolate combination therapies. US Patent 7,772,2090 B2, filed Jul. 11, 2007 and issued Aug. 10, 2010.

Edward C. Taylor is currently the A. Barton Hepburn Professor of Organic Chemistry Emeritus at Princeton University. He obtained his PhD from Cornell University in 1949, a National Academy of Sciences postdoctoral fellowship at the ETH in Zürich, Switzerland 1949–1950, and the duPont postdoctoral fellowship at the University of Illinois in Urbana 1950–1951. He was appointed to the faculty at the University of Illinois in 1951, and moved to Princeton University in 1954, where he has remained ever since. He is the author of >460 scientific research publications, the editor or co-editor of 88 volumes on organic chemistry and heterocyclic chemistry, and the author of three volumes on organic chemistry and heterocyclic chemistry. He has been the recipient of many honors, including the ACS Award for Creative Work in Synthetic Organic Chemistry, the Alfred Burger Award in Medicinal Chemistry (ACS), the Cope Scholar Award (ACS), the 5th International Award in Heterocyclic Chemistry, the ACS Heroes of Chemistry Award (together with Joe Shih and Homer Pearce of Lilly), the Thomas Alva Edison Award for Invention (for Alimta), and the NAS Prize for Chemistry in Service to Society. He has been awarded honorary degrees from Hamilton College and from Princeton University.

9
Perampanel: A Novel, Noncompetitive AMPA Receptor Antagonist for the Treatment of Epilepsy

Shigeki Hibi

9.1
Introduction

Traditional molecular targets for antiepileptic drug (AED) development have included voltage-gated sodium and calcium channels, γ-aminobutyric acid receptors, transporters, and enzymes [1, 2]. Despite the introduction of a number of second- and third-generation AEDs over the past two decades, one-third of patients still have poorly controlled epilepsy and continue to experience seizures [3]. Thus, there remains a need for novel treatment options with improved safety and efficacy profiles [4, 5].

In the brain, glutamate is the most abundant excitatory neurotransmitter [6] and two major families of glutamate receptor proteins have been identified. Metabotropic glutamate receptors (mGluRs) modulate neuro-excitation and synaptic transmission via second messenger pathways while ionotropic glutamate receptors (iGluRs) mediate fast synaptic responses to glutamate and are ligand-gated ion channels [7–9]. The iGluRs include α-amino-3-hydroxy-5-methyl-4-isoxazolepropionic acid (AMPA), kainate, and N-methyl-D-aspartic acid (NMDA) receptors, each subgroup being named after a class-selective agonist [8]. For over 20 years, the AMPA receptor has been investigated for its potential as a target for the treatment of epilepsy and other neurodegenerative diseases based on its direct role in glutamate-mediated excitatory neurotransmission and synaptic plasticity [10–13]. The AMPA receptor has also been specifically linked to temporal lobe epilepsy [14, 15], and it may have a key function in seizure-induced neuronal injury [16].

In 1988, the basic structure of some selective competitive non-NMDA receptor antagonists was discovered [17], and in 1990, the basic structure of noncompetitive AMPA receptor antagonists was identified from a 2,3-benzodiazepine muscle relaxant [18, 19]. Since then, several AMPA receptor antagonists have been in development as therapeutic agents, but most compounds have failed during the preclinical stages.

Successful Drug Discovery, First Edition. Edited by János Fischer and David P. Rotella.
© 2015 Wiley-VCH Verlag GmbH & Co. KGaA. Published 2015 by Wiley-VCH Verlag GmbH & Co. KGaA.

9.1.1
Competitive Receptor Antagonists

Quinoxalinedione derivatives (such as NBQX; Figure 9.1), were the first selective AMPA receptor antagonists to be identified, and these were found to compete with glutamate for its binding site on the AMPA receptors with high affinity. Unfortunately, these compounds were found to have both poor brain penetration and solubility at a neutral pH resulting in precipitation in the kidneys at therapeutic plasma concentrations [17, 20]. Safety issues have also limited the development of other competitive AMPA receptor antagonists. For ZK-200775, a clinical study involving patients who had experienced an acute ischemic stroke was terminated following strong depressant effects on their central nervous system (CNS) [21]. Tezampanel (Figure 9.1) [22, 23] was being investigated in a Phase II study for the treatment of acute migraine (ClinicalTrials.gov Identifier: NCT00567086).

9.1.2
Noncompetitive Receptor Antagonists

Meanwhile, noncompetitive AMPA receptor antagonists have actively been explored, driven by the belief that such drugs would exercise effectiveness independent of glutamate levels, without influencing normal glutamatergic activity [24]. Piriqualone (Figure 9.1) is categorized as a noncompetitive AMPA receptor

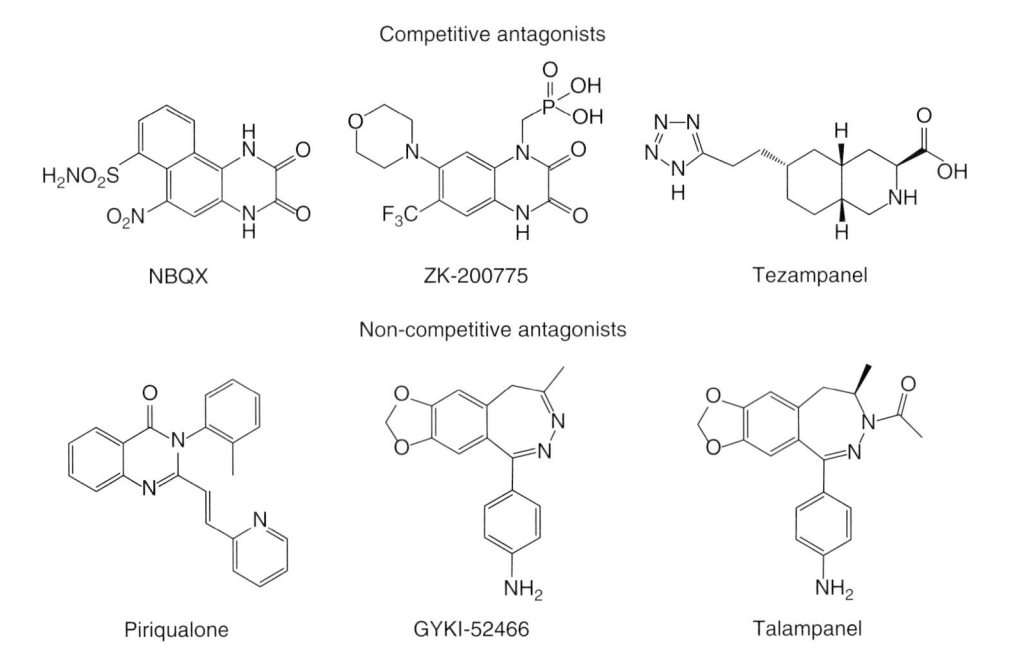

Figure 9.1 AMPA receptor antagonists.

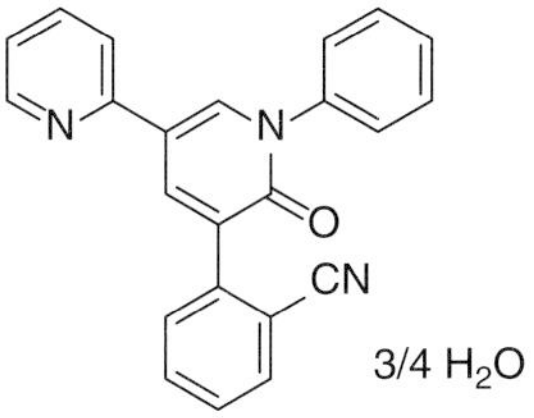

Figure 9.2 Perampanel.

antagonist [25]. GYKI-52466, a 2,3-benzodiazepine derivative, demonstrated a modest *in vitro* inhibitory effect but good oral bioavailability and *in vivo* efficacy, indicating effective blood–brain barrier (BBB) penetration [20]. A related compound, talampanel has been evaluated in patients with epilepsy, brain tumors, and amyotrophic lateral sclerosis [26–28]. Although efficacy was observed in epilepsy [26], there has been no further clinical development probably because talampanel has a half-life of 3–6 h [29], and multiple daily dosing was required in clinical trials [26, 28, 30].

Continued research identified the potential of targeting neuronal AMPA-type glutamate receptors, which have a key role in mediating fast excitatory neurotransmission. Here we describe a series of 1,3,5-triaryl-1*H*-pyridine-2-one derivatives as novel and noncompetitive antagonists of AMPA-type iGluRs and give an account of how they were developed for the treatment of epilepsy, a process finally resulting in the discovery of perampanel (Figure 9.2).

9.2
Seeds Identification by High Throughput Screening (HTS) Assays

The discovery process for perampanel started with a chemical library screen involving two HTS (high throughput screening) assays. The first assay, rat cortical neuron AMPA-induced cell death, allowed the identification of AMPA antagonist activity, while a binding assay with labeled AMPA allowed the detection and elimination of competitive AMPA antagonists. Positive compounds in HTS assays were further tested for their ability to inhibit AMPA-induced Ca^{2+} influx, to exclude neuro-protective compounds that did not act on the AMPA receptor target. As a result, we obtained the oxadiazinone scaffold (2,4-diphenyl-4*H*-[1,3,4]oxadiazin-5-one derivatives) as a latent seed compound for the development of noncompetitive AMPA receptor antagonists. Structural modifications of this scaffold eventually resulted in a drug candidate devoid of competitive antagonism, and led to the development of the first non-competitive AMPA receptor antagonist.

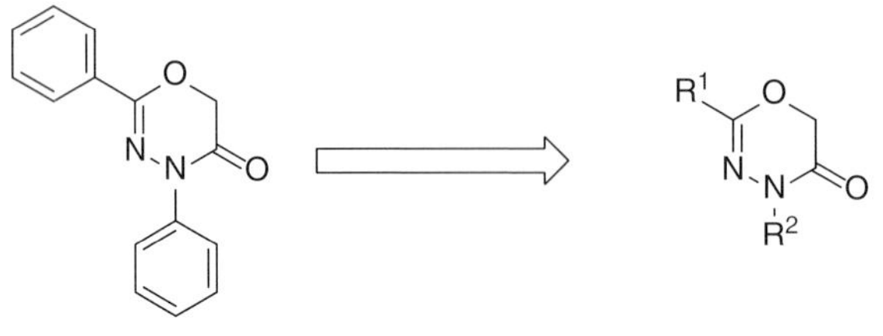

184 | *9 Perampanel: A Novel, Noncompetitive AMPA Receptor Antagonist for the Treatment of Epilepsy*

IC$_{50}$ = 9.17 µM
Conjugated aromaticity

IC$_{50}$ = 33.7 ->150 µM
Introduction of aliphatic substituents

Figure 9.3 SARs for the oxadiazinone derivatives in the AMPA-induced Ca^{2+} influx assay.

9.3
Structure and Activity Relationship (SAR) Study Starting from the Unique Structure of Seed Compounds

9.3.1
Introduction of Conjugated Aromaticity

We found the oxadiazinone scaffold (2,4-diphenyl-4*H*-[1,3,4]oxadiazin-5-one derivatives) as a latent seed compound for the development of noncompetitive AMPA receptor antagonists. The structure and activity relationships (SARs) for the oxadiazinone derivatives started with compounds having an aliphatic substituent at either R^1 or R^2 (Figure 9.3), which were tested in the AMPA-induced Ca^{2+} influx assay to assess *in vitro* activity. We then determined the effect of conjugated other aromaticity substituents. However, the aliphatic substituents reduced activity, suggesting that conjugated aromaticity would be required to maintain AMPA antagonism in these oxadiazinone derivatives.

9.3.2
Discovery of 1,3,5-Triaryl-1*H*-pyridin-2-one Template

For the next step, the oxadiazinone core structures were changed to improve both chemical and metabolic stability by modification of the hetero aromatic ring whilst keeping *in vitro* activity (Figure 9.4). For the six-membered hetero ring systems, the diazinone replaced the oxadiazinone to remove the oxygen atom and increase lipophilicity (Figure 9.4a), and the pyridone derivatives were replaced with oxadiazinone but still maintained weak *in vitro* activity in the Ca^{2+} influx assay (Figures 9.4b and c). When a comparison of metabolic stability was made between the pyridone derivatives, the acidic hydrogen was found to decrease the stability in human, mouse, and rat liver microsomes. To mask the acidic hydrogen in the 3,5-diphenyl-1,2-dihydro-pyridin-2-one derivative, another phenyl group was introduced to the pyridone (1,3,5-triphenyl-1,2-dihydropyridine-2-one), and this introduction surprisingly showed both good Ca^{2+} influx activity (IC$_{50}$ = 1.08 µM) and good metabolic stability. This modification was a critical key step for the invention of our drug candidate.

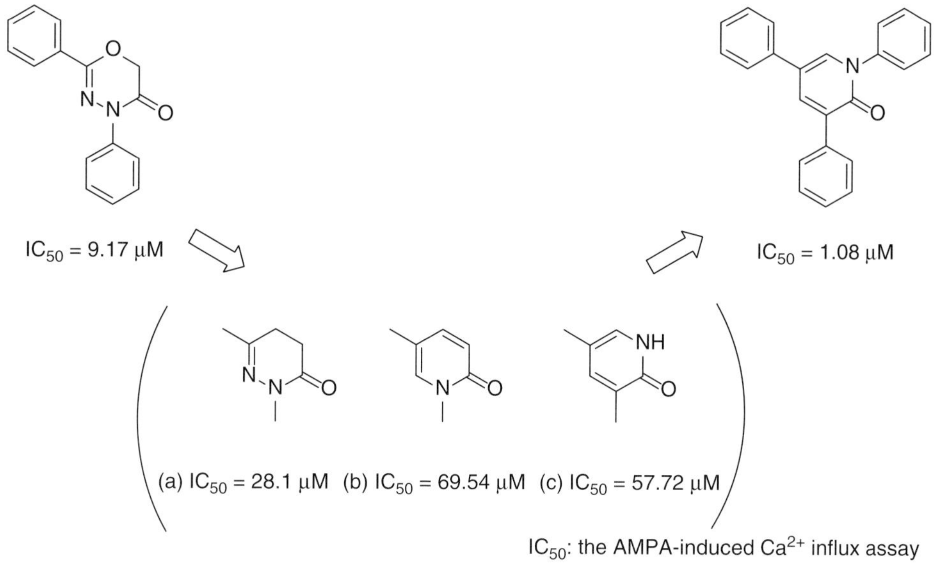

Figure 9.4 1,3,5-Triaryl-1*H*-pyridine-2-one derivatives.

9.3.3
Optimization of 1,3,5-Triaryl-1*H*-pyridin-2-one Derivatives

In order to clarify the contribution of each aromatic ring, additional optimization of the 1,3,5-triaryl-1*H*-pyridin-2-one template was conducted by focusing on manipulation of the individual aromatic rings located at positions 1, 3, and 5 of the pyridone ring. Firstly, the introduction of basicity via a 2-pyridine ring was assessed (Figure 9.5). In comparison with 1,3,5-triphenyl-1,2-dihydropyridine-2-one ($IC_{50} = 1.08\,\mu M$), 1,3-diphenyl-5-(2-pyridyl)-1,2-dihydropyridine-2-one ($IC_{50} = 0.32\,\mu M$) improved the activity to submicromolar levels. In addition,

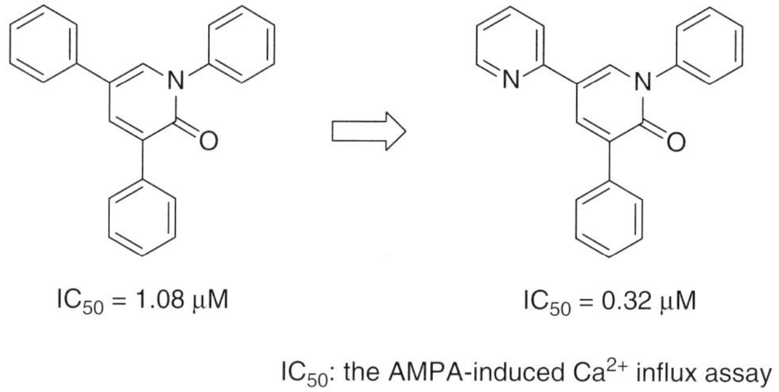

Figure 9.5 Introduction of 2-pyridyl moiety into 1,3,5-triaryl-1*H*-pyridine-2-one.

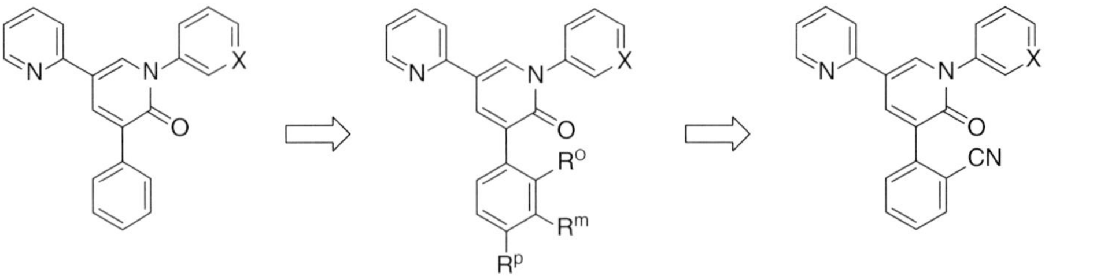

Figure 9.6 Introduction of −CN into the 3-phenyl of the pyridone ring.

IC$_{50}$ = 0.32 µM (X = C)
IC$_{50}$ = 0.44 µM (X = N)

R: halogen, −CN, alkoxy

IC$_{50}$ = 0.06 µM (X = C): perampanel
IC$_{50}$ = 0.21 µM (X = N)

IC$_{50}$: the AMPA-induced Ca^{2+} influx assay

a 2-pyridyl-substituent at position 5 showed improvement of *in vitro* and *in vivo* activities, resulting in the selection of a 2-pyridyl substituent at position 5 on the pyridone ring as the key aromatic ring in further optimization studies.

To determine if steric or electronic effects of the carbonyl group at position 2 on the pyridone ring affect *in vitro* activity, several substituents were introduced to the aromatic ring at position 3 on the pyridone ring in modification of 3-phenyl-5-(2-pyridyl)-1-phenyl-1,2-dihydropyridine-2-one as a template for SAR. In order to clarify the substituent effect (*ortho*, *meta*, and *para*), both bulk and electron density were investigated. To clarify the substituent effect by using the cyano group on the phenyl group, we synthesized *ortho*, *meta*, and *para* position substituted compounds (Figure 9.6). The *ortho* cyano-phenyl was shown to be the most appropriate, suggesting that both electron-withdrawing properties and steric hindrance, which induces a certain twist angle between the pyridone core and 3-aryl ring, are important.

Further investigations were carried out to confirm whether the 2-pyridine at position 5 and position 1 are optimized even in derivatives with a 2-cyano-phenyl substituent at position 3 of the pyridone. Several aryl rings were introduced at position 5 (Figure 9.7a) and comparing the substituent patterns of the pyridine rings, the 2-pyridyl derivative (perampanel) showed greater *in vitro* activity than the 3- and 4-pyridine isomers. Phenyl and thiophene or *ortho*-substituted phenyl groups decreased *in vitro* activity. In conclusion, the 2-pyridyl group was shown to confer the greatest *in vitro* potency. A final optimization at position 1 was executed by introduction of any substituted phenyl or pyridine ring, demonstrating that an unsubstituted phenyl substituent (perampanel) was best for *in vitro* activity (IC$_{50}$ = 0.06 µM) (Figure 9.7b) [31].

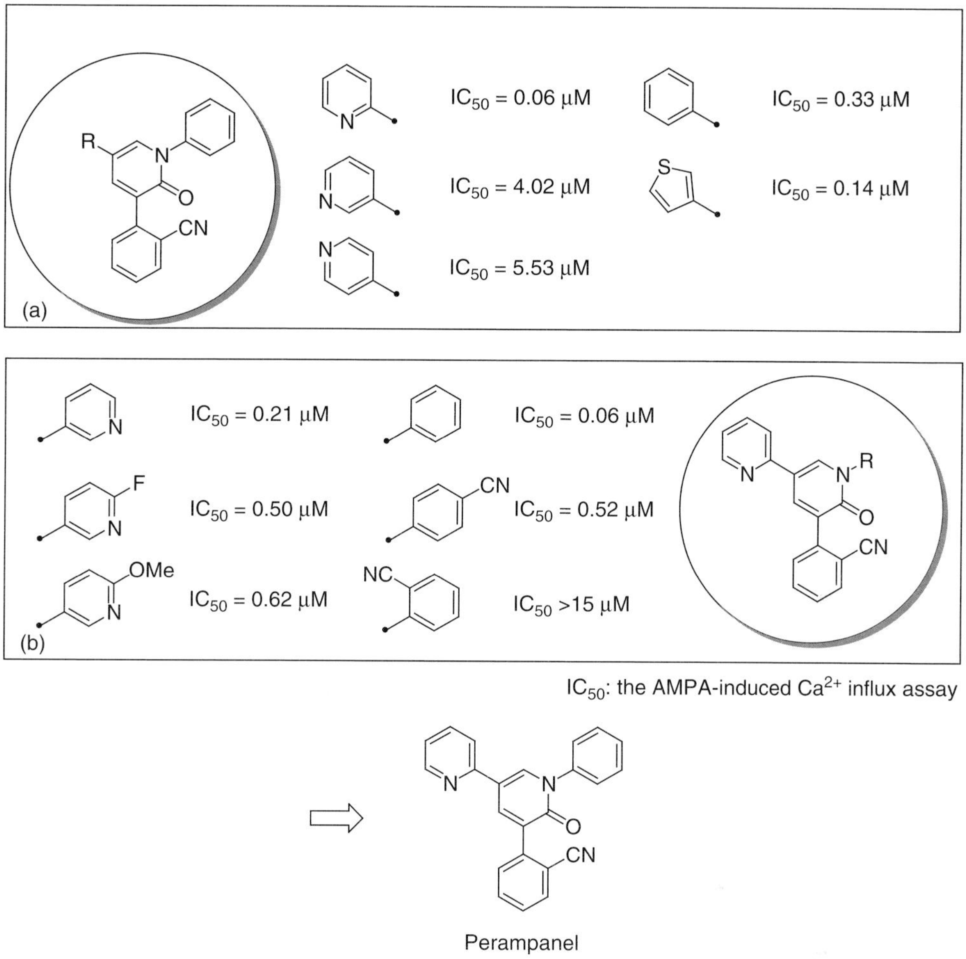

Figure 9.7 (a, b) SARs for the aryl group at positions 5 and 1 of the pyridone ring.

9.4
Pharmacological Properties of Perampanel; Selection for Clinical Development

9.4.1
The Pharmacological Evaluation of Perampanel

In pharmacological studies, perampanel had potent inhibitory effects both *in vitro* and *in vivo* (AMPA-induced Ca^{2+} influx assay: IC$_{50}$ of 60 nM (0.06 μM); AMPA-induced seizure model: minimum effective dose (MED) of 2 mg kg^{-1} po). In addition, binding assay results [32] indicated that perampanel has a similar binding site to other noncompetitive AMPA receptor antagonists. From radioligand binding assays, which included 63 physiologically relevant enzymes, ion channels, and neurotransmitter transporters, perampanel was also found to be highly selective for the AMPA receptor [33]. A number of preclinical seizure models were used to confirm the potent *in vivo* activity of oral perampanel (Table 9.1): audiogenic seizures in male DBA/2J mice $ED_{50} = 0.47$ mg kg^{-1}; maximal electroshock seizures (MES) in male ddY mice as an animal model of generalized tonic clonic seizure $ED_{50} = 1.6$ mg kg^{-1}; pentylenetetrazole (PTZ)-induced seizures in male ICR mice as an animal model of myoclonic or absence seizure $ED_{50} = 0.94$ mg kg^{-1}. In addition to efficacy in common seizure battery tests, oral perampanel showed efficacy in 6-Hz seizure test (32/44 mA), animal model of psychomotor seizures with ED_{50} values at 2.1 and 2.8 mg kg^{-1}, respectively. This broad spectrum activity distinguishes perampanel from sodium channel blocking AEDs (such as carbamazepine), which are weak or inactive in the PTZ and 6-Hz seizure tests. Many AEDs exhibit reduced potency in the 6-Hz test as the stimulus intensity is increased from 32 to 44 mA [34]. However, perampanel demonstrated nearly equal potencies at the two intensity levels, further reinforcing its distinctive profile. The activity of perampanel in the 6-Hz model was maintained and possibly augmented in the presence of carbamazepine, phenytoin, and valproate. These observations are consistent with the results of adjunctive clinical trials that have found perampanel to be efficacious in combination with these and other AEDs [35–38].

Perampanel was also studied in chronic epilepsy models. In rats with limbic seizure epilepsy induced by amygdala kindling, perampanel at a dose of 10 mg kg^{-1} *po* reduced the duration of the electrographic seizure discharge (after discharge) evoked by electrical stimulation, and at doses of 5 and 10 mg kg^{-1} caused a reduction in the behavioral seizure duration and seizure severity. However, as it was generally experienced with AMPA receptor antagonists [39], perampanel was inactive in an animal model (Genetic Absence Epilepsy Rat from Strasbourg) of spontaneous absence seizure.

Table 9.1 Comparison of the protective activity of perampanel in various acute seizure models with other antiepileptic drugs in mice and its potency to induce motor side effects.

Drug	Audiogenic	MES	PTZ	6-Hz	Rotarod
	ED_{50} (mg kg^{-1}), *po*				TD_{50} (mg kg^{-1})
Carbamazepine	6.1	21	>100	50 (32 mA)	ND
Sodium valproate	160	460	350	394 (32 mA)	ND
Perampanel	0.47	1.6	0.94	2.1 (32 mA)	1.8
				2.8 (44 mA)	

ED_{50}, estimated dose at which 50% of animals are protected; MES, maximal electroshock test, *po*, orally; PTZ, pentylenetetrazol seizure test; 6 Hz, 6 Hz seizure test (32 or 44 mA stimulation intensity); TD_{50}, estimated dose at which 50% of animals exhibit motor impairment in the rotarod test. Drugs were dosed orally 30–60 min prior to testing. In rats, the TD_{50} value of perampanel is 9.14 mg kg^{-1}.

9.4.2
The Pharmacokinetic Evaluation of Perampanel

Extensive metabolism of 2,3-benzodiazepine AMPA receptor antagonists and poor BBB penetration of competitive AMPA receptor antagonist have been an issue for potential drugs in development. One reason for the selection of the basic structure of perampanel (1,3,5-triaryl-1*H*-pyridine-2-one structure) was its metabolic stability in human liver microsome assays. The pharmacokinetic properties of perampanel were evaluated in rats after *iv* and *po* administration at a dose of 1 mg kg^{-1}. The results indicated a time to reach maximal plasma concentration of 0.50 h, a half-life of 2.37 h, a plasma clearance (CL) of 1.82 l h^{-1} kg^{-1}, and an oral bioavailability of 64.3% [31]. *In vitro* metabolic stability (intrinsic clearance (CLint)) with *in vitro–in vivo* extrapolation *(IVIVE)* indicated that hepatic metabolism is the main elimination route for perampanel. Perampanel was predicted by *IVIVE* to have a resistance to liver metabolic enzymes in humans since the CLint of perampanel in human liver microsomes was low (0.009 μl min^{-1} mg^{-1} protein). Good oral bioavailability was verified in rats and monkeys using radiolabeled perampanel, thus perampanel was expected to have high oral availability with good absorption and minimal hepatic extraction in human. Furthermore, the half-life of perampanel in humans was predicted to be long enough for once-daily dosing by allometric scaling of volume of distribution (Vd) and *IVIVE*. The concentration ratios of perampanel for brain: plasma or CSF: unbound plasma, indicated good brain penetration (concentration ratios in mice and rats ranged from 1.06 to 1.14) [31]. In addition, perampanel was not found to be a substrate for the efflux transporters p-glycoprotein and breast cancer resistant protein expressed on the BBB.

9.5
Clinical Development of Perampanel

9.5.1
Phase I

In humans, oral perampanel is rapidly and almost completely absorbed, with low systemic CL and high relative bioavailability. From two Phase I clinical studies, the half-life values for perampanel ranged from 52 to 129 h in the single-dose study and 66–90 h in the multiple-dose study [40]. Due to perampanel's long half-life, the plasma concentration, when perampanel was administered QD (quaque die) once daily, accumulated gradually and steady state was achieved after 2 weeks of repeated dose. Saccade eye velocity and Bond and Lader sedation subscale assessments were sensible measures of the CNS depressant effect of perampanel. Repeated dose study results indicated that perampanel more than 2 mg daily showed significant CNS depressive effect at first dose but severe somnolence was only observed in one patient at 4 mg after 2 weeks repeated dose [40]. It is suggested that gradual accumulation of plasma concentration caused self-titration and helped the development of tolerance to the sedative effect during repeated dosing.

9.5.2
Phase II and Phase III

Based on favorable preclinical and pharmacokinetic data, a global clinical development program investigated the effect of perampanel in patients with refractory partial-onset seizures who were receiving 1–3 concomitant AEDs. In the two randomized, double-blind, placebo-controlled Phase II trials, perampanel was titrated up to 4 mg twice daily or 12 mg once daily [41]. Three randomized, double-blind, placebo-controlled Phase III trials were also carried out and the perampanel dose was titrated up to 8 or 12 mg once daily [35–37]. Compared with placebo, treatment with perampanel (4–12 mg day^{-1}) was associated with significant reductions in the frequency of partial-onset seizures in these three pivotal Phase III trials. In addition, treatment was associated with an acceptable and consistent safety profile. Two long-term extension trials were also carried out to investigate the long-term efficacy and safety of perampanel [42, 43], and interim analyses have so far shown consistent results with the Phase III trials.

Based on the Phase III data, new drug and marketing authorization applications were submitted in 2011 and were approved in 2012. Perampanel is now approved in the European Union, the United States, Canada, and Switzerland as an adjunctive therapy for the treatment of partial-onset seizures with or without secondarily generalized seizures in patients with epilepsy aged 12 years and older (18 years and older in Canada) [44, 45].

9.6
Conclusion

Perampanel, which has been developed through a process of systematic optimization following HTS, exhibited extremely potent *in vitro* and *in vivo* effects. The unique 1,3,5-triaryl-1*H*-pyridine-2-one structure led to good drug metabolism and pharmacokinetic properties, overcoming the weaknesses of some earlier AMPA receptor antagonists. The discovery of perampanel and its progression through Phases I to III of clinical development have validated the clinical utility of an orally active, highly selective, noncompetitive AMPA receptor antagonist. Perampanel is now the first AMPA receptor antagonist to be approved in an epilepsy indication.

List of Abbreviations

AED	antiepileptic drug
CLint	intrinsic clearance
mGluRs	metabotropic glutamate receptors
iGluRs	ionotropic glutamate receptors

AMPA	α-amino-3-hydroxy-5-methyl-4-isoxazolepropionic acid
NMDA	N-Methyl-D-aspartic acid
BBB	blood–brain barrier
HTS	high throughput screening
MED	minimum effective dose
MES	maximal electroshock seizures
PTZ	pentylenetetrazole
IVIVE	in vitro–in vivo extrapolation
QD	quaque die

References

1. Rogawski, M.A. and Loscher, W. (2004) The neurobiology of antiepileptic drugs. *Nat. Rev. Neurosci.*, **5**, 553–564.
2. Landmark, C.J. (2007) Targets for antiepileptic drugs in the synapse. *Med. Sci. Monit.*, **13**, RA1–RA7.
3. Perucca, E., French, J., and Bialer, M. (2007) Development of new antiepileptic drugs: challenges, incentives, and recent advances. *Lancet Neurol.*, **6**, 793–804.
4. Ngugi, A.K., Bottomley, C., Kleinschmidt, I. et al. (2010) Estimation of the burden of active and life-time epilepsy: a meta-analytic approach. *Epilepsia*, **51**, 883–890.
5. French, J.A. (2007) Refractory epilepsy: clinical overview. *Epilepsia*, **48** (Suppl 1), 3–7.
6. Meldrum, B.S. (2000) Glutamate as a neurotransmitter in the brain: review of physiology and pathology. *J. Nutr.*, **130**, 1007S–1015S.
7. Mayer, M.L. and Armstrong, N. (2004) Structure and function of glutamate receptor ion channels. *Annu. Rev. Physiol.*, **66**, 161–181.
8. Watkins, J.C. and Jane, D.E. (2006) The glutamate story. *Br. J. Pharmacol.*, **147** (Suppl. 1), S100–S108.
9. Niswender, C.M. and Conn, P.J. (2010) Metabotropic glutamate receptors: physiology, pharmacology, and disease. *Annu. Rev. Pharmacol. Toxicol.*, **50**, 295–322.
10. Liu, S.Q. and Cull-Candy, S.G. (2000) Synaptic activity at calcium-permeable AMPA receptors induces a switch in receptor subtype. *Nature*, **405**, 454–458.
11. Rao, V.R. and Finkbeiner, S. (2007) NMDA and AMPA receptors: old channels, new tricks. *Trends Neurosci.*, **30**, 284–291.
12. Seeburg, P.H. (1993) The TINS/TiPS lecture. The molecular biology of mammalian glutamate receptor channels. *Trends Neurosci.*, **16**, 359–365.
13. Tanaka, H., Grooms, S.Y., Bennett, M.V., and Zukin, R.S. (2000) The AMPAR subunit GluR2: still front and center-stage. *Brain Res.*, **886**, 190–207.
14. Eid, T., Kovacs, I., Spencer, D.D., and de Lanerolle, N.C. (2002) Novel expression of AMPA-receptor subunit GluR1 on mossy cells and CA3 pyramidal neurons in the human epileptogenic hippocampus. *Eur. J. Neurosci.*, **15**, 517–527.
15. Hosford, D.A., Crain, B.J., Cao, Z. et al. (1991) Increased AMPA-sensitive quisqualate receptor binding and reduced NMDA receptor binding in epileptic human hippocampus. *J. Neurosci.*, **11**, 428–434.
16. Langer, M., Brandt, C., Zellinger, C., and Loscher, W. (2011) Therapeutic window of opportunity for the neuroprotective effect of valproate versus the competitive AMPA receptor antagonist NS1209 following status epilepticus in rats. *Neuropharmacology*, **61**, 1033–1047.
17. Honore, T., Davies, S.N., Drejer, J. et al. (1988) Quinoxalinediones: potent competitive non-NMDA glutamate receptor antagonists. *Science*, **241**, 701–703.
18. Tarnawa, I., Farkas, S., Berzsenyi, P. et al. (1989) Electrophysiological studies with a 2,3-benzodiazepine muscle relaxant: GYKI 52466. *Eur. J. Pharmacol.*, **167**, 193–199.
19. Tarnawa I., Farkas S., Berzsenyi P. et al. Reflex inhibitory action of a non-NMDA

20. Weiser, T. (2005) AMPA receptor antagonists for the treatment of stroke. *Curr. Drug Targets CNS Neurol. Disord.*, **4**, 153–159.
21. Walters, M.R., Kaste, M., Lees, K.R. et al. (2005) The AMPA antagonist ZK 200775 in patients with acute ischaemic stroke: a double-blind, multicentre, placebo-controlled safety and tolerability study. *Cerebrovasc. Dis.*, **20**, 304–309.
22. Ornstein, P.L., Arnold, M.B., Augenstein, N.K., Lodge, D., Leander, J.D., and Schoepp, D.D. (1993) (3SR,4aRS,6RS, 8aRS)-6-[2-(1H-tetrazol-5-yl)ethyl]decahydroisoquinoline-3-carboxylic acid: a structurally novel, systemically active, competitive AMPA receptor antagonist. *J. Med. Chem.*, **36**, 2046–2048.
23. Schoepp, D.D., Lodge, D., Bleakman, D., Leander, J.D., Tizzano, J.P., Wright, R.A., Palmer, A.J., Salhoff, C.R., and Ornstein, P.L. (1995) *In vitro* and *in vivo* antagonism of AMPA receptor activation by (3S,4aR,6R,8aR)-6-[2-(1(2)H-tetrazole-5-yl) ethyl] decahydroisoquinoline-3-carboxylic acid. *Neuropharmacolgy*, **34**, 1159–1168.
24. De Sarro, G., Gitto, R., Russo, E., Ibbadu, G.F., Barreca, M.L., De Luca, L., and Chimirri, A. (2005) AMPA receptor antagonists as potential anticonvulsant drugs. *Curr. Top. Med. Chem.*, **5**, 31–42.
25. Welch, W.M., Ewing, F.E., Huang, J., Menniti, F.S., Pagnozzi, M.J., Kelly, K., Seymour, P.A., Guanowsky, V., Guhan, S., Guinn, M.R., Critchett, D., Lazzaro, J., Ganong, A.H., DeVries, K.M., Staigers, T.L., and Chenard, B.L. (2001) Atropisomeric quinazolin-4-one derivatives are potent noncompetitive alpha-amino-3-hydroxy-5-methyl-4-isoxazolepropionic acid (AMPA) receptor antagonists. *Bioorg. Med. Chem. Lett.*, **11**, 177–181.
26. Chappell, A.S., Sander, J.W., Brodie, M.J. et al. (2002) A crossover, add-on trial of talampanel in patients with refractory partial seizures. *Neurology*, **58**, 1680–1682.
27. Grossman, S.A., Ye, X., Chamberlain, M. et al. (2009) Talampanel with standard radiation and temozolomide in patients with newly diagnosed glioblastoma: a multicenter phase II trial. *J. Clin. Oncol.*, **27**, 4155–4161.
28. Pascuzzi, R.M., Shefner, J., Chappell, A.S. et al. (2010) A phase II trial of talampanel in subjects with amyotrophic lateral sclerosis. *Amyotroph. Lateral Scler.*, **11**, 266–271.
29. Langan, Y.M., Lucas, R., Jewell, H., Toublanc, N., Schaefer, H., Sander, J.W., and Patsalos, P.N. (2003) Talampanel, a new antiepileptic drug: single- and multiple-dose pharmacokinetics and initial 1-week experience in patients with chronic intractable epilepsy. *Epilepsia*, **44**, 46–53.
30. Iwamoto, F.M., Kreisl, T.N., Kim, L., Duic, J.P., Butman, J.A., Albert, P.S., and Fine, H.A. (2010) Phase 2 trial of talampanel, a glutamate receptor inhibitor, for adults with recurrent malignant gliomas. *Cancer*, **116**, 1776–1782.
31. Hibi, S., Ueno, K., Nagato, S., Kawano, K., Ito, K., Norimine, Y., Takenaka, O., Hanada, T., and Yonaga, M. (2012) Discovery of 2-(2-Oxo-1-Phenyl-5-Pyridin-2-yl-1,2-Dihydropyridin-3-yl)Benzonitrile (Perampanel): a novel, noncompetitive α-Amino-3-Hydroxy-5-Methyl-4-Isoxazolepropanoic acid (AMPA) receptor antagonist. *J. Med. Chem.*, **55**, 10584–10600.
32. Hanada, T., Hashizume, Y., Tokuhara, N., Takenaka, O., Kohmura, N., Ogasawara, A., Hatakeyama, S., Ohgoh, M., Ueno, M., and Nishizawa, Y. (2011) Perampanel: a novel, orally active, non-competitive AMPA-receptor antagonist that reduces seizure activity in rodent models of epilepsy. *Epilepsia*, **52**, 1331–1340.
33. Tokuhara, N., Hatakeyama, S., Amino, H., Hanada, T., and Nishizawa, Y. (2008) Pharmacological profile of perampanel: a novel, non-competitive selective AMPA receptor antagonist. Abstract P02.112 presented at the 60th Annual Meeting of the American Academy of Neurology, Chicago, IL, April 12–19, 2008.
34. Barton, M.E., Klein, B.D., Wolf, H.H., and White, H.S. (2001) Pharmacological

characterization of the 6 Hz psychomotor seizure model of partial epilepsy. *Epilepsy Res.*, **47**, 217–227.
35. French, J.A., Krauss, G.L., Biton, V., Squillacote, D., Yang, H., Laurenza, A., Kumar, D., and Rogawski, M.A. (2012) Adjunctive perampanel for refractory partial-onset seizures: randomized phase III study 304. *Neurology*, **79**, 589–596.
36. French, J.A., Krauss, G.L., Steinhoff, B.J., Squillacote, D., Yang, H., Kumar, D., and Laurenza, A. (2013) Evaluation of adjunctive perampanel in patients with refractory partial-onset seizures: results of randomized global phase III study 305. *Epilepsia*, **54**, 117–125.
37. Krauss, G.L., Serratosa, J.M., Villanueva, V., Endziniene, M., Hong, Z., French, J., Yang, H., Squillacote, D., Edwards, H.B., Zhu, J., and Laurenza, A. (2012) Randomized phase III study 306: adjunctive perampanel for refractory partial-onset seizures. *Neurology*, **78**, 1408–1415.
38. Steinhoff, B.J., Ben-Menachem, E., Ryvlin, P., Shorvon, S., Kramer, L., Satlin, A., Squillacote, D., Yang, H., Zhu, J., and Laurenza, A. (2013) Efficacy and safety of adjunctive perampanel for the treatment of refractory partial seizures: a pooled analysis of three phase III studies. *Epilepsia*, **54**, 1481–1489.
39. Kaminski, R.M., Van Rijn, C.M., Turski, W.A., Czuczwar, S.J., and Van Luijtelaar, G. (2001) AMPA and GABA(B) receptor antagonists and their interaction in rats with a genetic form of absence epilepsy. *Eur. J. Pharmacol.*, **430**, 251–259.
40. Templeton, D. (2009) Pharmacokinetics of perampanel, a highly selective AMPA-type glutamate receptor antagonist. *Epilepsia*, **50** (Suppl. 11), 98 (abstract 1.199).
41. Krauss, G.L., Bar, M., Biton, V., Klapper, J.A., Rektor, I., Vaiciene-Magistris, N., Squillacote, D., and Kumar, D. (2012) Tolerability and safety of perampanel: two randomized dose-escalation studies. *Acta Neurol. Scand.*, **125**, 8–15.
42. Rektor, I., Krauss, G.L., Bar, M., Biton, V., Klapper, J.A., Vaiciene-Magistris, N., Kuba, R., Squillacote, D., Gee, M., and Kumar, D. (2012) Perampanel study 207: long-term open-label evaluation in patients with epilepsy. *Acta Neurol. Scand.*, **126**, 263–269.
43. Krauss, G.L., Perucca, E., Ben-Menachem, E., Kwan, P., Shih, J.J., Squillacote, D., Yang, H., Gee, M., Zhu, J., and Laurenza, A. (2012) Perampanel, a selective, noncompetitive a-amino-3-hydroxy-5-methyl-4-isoxazolepropionic acid receptor antagonist, as adjunctive therapy for refractory partial-onset seizures: interim results from phase III, open-label extension study 307. *Epilepsia*, **54**, 126–134.
44. European Medicines Agency Fycompa Summary of Product Characteristics 2012, http://www.ema.europa.eu/docs/en_GB/document_library/EPAR_-_Product_Information/human/002434/WC500130815.pdf (accessed 8 August 2014).
45. Food and Drug Administration Fycompa Prescribing Information 2012, http://www.accessdata.fda.gov/drugsatfda_docs/label/2012/202834lbl.pdf (accessed 8 August 2014).

Shigeki Hibi majored in organic chemistry at Tsukuba University, Japan, and went on to complete his master's degree in 1987. He then joined Eisai Tsukuba Research, working for 17 years as a medicinal chemist where he focused on immunology, bone, gastrointestinal, and CNS projects. In 1985 he completed his PhD in Pharmacy at Kyushu University, Japan. In 2004 he moved into global project management to specialize in project management for CMC, nonclinical safety, and clinical development, planning for global submission and Asian clinical development. In 2010 he became Senior Director of Strategic Operations for Eisai's Product Creation Headquarters.

10
Discovery and Development of Telaprevir (Incivek™) – A Protease Inhibitor to Treat Hepatitis C Infection

Bhisetti G. Rao, Mark Murcko, Mark J. Tebbe, and Ann D. Kwong

10.1
Introduction

Hepatitis C is an infection caused by a virus that attacks the liver and leads to inflammation [1]. Most people infected with the hepatitis C virus (HCV) have no symptoms. In fact, most people do not know they have the hepatitis C infection until liver damage shows up, decades later, during routine medical tests. The infection that continues over many years can cause significant complications, such as cirrhosis of the liver (scarring of the liver tissue). HCV infection leads to liver cancer in a small number of people with liver cirrhosis and ultimately to liver failure. It is, therefore, a leading cause of liver transplantation in the United States [2]. It is estimated that more than 150 million individuals are infected with HCV in the world. It is a serious disease and a hidden epidemic causing about 350 000 deaths worldwide annually [3].

The HCV genome encodes a polyprotein of structural and non-structural (NS) proteins [4]. The polyprotein is proteolytically processed into 10 individual proteins by host cell and viral proteases. HCV NS3/4A serine protease (NS3/4A) is a noncovalent heterodimer of the N-terminal, ~180-residue portion of the 631-residue NS3 protein with the NS4A co-factor. NS3/4A cleaves the polyprotein sequence at four specific regions releasing the NS proteins 4A, 4B, 5A, and 5B [5]. Therefore, NS3/4A protease activity is essential for viral replication and has been recognized to be an attractive drug target. Protease inhibition can impact early steps of viral replication by blocking the release of the NS proteins and formation of the HCV replicase complex. Scientists at Chiron identified HCV in 1987 [6], but progress toward finding direct antiviral agents for treatment of this disease were slow to start due to many factors including lack of crystal structures of HCV targets and robust primary and secondary assays for lead discovery and optimization. In a recent review, Kwong *et al.* [7] described the important research and commercial milestones in discovery and development of telaprevir. The protease inhibitor project that led to discovery of telaprevir was initiated at Vertex Pharmaceuticals in 1994. The following sections describe how the major challenges of

Successful Drug Discovery, First Edition. Edited by János Fischer and David P. Rotella.
© 2015 Wiley-VCH Verlag GmbH & Co. KGaA. Published 2015 by Wiley-VCH Verlag GmbH & Co. KGaA.

discovery and development were addressed by scientists at Vertex and its partners in a successful collaboration.

10.1.1
Crystal Structure of NS3/4A Protease

The small size of the protein and its dependence on Zn and a cofactor peptide for stability and activity caused many hurdles in expression and crystallization of the protein. The scientists at Vertex Pharmaceuticals were successful in overcoming these hurdles and solved the first crystal structure of the active form of the NS3/4A protease in 1996 [8]. The protease has a chymotrypsin-like fold composed of two beta barrels with the catalytic triad (consisting of Ser-139, His-57, and Asp-81) located at the domain-interface. The beta strand of the NS4A peptide forms a part of the N-terminal beta barrel of the NS3 protein, and helps the formation of a competent catalytic triad, explaining the role of the cofactor in protease activation. The role of the structural Zn in the maintenance of the protease tertiary structure was also evident in its coordination of the side chains of four residues (Cys-97, Cys-99, Cys-145, and His-149) that are distal to the active site. The crystal structure opened up a great opportunity for structure-based drug design, but also presented a major challenge for designing potent inhibitors by revealing that the active site was flat and open without any deep binding pockets. Consistent with the structure of the active site, the minimal length of the enzyme's natural substrates were long, encompassing 10 residues spanning the S6 to S4′ subsites of the enzyme. Designing inhibitors with required affinity and drug-like features (e.g., cellular permeability, acceptable DMPK) appeared daunting [9].

10.1.2
Assays

An additional initial hurdle in tackling the target was the lack of robust enzymatic and cellular assays to drive screening, characterization, and optimization of inhibitors. The initial assay used to evaluate compounds as inhibitors of HCV protease was a spectrophotometric assay that followed the cleavage of *p*-nitroanilide (*p*NA) from the C-terminus of a hexapeptide substrate [10]. This assay provided significant throughput and the kinetic advantage of continuously monitoring product release. However, the *p*NA assay is limited by two factors: a high enzyme concentration that limits the overall sensitivity of the assay and a high concentration of substrate (~1 mM) that can limit the solubility of compounds of interest. To combat these limitations an high-performance liquid chromatography (HPLC) assay was developed based on the detection of products generated from the cleavage of an unlabeled peptide with high homology to a naturally occurring cleavage site within the HCV polypeptide [10]. The HPLC-based assay gave good sensitivity along with excellent data quality. Coupling the HPLC to a microplate-compatible autosampler allowed maintenance of sufficient throughput to use this assay as the primary assay for initial compound evaluation.

Compounds that showed significant potency in the above enzyme assays were further evaluated in cells using a 2-day replicon assay [11]. Huh-7 cells harboring a subgenomic replicon of the HCV subtype 1b Con1 strain were incubated with the compound for 48 h, at which point the level of HCV RNA was quantitated using quantitative real time reverse transcription–polymerase chain reaction (QRT–PCR).

10.2
Discussion

Attempts to find chemical starting points utilizing high-throughput screening against the HCV NS3/4A protease proved futile and did not provide any useful hits. Therefore, a design strategy was devised that combined the use of the protease crystal structure and binding information of the natural substrates to generate chemical starting points. These starting points could then be optimized using structure-based drug design cycles employing the crystal structures of the protease-inhibitor complexes at different stages of optimization. Further, it was decided to pursue compounds that would distribute preferentially to the liver, as the liver is the primary site of HCV infection. Targeting a drug to the liver had the advantage of potentially limiting systemic toxicity by lowering systemic exposure. Therefore, the compound level in the liver of rodents was regularly measured in addition to evaluating *in vitro* potency derived from the enzyme and HCV replicon assays. Compounds with higher potency and better liver exposure were prioritized for advancement. The details of this strategy have been described previously [12–15].

10.2.1
Substrate-Based Inhibitor Design

The key information used in pursuing a structure-based inhibitor design approach for the HCV NS3/4A protease in the initial stages was the model of its complex with a decamer substrate peptide (Figure 10.1). The natural NS5A-5B cleavage site was chosen as the 10-mer substrate starting point for inhibitor design. The 10-amino acid long segment (EDVVCC-SMSY) of the substrate spans the S6 to S4′ subsites of the enzyme. At this point, only the crystal structure of the apo enzyme was available. Therefore, a study of the effect of truncations from the N- and C-terminus of the decapeptide on Ki in an HCV protease enzyme assay was performed (Landro *et al.*) and a model of the 10-amino acid substrate bound to the active site of the NS3/4A protease active site was built [10]. The peptide truncation study revealed that an 80-fold decrease in affinity was observed when a single amino acid was deleted from the N-terminus and a 200-fold decrease in affinity was observed when two amino acids were truncated from the C-terminus. These results were consistent with the crystal structure of a peptide substrate co-complexed with the HCV protease catalytic domain, which revealed that substrate

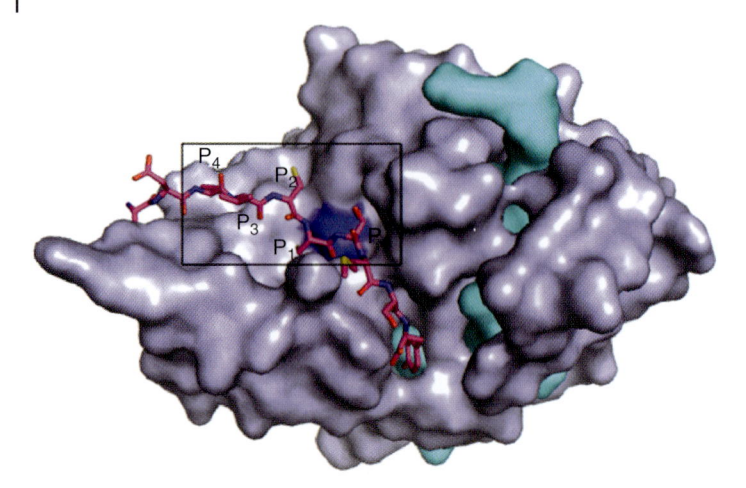

Figure 10.1 Model of NS5A-5B substrate (dark sticks) bound to the active site of NS3/NS4A protease. The NS3 protease is shown as a gray surface and the NS4A peptide is shown in cyan. The dark blue patch on the protein surface is the location of the active site Ser-139. The box highlights different hydrophobic subsites that are filled by an inhibitor that spans P4 to P_1'. The PyMOL Molecular Graphics System (Schrödinger LLC) was used to generate Figures 10.1–10.3.

binding was spread over a large surface area (Figure 10.1). The model revealed that the binding site for the substrate had limited topology that would allow effective, high-energy interactions to be made except for some shallow pockets at S1 and S4'. The other subsites showed only small indentations on the enzyme surface. The six residues of the nonprime side make several hydrogen bonds and hydrophobic interactions whereas only the P4' Tyr side chain on the prime side makes strong interactions with the enzyme. Therefore, it was not surprising that the removal of the P4' residue resulted in a dramatic loss of activity (∼100-fold), and subsequent truncations at P3' and P2' had little additional loss. These results suggested that S3' and S2' were not sites where optimization would lead to significant gains in affinity. On the nonprime side, removal of the P6 and P5 acidic residues resulted in significant decreases in binding affinities. Additionally, truncation of the hydrophobic residues at P4 and P3 led to further loss in affinity, suggesting that increased binding could be gained through hydrophobic contacts at P3 and P4. These results suggested rather large hydrophobic inhibitors spanning P4 to P1/P1' would be needed to achieve adequate affinity and that achieving good physicochemical properties consistent with good oral bioavailability was going to be a challenge.

To gain binding with smaller size inhibitors, an electrophilic group (i.e., warhead in the serine protease literature) was added to the P1 residue to engage the catalytic serine side chain to form a reversible, covalent bond between the enzyme and the inhibitor. The resulting enzyme–inhibitor complex is a transition state-like tetrahedral intermediate that mimics amide bond hydrolysis and is stabilized through several ionic and hydrogen bond interactions. The resulting covalent and reversible inhibitors were 10–1000 times more potent than inhibitors that relied solely on noncovalent interactions. This strategy appeared viable as a way to gain back the affinity lost by necessary truncation of the natural substrate. Therefore, it was decided to pursue covalent, reversible inhibitors spanning the S4 to S1′ subsites of the enzyme, as shown in Figure 10.1. The aldehyde functionality was initially chosen as the warhead moiety. However, it was quickly determined that the nucleophilic Cys at P1 was not compatible with the aldehyde warhead, and it was replaced by Abu (2-aminobutanoic acid). The P2 Cys was replaced by benzyloxy-proline as proline is an observed P2 residue of the HCV NS4B-5A substrate. The benzyloxy-proline modeled well into the open S2 pocket of the protease. This simple design of a hexapeptide aldehyde inhibitor (**2**) achieved sub-μM potency (Ki = 0.89 μM) (Scheme 10.1). The P6 and P5 residues were then replaced with a heterocyclic cap, which led to the tetrapeptide aldehyde **3** (Ki = 12 μM) as the starting point for further optimization. This substitution reduced the size of the inhibitor and eliminated two negatively charged side chains but lost potency by about 14-fold [16].

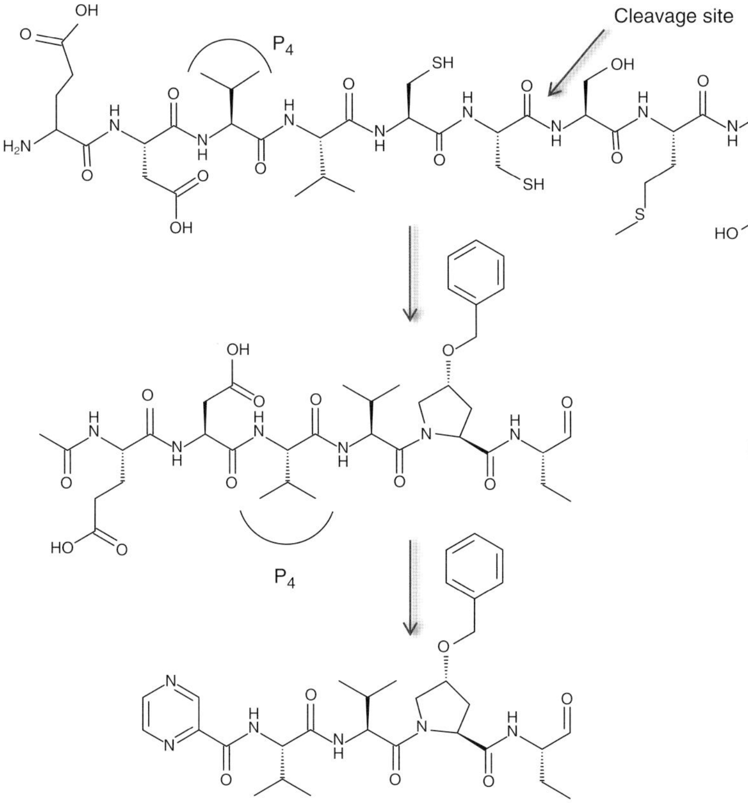

1
NS5A-5B substrate

2
Ki = 0.89 μM

3
Ki = 12 μM

Scheme 10.1 Truncation of a substrate-based hexapeptide aldehyde to a tetrapeptide aldehyde inhibitor. Accelrys Draw 4.1 was used to draw chemical structures in Schemes 10.1–10.4.

10.2.2
Structure-Based Inhibitor Optimization

To improve potency of the lead tetrapeptide aldehyde inhibitor, the side chains were examined guided by the shape and the electrostatic characteristics of the binding subsites that these side chains occupied. Crystal structure based docking studies were employed to select the preferred side chains at each subsite for synthesis. Parallel synthesis methods were utilized extensively to conduct multi-site optimization as needed. Early optimization focused on the P2 residue where optimization of interactions with residues in the S2 pocket could potentially lead to gains in inhibitor affinity. There are three key residues of this pocket: Arg-155, Asp-81, and His-83. The latter two residues are part of the catalytic triad and provide a rigid, flat surface for stacking interactions. However, the long, flexible side chain of Arg-155 can adopt many conformations and greatly affects the shape of the S2 pocket. In most of the conformations, Arg-155 tends to maintain ionic and hydrogen bond interactions with the neighboring Asp-81 and Asp-168 side chains. Based on the chemistry available for the elaboration of a 4-hydroxyproline starting material, ethers, esters, and carbamates were investigated (Scheme 10.2). Crystal structures of key inhibitors from each class revealed that the orientation of the P2 proline 4-substituent is influenced by the linker geometry. The aromatic rings of these substituents make stacking interactions with the Arg-155 side chain that is in an extended conformation in all cases. The shape and the orientation of the bicyclic group appear to be critical for the substituent's interaction with Arg-155. Esters such as **5** (Ki = 0.4 µM) with planar rigid geometry were found to be generally more potent than the flexible ethers (e.g., **4**, Ki = 1.7 µM). As esters are hydrolytically unstable, carbamates were investigated as an alternative to esters, with varying degrees of success. Replacement of the 2-naphthyl ester with the isosteric tetrahydroisoquinolyl (THIQ) carbamate led to compound **6**, which showed similar affinity for the enzyme (Ki = 0.89 µM) although it showed a dramatic change in the orientation of the P2-substituent [16] (Figure 10.2).

After finding a potent and stable P2 group, the optimization moved to P1 and the warhead. The S1 pocket is a well-defined specificity subsite of the protease.

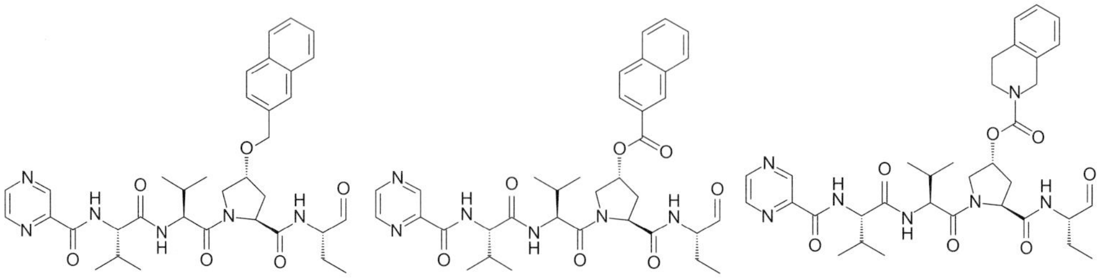

Scheme 10.2 P2 variation of aldehyde inhibitors.

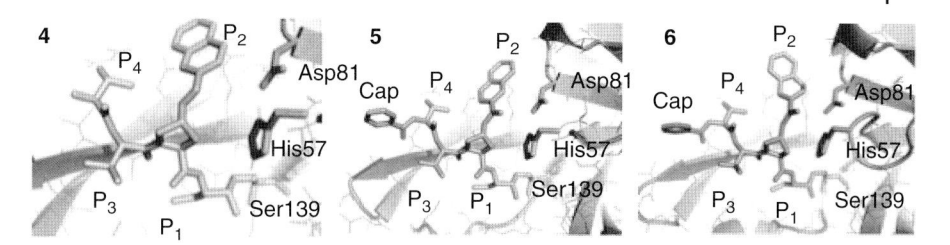

Figure 10.2 Crystal structure of inhibitors 4–6. The inhibitor and the catalytic triad (Ser-139, His-57, and Asp-81) are shown as sticks. The protein is shown in lines as well as in ribbon representation. The naphthyl of the P2 group in compounds 4 and 5 make the same contacts with the binding site; however, compound 5 is more potent due to its rigid ester linker. The THIQ group linked as a carbamate in compound 6 is twisted and makes less optimal contacts and is, therefore, slightly weaker than compound 4. The catalytic Ser-139 makes a covalent interaction with the inhibitor aldehyde warhead. The His-57 and Asp-81 residues are also shown in stick diagram. (The P4-cap of compound 4 was not defined in the X-ray structure.)

It is a small pocket lined by Phe-154, Leu-135, and Ala-157 and prefers residues with small side chains. In fact, Cys appears to be the consensus residue of the protease substrates involved in trans cleavage. The Cys side chain makes a classical thiol–aromatic interaction with Phe-154. Exploration of SAR at this position showed preference for small hydrophobic side chains such as ethyl, propyl, and trifluoroethyl. Incorporation of an oxygen atom as well as disubstitution at the geminal position led to loss of affinity. The propyl side chain was selected as an optimal P1 side chain due to its good affinity and synthetic accessibility [16].

Aldehyde-based inhibitors were useful in rapidly exploring structure activity relationships of various sites on the inhibitor, but aldehydes perform poorly in drugs for numerous reasons (e.g., they are readily oxidized to carboxylic acids *in vivo*). To find a potent and acceptable replacement, a number of warheads (e.g., carboxylic acid, trifluoromethyl ketones, chloromethyl ketones) commonly used in serine protease inhibitors were investigated. Successful replacement of the aldehyde was accomplished with an alpha-ketoamide, which resulted in greater than 10-fold improvements in binding affinity (*vide infra*, 3 in Scheme 10.1 and 7 in Scheme 10.3). Replacement of the benzyloxy group by a THIQ carbamate 8 boosted the potency further by about fourfold. X-ray structure analysis of alpha-ketoamide based inhibitors such as 7 showed an unexpected arrangement of the tetrahedral intermediate [17]. Instead of occupying the oxyanion hole formed by the main-chain NH groups of Ser-139 and Gly-137, the oxyanion of the inhibitor–enzyme complex pointed toward His-57 of the catalytic triad and the oxygen from the non-electrophilic carbonyl is directed to occupy the oxyanion hole. The same binding mode was observed for telaprevir (Figure 10.3).

The ketoamide warhead provided a handle to extend the inhibitor into the S1′ pocket of the enzyme and gain additional affinity. Introducing a carboxylic acid in the prime side and a more optimal propyl side chain at P1 provided substantial

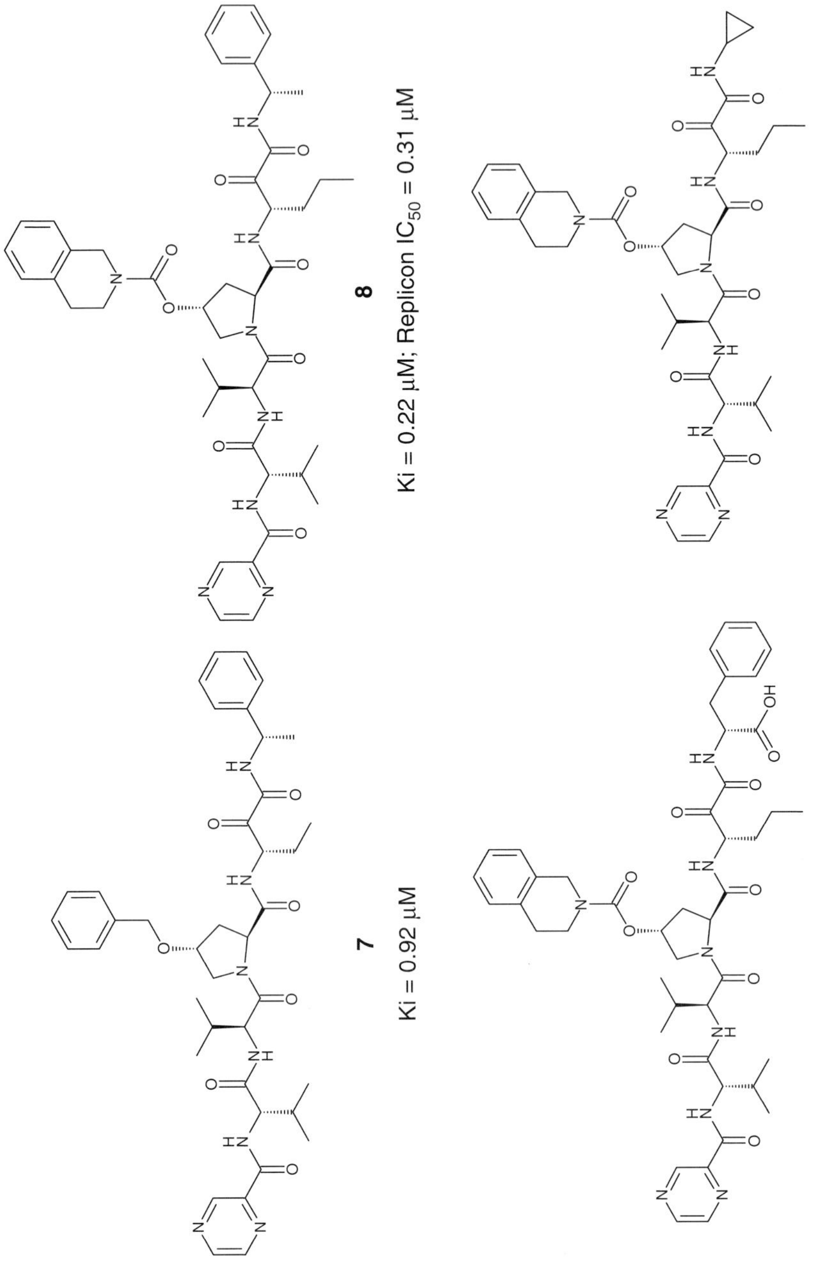

202 | *10 Discovery and Development of Telaprevir (Incivek™)*

7 Ki = 0.92 μM

8 Ki = 0.22 μM; Replicon IC$_{50}$ = 0.31 μM

9 Ki = 0.026 μM; Replicon IC$_{50}$ > 10 μM

10 Ki = 0.42 μM; Replicon IC$_{50}$ = 2.3 μM

Scheme 10.3 Optimization of ketoamide inhibitors 7–9.

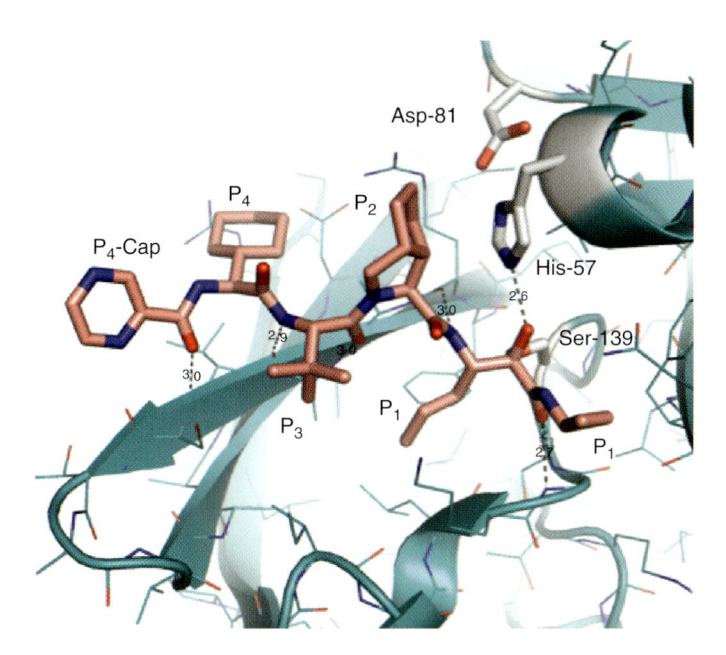

Figure 10.3 Crystal structure of telaprevir complexed with NS3 protease and NS4A peptide cofactor [41] (PDB code:3SV6). The inhibitor is shown in stick diagram with the color scheme: C (salmon), N (blue), and O (red). The active site triad, Ser-139, His-57, and Asp-81 are also shown as sticks with C in gray. The rest of the protein is shown with C in cyan. The secondary structure of the protein is shown in ribbon diagram. Telaprevir makes four hydrogen bonds between the inhibitor main-chain and the protein main-chain. The side chains of the inhibitor fill the S4, S3, S2, S1, and S1' binding pockets. The Ser-139 side chain makes a covalent interaction with the ketoamide warhead.

improvements in enzyme potency (**9**, Ki = 26 nM, Scheme 10.3). Many carboxyl group based analogs showed low nanomolar affinity, but these charged inhibitors displayed no potency in the replicon assay, presumably due to their lack of cell membrane penetration. These results prompted exploration of P1 side chain assisted by docking analysis of a virtual library of ketoamides with small, neutral groups. Of the best-scoring groups, several small hydrophobic groups displayed encouraging enzyme and cell potency. Of these, the cyclopropyl analog (**10**, Scheme 10.3) with the smallest size showed a Ki of 420 nM.

Compound **8** showed submicromolar potency in both enzyme and replicon assays. Therefore, considerable effort and time was spent on the optimization of the THIQ carbamate proline alpha ketoamide subclass, and over 300 compounds were prepared, leading to compounds with nanomolar potency in the replicon assay. More optimal P3 (*tert*-butyl-glycine) and P4 (cyclohexyl-alanine) substituents were discovered during this exploration. However, the potency improvements came at the cost of increasing molecular weight and deteriorating physicochemical properties. Not surprisingly, these compounds displayed poor

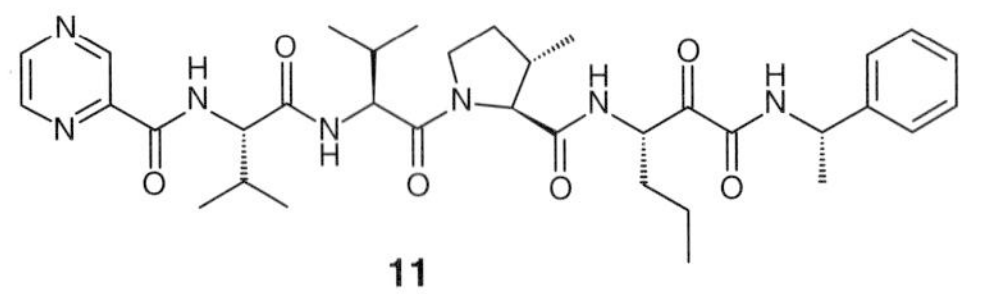

11

Ki = 1.4 μM

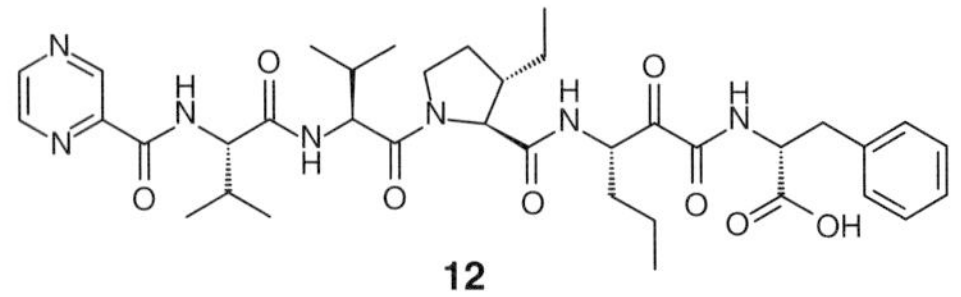

12

Ki = 0.22 μM

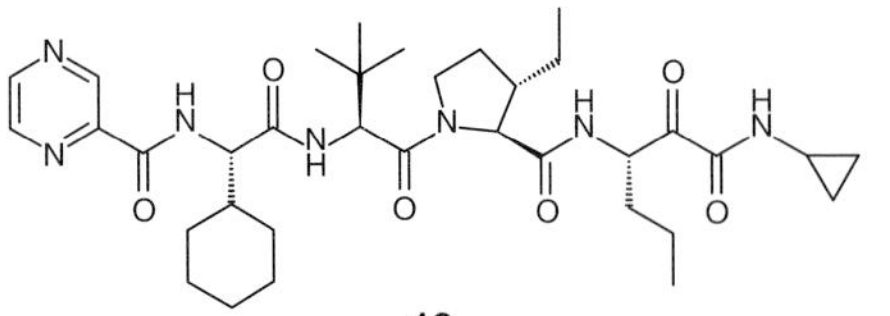

13

Ki = 0.15 μM

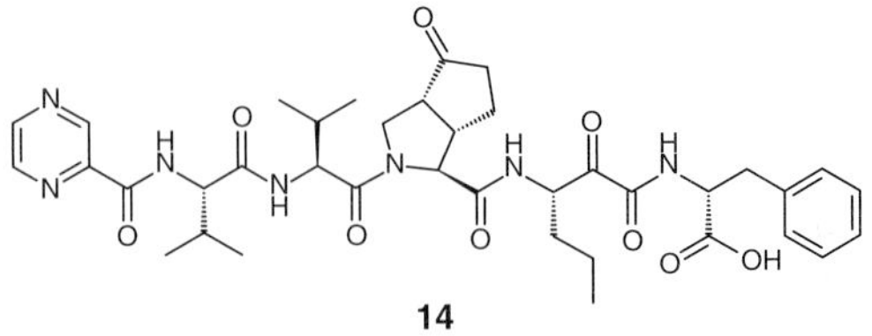

14

Ki = 0.04 μM

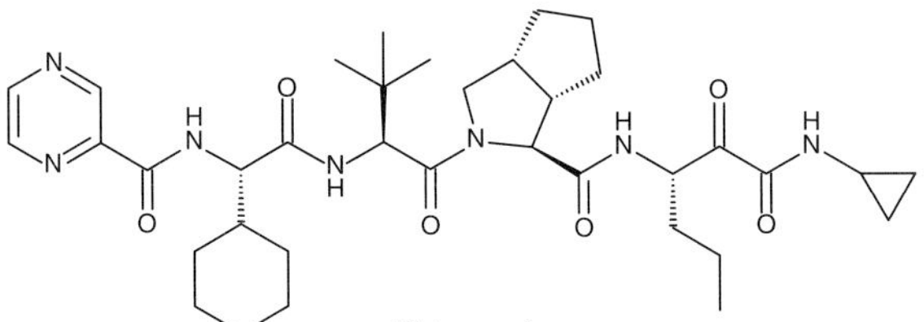

Telaprevir

Ki = 0.044 μM

Scheme 10.4 Key compounds in the progression toward telaprevir.

PK profiles giving low liver and plasma exposures upon single PO dose administration in mice. At this time, the P2 site was revisited to investigate proline substitution other than 4-hydroxyproline-based analogs [18]. Re-examination of the crystal structure of hydroxyproline-based inhibitor **7** led to the hypothesis that proline-based P2s bearing a 1–4 carbon substituent at the 3 position on the alpha face could result in displacement of a putative water molecule in the enzyme active site, thus leading to improved binding affinity. To test this hypothesis, compound **11** (Scheme 10.4) was prepared and showed a Ki of 1.4 μM. Subsequent crystal structure examination disproved the water molecule displacement hypothesis, yet the reasonable activity of compound **11** led to further exploration of P2 groups having this substitution pattern. Changing 3-methyl to 3-ethyl showed dramatic improvement in potency both in the charged series (**12**) and neutral series (**13**) as can be seen in Scheme 10.4. A more dramatic improvement in the potency of neutral series compound **13** is assisted by more optimal P3 and P4 substituents. Furthermore, the 3-alkyl proline-based inhibitors exhibited a PK profile that resulted in significantly improved liver exposure following single PO dose administration in mice providing further incentive for optimization. Exploration of 3-alkyl substituents eventually led to the bicyclic ketone **14** (Ki = 40 nM). Reduction of the ketone moiety to the bicyclic carbocycle, coupled with optimized substituents at P1′, P1, P3, and P4 led to the discovery of telaprevir (Ki = 44 nM) [19].

Telaprevir was selected for advancement based on its potency and its good liver exposure in rodents. The detailed steps of synthesis of telaprevir are reported in a recent publication [15]. The crystal structure of telaprevir complexed with the NS3 protease catalytic domain and an NS4A peptide cofactor (Figure 10.3) shows that telaprevir forms a covalent, tetrahedral intermediate with the nucleophilic active site serine of HCV protease. This observation led to the investigation of the kinetic nature of telaprevir's inhibition of the HCV protease using a continuous fluorescent depsipeptide cleavage assay [20]. Progress curve analysis suggests that telaprevir forms an initial, weakly bound complex with the HCV protease that then rearranges to a more tightly bound form. Under steady-state conditions the inhibition constant for telaprevir inhibition of HCV protease (K_i^*) is 7 nM. As expected for a reversible, covalent mechanism of action, it was found that telaprevir is a slowly dissociating inhibitor of the HCV protease with an observed half-life of the tightly bound complex of 58 min. The slow-binding nature of telaprevir inhibition may provide a therapeutic advantage over compounds of similar potency, which act at diffusion-controlled rates [21].

Telaprevir has an IC_{50} of 354 nM in the 2-day replicon assay and, the CC_{50}, (determined in parallel with the IC_{50}) of 83 μM, yielding a selectivity index (CC_{50}/IC_{50}) of 230 [22]. Using a more stringent assay, it was shown that the replicon cells incubated with 7 μM telaprevir (20 × 2-day replicon IC_{50}) for 3, 6, or 9 days before HCV RNA quantitation showed a continued drop in the number of HCV RNA copies per cell, resulting in a near 4-log drop in RNA levels after 9 days of treatment. To obviate the possibility that the results from the subtype 1b subgenomic replicon assay were artifactual, telaprevir was tested in a subtype 1a infectious assay system using primary cells. The IC_{50} for telaprevir in this assay was determined to be 280 nM, consistent with the inhibition observed in the 2-day replicon assay (354 nM). Telaprevir has the S configuration at its P1 chiral center and the R-diastereomer of telaprevir is nearly 30-fold less potent in the enzyme assay. However, when the antiviral activities of the two diastereomers and their mixture were tested in the standard 48-h HCV replicon cell assay, their IC_{50} were surprisingly similar to each other [13]. These results suggest that the compounds might have epimerized at the P1 position in the replicon cells over the 48-h incubation time, ending up with the same equilibrium mixture regardless of the chiral composition of the starting compound. Telaprevir exhibited a favorable PK profile in several animal species, and demonstrated potent inhibition of HCV protease activity in a mouse model [13]. Animal PK studies showed that the concentrations of telaprevir in the liver relative to plasma is significantly higher at all time points tested (0.5, 1, 2, 4, and 8 h). The average (0–8 h) concentration of telaprevir was $9.82 \pm 5.00\ \mu g\ g^{-1}$ in liver versus $0.28 \pm 0.19\ \mu g\ ml^{-1}$ in plasma of rats, resulting in a liver-to-plasma ratio of 35 to 1. These data for telaprevir satisfied our design requirements for an oral drug candidate to target a liver disease.

In summary, preclinical studies demonstrated that telaprevir is a covalent, reversible, tight binding inhibitor of the HCV NS3/4A protease with good antiviral activity in both HCV replicon cells and human fetal hepatocytes infected with HCV-positive patient sera. In addition, telaprevir exhibited a favorable pharmacokinetic profile in several animal species and demonstrated potent anti-HCV protease activity in an NS3/4A protease mouse model. These results are commensurate with the properties expected for a clinically useful drug.

10.2.3
Pre-Clinical Development

Telaprevir was found to have challenging physicochemical properties for a drug intended for oral delivery. The high molecular weight, high logP, and high polar surface area (PSA) makes telaprevir fall well outside the range for marketed orally delivered drugs that are known to be well absorbed [23]. Also, telaprevir is highly crystalline, with a high melting point (mp) of 246 °C, contributing to the extremely poor water solubility (S_{aq}) of crystalline material (4.7 µg ml^{-1}). Preliminary formulation attempts showed that solutions, crystalline suspensions, and even a nanosuspension resulted in very low oral exposure of telaprevir. It was found that formulating telaprevir as an amorphous spray-dried dispersion substantially improved the kinetic solubility (S_{aq}^{kin}) in aqueous media, and thereby improved bioavailability. The early formulation used in toxicology and phase 1 studies was less physically stable than the final formulation. Stabilizing the dispersion led to substantially improved exposure, allowing the safety margin for telaprevir to be confidently established [7].

10.2.4
Clinical Development

The results from different phases of clinical trials of telaprevir have been recently described [7, 24, 25].

A phase I dose-ranging study of telaprevir monotherapy found that the majority of subjects had marked declines in HCV RNA over the 14-day dosing period (median change of −4.41 log$_{10}$ in the 750 mg every 8 h arm) [26]. However, several patients had viral breakthrough during the dosing period. The rapid emergence of viral resistance mutants to telaprevir monotherapy likely reflects the large number of viral quasispecies in the infected host that results from the highly error-prone replication mechanism of the virus. Nevertheless, these resistant variants remain sensitive to the broad antiviral actions of interferon [27].

In a Phase II clinical trial PROVE-1 (The Protease Inhibition for Viral Evaluation 1), which included patients solely from the United States, the observed rate of SVR (*Sustained Virologic Response*) was 35–67% in the telaprevir (750 mg tid) plus PEGylated interferon, and ribavirin treatment arms, compared with 41% in the control arm [28]. In the PROVE-2 trial, performed in Europe, treatment with telaprevir in combination with PEGylated interferon and ribavirin for 12 weeks, followed by a further 12 weeks of treatment with PEGylated interferon and ribavirin resulted in an SVR rate of 69% [29]. The PROVE-3 study looked at the use of telaprevir in patients who failed previous antiviral therapy [30]. The necessity of including ribavirin with telaprevir and peginterferon (pegIFN) was demonstrated in the PROVE-2 trial: SVR of 36% in the telaprevir and pegIFN 12-week arm compared with 60% in the telaprevir, pegIFN, and ribavirin 12-week arm. Not only did the SVR rates increase when ribavirin was included, but the rates of viral breakthrough and relapse were also significantly lower in the groups that received ribavirin. Therefore, all phase 3 trials assessed the use of telaprevir in combination with both pegIFN α-2a and ribavirin.

Two Phase III clinical trials (ILLUMINATE and ADVANCE) in treatment-naïve HCV infected patients and one trial (REALIZE) in treatment experienced patients were performed [31–33]. These Phase III trials confirmed the clinical benefit of telaprevir demonstrated in Phase II trials. In treatment-naïve patients, the data suggest that after the initial 12 weeks of treatment with telaprevir, pegIFN, and ribavirin, 12 additional weeks of pegIFN and ribavirin (total of 24 weeks of therapy) result in higher SVR rates than the current standard of therapy. The data from the ADVANCE trial also suggest that a longer initial duration of 12 weeks with telaprevir plus pegIFN plus ribavirin results in lower virology failure rates compared with 8 weeks. The ADVANCE study achieved an SVR rate of 79%.

Telaprevir is generally well tolerated, but with significant side effects [34–36]. More than 2500 individuals received telaprevir during different phases of clinical trials. The most significant adverse reactions have been rash, pruritus, anemia, and gastrointestinal disturbance. Rash is the most notable adverse event. Different grades of rash were reported in 56% of patients receiving telaprevir compared with 34% of patients receiving pegIFN and ribavirin alone. Telaprevir rash is typically eczematous, and 90% of reported cases have been of mild/moderate severity. Severe rash (requiring discontinuation of telaprevir) was reported in 4% of subjects receiving telaprevir regimens in the registration studies, compared with <1% of patients receiving pegIFN. The incidence of anemia in the same clinical studies was 36% for telaprevir combination treatment compared with 17% for pegIFN and ribavirin. In the majority of patients who do not achieve SVR following telaprevir-containing treatment, viral variants conferring telaprevir resistance can be detected [37]. Although wild-type virus gradually replaces these variants over months following treatment [38], the implications of these variants on future treatment with telaprevir or other protease inhibitors are currently unknown.

10.3 Summary

Telaprevir (Incivek™) – an NS3/4A protease inhibitor was approved in May 2011 for treatment of HCV infection. It is the culmination of more than a dozen years of intense research and development that involved several innovative approaches including structure-based drug design, assay development, formulation methods, and clinical trial design. Another protease inhibitor, boceprevir, was also approved in the same month. Commercial launch of boceprevir and telaprevir ushered in a new era of HCV treatment by markedly improving the cure rate from below 50% on standard therapy to about 75%. For more than 20 years before this advance, the standard treatment involved a grueling 1-year regimen of pills (ribavirin) and injections (interferon) that caused nausea, fever, and headaches. Telaprevir recorded sales of $1.56 billion in its first four quarters in the market, reaching blockbuster status and the top spot as the fastest drug launch ever. The drug treated more than 100 000 patients around the world generating more than $2 billion in revenue in ~2.5 years until October of 2013.

However, the triple combination therapy of telaprevir, ribavirin, and interferon has several shortcomings such as high pill burden, a TID dosing schedule, a requirement for dosing with a fat-containing meal, and notable adverse effects that limit the use of the therapy in some groups of patients [35]. There is a need for improved therapies as the progression of HCV infected patients toward advanced liver disease is projected to grow [39]. The new therapies should be able to target all genotypes of HCV with regimens that are all oral, once or twice daily dosing, and requiring no interferon [40]. In November 2013, a third protease inhibitor, simeprevir was approved, which requires only once-daily dosing. Two weeks later in December of 2013, a more remarkable and a major advance in treatment of HCV was announced with the approval of HCV polymerase nucleoside inhibitor – sofosbuvir in combination with pegylated interferon and ribavirin for genotype 1 HCV and sofosbuvir plus ribavirin for HCV genotypes 2, 3, and 4 – a once-daily pill that does not require interferon and has a higher cure rate. Although telaprevir had a very successful launch, the success was short lived due to quick follow-up launches of more effective therapies. Telaprevir and boceprevir were the first-in-class of direct acting anti-viral treatments that reached out to help the HCV patients in urgent need, and paved the way for more effective drugs. More combinations of drugs in all oral regimens targeted to cure all genotypes of HCV are undergoing clinical trials and the future looks very promising for HCV patients worldwide.

List of Abbreviations

CC_{50}	50% cytotoxic concentration
CDC	Center for Disease Control
HCV	hepatitis C virus
HPLC	high-performance liquid chromatography
Huh-7	well-differentiated hepatocyte derived cellular carcinoma cell line
IC_{50}	half maximal inhibitory concentration
Ki	inhibition constant
LLC	Limited Liability Company
NS	nonstructural

pegIFN	peginterferon
pNA	*p*-nitroanilide
QRT–PCR	quantitative real time reverse transcription–polymerase chain reaction
SVR	sustained virologicresponse
THIQ	tetrahydroisoquinolyl
TID	three times a day

References

1. Purcell, R.H. (1997) The hepatitis C virus: overview. *Hepatology*, **26** (Suppl. 1), 11S–14S.
2. Saito, I.T. *et al.* (1990) Hepatitis C virus infection is associated with the development of hepatocellular carcinoma. *Proc. Natl. Acad. Sci. U.S.A.*, **87**, 6547–6549.
3. Perz, J.F., Armstrong, G.L., Farrington, L.A., Hutin, Y.J., and Bell, B.P. (2006) The contribution of hepatitis B virus and hepatitis C virus infections to cirrhosis and primary liver cancer worldwide. *J. Hepatol.*, **45**, 529–38.
4. Lindenbach, B.D. and Rice, C.M. (2005) Unravelling hepatitis C virus replication: from genome to function. *Nature*, **436**, 933–938.
5. Lindenbach, B.D. and Rice, C.M. (2001) in Fields Virology, 4th edn (eds D.M. Knipe, P.M. Howley, and D.E. Griffin), Lippincott Williams & Wilkins, Philadelphia, PA, pp. 991–1041.
6. Choo, Q.L. *et al.* (1991) Genetic organisation and diversity of the hepatitis C virus. *Proc. Natl. Acad. Sci. U.S.A.*, **88**, 2451–2455.
7. Kwong, A.D., Kauffman, R.S., Hurter, P., and Mueller, P. (2011) Disocvery and development of telaprevir: an NS3-4A protease inhibitor for treating genotype 1 chronic hepatitis C virus. *Nat. Biotech.*, **29**, 993–1003.
8. Kim, J.L. *et al.* (1996) Crystal structure of the hepatitis C Virus NS3 protease domain complexed with a synthetic NS4A cofactor peptide. *Cell*, **87**, 343–355.
9. Perni, R.B. (2000) NS3·4A protease as a target for interfering with hepatitis C virus replication. *Drug News Perspect.*, **13**, 69–77.
10. Landro, J.A. *et al.* (1997) Mechanistic role of an NS4A peptide cofactor with the truncated NS3 protease of hepatitis C virus: elucidation of the NS4A stimulatory effect via kinetic analysis and inhibitor mapping. *Biochemistry*, **36**, 9340–9348.
11. Lohmann, V. *et al.* (1999) Replication of subgenomic hepatitis C virus RNAs in a hepatoma cell line. *Science*, **285**, 110–113.
12. Lin, C. *et al.* (2006) Discovery and development of VX-950, a novel, covalent and reversible inhibitor of hepatitis C virus NS3·4A serine protease. *Infect. Disord. Drug Targets*, **6**, 3–16.
13. Perni, R.B. *et al.* (2006) Preclinical profile of VX-950, a potent, selective, and orally bioavailable inhibitor of hepatitis C virus NS3-4A serine protease. *Antimicrob. Agents Chemother.*, **50**, 899–909.
14. Mani, N., Rao, B.G., Kieffer, T.L., and Kwong, A.D. (2011) Recent progress in the development of HCV protease inhibitors, in Antiviral Drug Strategies (ed E. De Clercq), Wiley-VCH Verlag GmbH & Co. KGaA, Weinheim.
15. Grillot, A.-L., Farmer, L.J., Rao, B.G., Taylor, W.P., Weisberg, I.S., Jacobson, I.M., Perni, R.B., and Kwong, A.D. (2011) Discovery and development of telaprevir, in Antiviral Drugs: From Basic Discovery through Clinical Trials (ed W.M. Kazmierski), John Wiley & Sons, Inc, Hoboken, NJ.
16. Perni, R.B. *et al.* (2003) Inhibitors of hepatitis C virus NS3·4A protease 1. Non-charged tetrapeptide variants. *Bioorg. Med. Chem. Lett.*, **13**, 4059–4063.

17. Perni, R.B. et al. (2004) Inhibitors of hepatitis C virus NS3·4A protease 2. Warhead SAR and optimization. *Bioorg. Med. Chem. Lett.*, **14**, 1441–1446.
18. Perni, R.B. et al. (2004) Inhibitors of hepatitis C virus NS3·4A protease. Part 3: P2 proline variants. *Bioorg. Med. Chem. Lett.*, **14**, 1939–1942.
19. Yip, Y. et al. (2004) Discovery of a novel bicycloproline P2 bearing peptidyl α-ketoamide LY514962 as HCV protease inhibitor. *Bioorg. Med. Chem. Lett.*, **14**, 251–256.
20. Taliani, M. et al. (1996) A continuous assay of hepatitis C virus protease based on resonance energy transfer depsipeptide substrates. *Anal. Biochem.*, **240**, 60–67.
21. Copeland, R.A., Pompliano, D.L., and Meek, T.D. (2006) Drug-target residence time and its implications for lead optimization. *Nat. Rev. Drug Discov.*, **5**, 730–739.
22. Lin, K. et al. (2006) VX-950, a Novel Hepatitis C Virus (HCV) NS3-4A protease inhibitor, exhibits potent antiviral activities in HCV replicon cells. *Antimicrob. Agents Chemother.*, **50**, 1813–1822.
23. Egan, W.J., Merz, K.M. Jr.,, and Baldwin, J.J. (2000) Prediction of drug absorption using multivariate statistics. *J. Med. Chem.*, **43**, 3867–3877.
24. Kim, J.J., Culley, C.M., and Mohammad, R.A. (2012) Telaprevir - an oral protease inhibitor for hepatitis C virus infection. *Am. J. Health Syst. Pharm.*, **69**, 19–33.
25. Klibanov, O.M. et al. (2011) Telaprevir: a novel NS3/4 protease inhibitor for the treatment of hepatitis C. *Pharmacotherapy*, **31**, 951–974.
26. Reesink, H.W. et al. (2006) Rapid decline of viral RNA in hepatitis C patients treated with VX-950: a phase 1b, placebo-controlled, randomized study. *Gastroenterology*, **131**, 997–1002.
27. Lin, C. et al. (2004) *In vitro* resistance studies of hepatitis C virus serine protease inhibitors, VX-950 and BILN 2061: structural analysis Indicates different resistance mechanisms. *J. Biol. Chem.*, **279**, 17508–17514.
28. Forestier, N.H. et al. (2007) Antiviral activity of telaprevir (VX-950) and peginterferon alfa-2a in patients with hepatitis C. *Hepatology*, **46**, 640–648.
29. McHutchison, J.G. et al. (2009) Telaprevir with peginterferon and ribavirin for chronic HCV genotype 1 infection. *N. Engl. J. Med.*, **360**, 1827–1838.
30. Hézode, C. et al. (2009) Telaprevir and peginterferon with or without ribavirin for chronic HCV infection. *N. Engl. J. Med.*, **360**, 1839–1850.
31. Jacobson, I.M., McHutchison, J.G., Dusheiko, G. et al. (2011) Telaprevir for previously untreated chronic hepatitis C virus infection. *N. Engl. J. Med.*, **364**, 2405–16.
32. Sherman, K.E., Flamm, S.L., Afdhal, N.H. et al. (2010) Telaprevir in combination with peginterferon alfa2a and ribavirin for 24 or 48 weeks in treatment-naïve genotype 1 HCV patients who achieved an extended rapid viral response: final results of phase 3 ILLUMINATE study (abstract LB-2). Presented at 61st Annual Meeting of the American Association for the Study of Liver Diseases, Boston, MA, October 2010.
33. Zeuzem, S., Andreone, P., Pol, S. et al. (2011) Telaprevir for retreatment of HCV infection. *N. Engl. J. Med.*, **364**, 2417–28.
34. Cunningham, M. and Foster, G.R. (2012) Efficacy and safety of telaprevir in patients with genotype 1 hepatitis C infection. *Therap. Adv. Gastroenterol.*, **5**, 139–151.
35. Thompson, A.J. and Patel, K. (2012) New agents for the treatment of hepatitis C virus – focus on telaprevir. *Virus Adapt. Treat.*, **4**, 75–84.
36. Hézode, C. (2012) Boceprevir and telaprevir for the treatment of chronic hepatitis C: safety management in clinical practice. *Liver Int.*, **32** (s1), 32–38.
37. Sarrazin, C., Kieffer, T.L., Bartels, D., Hanzelka, B., Muh, U., Welker, M. et al. (2007) Dynamic hepatitis C virus genotypic and phenotypic changes in patients treated with the protease inhibitor telaprevir. *Gastroenterology*, **132**, 1767–1777.
38. Sullivan, J.C., De Meyer, S., Bartels, D.J., Dierynck, I., Zhang, E., Spanks, J. et al. (2011) Evolution of treatment-emergent

resistant variants in telaprevir phase 3 clinical trials. *J. Hepatol.*, **54**, S4.
39. Zalesak, M., Francis, K., Gedeon, A., Gillis, J., Hvidsten, K. *et al.* (2013) Current and future disease progression of the chronic HCV population in the United States. *PLoS One*, **8**, e63959.
40. Garber, K. (2011) Hepatitis C: move over interferon – beyond telaprevir and boceprevir. *Nat. Biotech.*, **29**, 963–966.
41. Romano, K.P. *et al.* (2012) The molecular basis of drug resistance against hepatitis C virus NS3/4A protease inhibitors. *PLoS Pathog.*, **8** (7), e1002832.

B. Govinda Rao, PhD, is a Computational Chemist with 23+ years of experience in the discovery and development of drugs. He is currently a Principal Investigator at Biogen Idec. He was an independent consultant at the time of writing this chapter. Before this, he was a Research Fellow at Vertex Pharmaceuticals from 1990 to 2013 where he led molecular modeling and drug design efforts on several drug discovery projects. He is a co-inventor of two marketed AIDS drugs, Agenerase™ (amprenavir) and Lexiva™ (fosamprenavir), and of one Hepatitis C drug, Incivek™ (Telaprevir) and contributed to discovery of several second-generation protease inhibitors including VX-175 (GW433908), VX-500, VX-813, and VX-985. He is a named inventor on 25 US patents. He has published 57 research papers, review articles, and five book chapters. He got his PhD from the Indian Institute of Science, Bangalore, India, and worked as postdoctoral fellow at Scripps Research Institute before joining Vertex.

Mark Murcko, PhD, is the Principal at Disruptive Biomedical, LLC, a Professor of Practice at both MIT and Northeastern, and an independent consultant. He currently serves on numerous scientific advisory boards and corporate boards of directors for a diverse range of companies in the biomedical space. Until November 2011, Mark was Chief Technology Officer and Chair of the Scientific Advisory Board of Vertex Pharmaceuticals. He is a co-inventor of the HCV protease inhibitor Incivek™ (telaprevir), as well as Agenerase™ (amprenavir) and Lexiva™ (fosamprenavir). Prior to Vertex, Mark worked at Merck Sharpe and Dohme, where he helped discover several clinical candidates including inhibitors of the enzyme carbonic anhydrase for the treatment of glaucoma. He has served on the editorial boards of many scientific publications, was the co-organizer of the 2008 ACS National Medicinal Chemistry Symposium, and served as the Chair of the 2013 Gordon Research Conference in Medicinal Chemistry. He is a co-inventor on more than 50 issued and pending patents, has co-authored more than 85 scientific articles, and has delivered more than 160 invited lectures.

Mark J. Tebbe, PhD, is a Medicinal Chemist and Drug Hunter with over 20 years of experience and expertise in the areas of cancer, infectious diseases, endocrine, and metabolism. Dr Tebbe held positions of increasing responsibility within Eli Lilly including head of medicinal and computational chemistry in Research Triangle Park and Lilly's site in Hamburg, Germany. Mark has also worked in biotech (VP of Medicinal Chemistry at Forma Therapeutics), and he is currently consulting in the biotechnology and pharmaceutical industries. He is a co-inventor of Incivek™ (Teleprevir). He obtained his undergraduate degree from the University of Notre Dame and his PhD in organic chemistry from Stanford University.

Ann D. Kwong, PhD, is an Industry Leader in Antiviral Drug Discovery with more than 20 years' experience in developing successful drug candidates. She created the Infectious Diseases group at Vertex Pharmaceuticals. Her group played a leading role in the research, development, and commercialization of telaprevir (INCIVEK™), a HCV protease inhibitor with the best drug launch in history (over $1B in sales in <1 year). She is a founding member of HCV DRAG (HCV Drug Development Advisory Group), a consortium of industry leaders, clinical trial leaders, community representatives, and FDA and EMA regulators who work together to optimize HCV drug development. Since leaving Vertex, she has developed the concept of *InnovaTID*, which stands for *Innovative Thinking for Innovative Drugs*, and started two organizations: *InnovaTID Pharmaceuticals and The InnovaTID Institute for Drug Creation*.

11
Antibody–Drug Conjugates: Design and Development of Trastuzumab Emtansine (T-DM1)

Sandhya Girish, Gail D. Lewis Phillips, Fredric S. Jacobson, Jagath R. Junutula, and Ellie Guardino

11.1
Introduction

Chemotherapy has been a mainstay of cancer treatment for several decades. Unfortunately, the cytotoxic agents administered in chemotherapy suffer from a major drawback: they are relatively nonselective for tumor cells. As a result, cytotoxic agents are often poorly tolerated, with many patients suffering severe toxicities, which frequently include bone marrow and immune suppression. Furthermore, dose reductions are often needed to manage these toxicities and can contribute to suboptimal antitumor efficacy.

These limitations have led to the development of a novel class of agents called "antibody–drug conjugates" or "ADCs," which consist of a chemotherapeutic agent covalently linked to a monoclonal antibody that is selective for a tumor-specific antigen [1]. The chemotherapeutic component of an ADC is anticipated to be nontoxic while in circulation, owing to its inability to permeate most cells while attached to the antibody. Binding of the antibody component to antigens on target cells leads to receptor-mediated endocytosis of the ADC. Following internalization, the linker between the antibody component and the chemotherapeutic agent is cleaved, in many cases in the lysosome [2], leading to the release of the cytotoxic chemotherapeutic component within the target cells. This therapeutic strategy is expected to deliver cytotoxic agents selectively to tumor cells, increase the potency and efficacy of therapy, and limit toxic effects in nontarget cells [1].

More than 20 ADCs are now in clinical development for a variety of malignancies, including breast, ovarian, lung, and prostate cancers, as well as non-Hodgkin's lymphoma and leukemia [3–5]. These ADCs utilize a variety of cytotoxic components and are targeted at diverse tumor-specific antigens [6, 7]. As of this writing, three ADCs have received approval from the US Food and Drug Administration (FDA): gemtuzumab ozogamicin (Mylotarg®) for acute myelogenous leukemia in 2000, brentuximab vedotin (Adcetris®) for anaplastic

Successful Drug Discovery, First Edition. Edited by János Fischer and David P. Rotella.
© 2015 Wiley-VCH Verlag GmbH & Co. KGaA. Published 2015 by Wiley-VCH Verlag GmbH & Co. KGaA.

large-cell lymphoma and Hodgkin's lymphoma in 2011, and trastuzumab emtansine (T-DM1; Kadcyla®) for metastatic breast cancer in 2013. Gemtuzumab ozogamicin was subsequently withdrawn from the market in 2010 because of safety concerns and a lack of clinical benefit observed in post-marketing testing [8]. This chapter focuses on the discovery, design, and development of T-DM1 for metastatic breast cancer, with particular attention to the scientific challenges posed by this new class of therapeutic agents for solid tumors, which includes both small molecule and biologic large molecule components.

11.2
Molecular Design of T-DM1

The human epidermal growth factor receptor 2 (HER2) is overexpressed in 15–20% of human breast cancers [9–11]. Before the availability of HER2-directed therapies, HER2 expression was associated with poor clinical outcomes [9, 12–14]. In 1998, trastuzumab (Herceptin®), a humanized monoclonal antibody against HER2, became the first HER2-targeted therapy to gain approval from the FDA. The addition of trastuzumab to standard chemotherapy was found to delay disease progression and extend survival in patients with metastatic breast cancer compared with chemotherapy alone [15]. Despite these improvements in clinical outcomes, some tumors do not respond or become refractory to therapy, and disease progression eventually ensues.

The molecular structure of T-DM1 (Figure 11.1) consists of three primary components: trastuzumab, the thioether linker maleimidomethyl-cyclohexane-1-carboxylate (MCC), and the cytotoxic agent DM1. Among ADCs in Phase III development, a unique feature of T-DM1 is that it retains the antitumor activity of

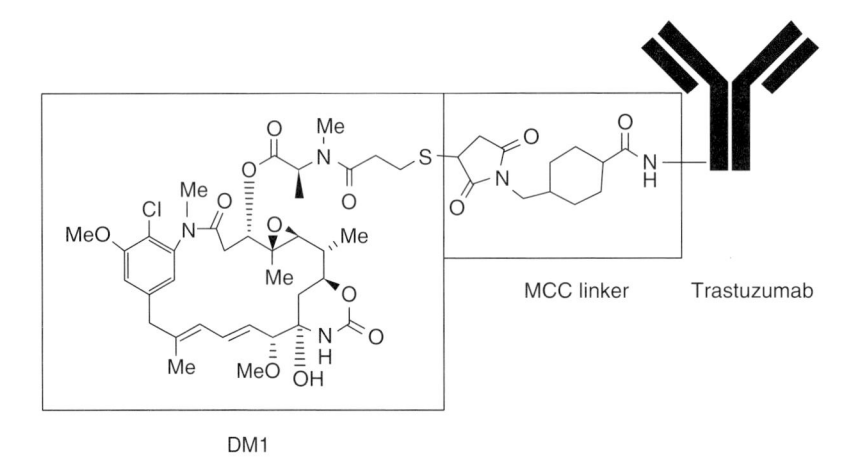

Figure 11.1 Molecular structure of T-DM1 comprising DM1 (cytotoxic component), MCC (covalent linker), and trastuzumab (anti-HER2 monoclonal antibody).

its antibody component, trastuzumab, in addition to targeting the cytotoxic DM1 to HER2-overexpressing cells. Trastuzumab exerts antitumor activity through multiple mechanisms. Trastuzumab binding to HER2 results in the disruption of downstream intracellular signaling via the phosphoinositide 3-kinase (PI3K) pathway, thereby inhibiting tumor growth [16]. Trastuzumab also impedes tumor growth by inhibiting proteolytic cleavage of the extracellular domain (ECD) of HER2, thus preventing formation of a constitutively active, truncated receptor [16]. In addition, the Fc portion of trastuzumab binds Fcγ receptors on immune cells resulting in tumor cell death through antibody-dependent cellular cytotoxicity (ADCC) [16, 17]. T-DM1 binds HER2 with an affinity similar to unconjugated trastuzumab [16]; studies with cell lines demonstrate that T-DM1 retains the mechanisms of action of trastuzumab [16].

DM1, the cytotoxic component of T-DM1, is a derivative of the microtubule depolymerizing agent maytansine [18]. DM1 induces mitotic arrest and triggers apoptosis in a manner similar to vinorelbine and other vinca alkaloids [19, 20]. Early ADCs using conventional chemotherapeutic drugs conjugated to antibodies had shown preclinical promise, but they have proved ineffective in clinical trials, likely because they failed to achieve sufficient cytotoxic drug concentrations within target cells at the maximum tolerated doses (MTDs). Achieving sufficient intracellular concentrations presents challenges owing to the barriers of receptor binding, internalization, and rapid proteolytic degradation and/or nonspecific deconjugation of the cytotoxic drug. Therefore, ADCs may require cytotoxic drugs of particularly high potency. Maytansine was known as a potent cytotoxic agent for several decades, but its nonselective toxicity curtailed clinical development. DM1 exhibits *in vitro* potency 3–10 times greater than maytansine [18], and 25–500 times greater than vinblastine or taxanes [16, 21].

An important objective in the design of T-DM1 was to optimize the stability of the covalent linkage between the cytotoxic and antibody components. A more stable linker was expected to minimize systemic exposure to DM1 by preventing nonspecific deconjugation, but it might also affect efficacy by limiting drug release following internalization. Two classes of covalent linkers were evaluated during the early, nonclinical development of T-DM1 (Figure 11.2): disulfide linkers capable of being cleaved through thioreduction and nonreducible thioether linkers resistant to cleaving. The thioether linker, MCC, yielded optimal results in rat toxicity and mouse efficacy studies. Compared with disulfide linkers, T-DM1 with the MCC linker exhibited longer time in circulation (suggesting a lower rate of deconjugation), reduced toxicity, and increased antitumor potency [22].

T-DM1 is made up of a mixture of drug-loaded species (from 0 to 8 drugs per antibody), with the average being 3.5 DM1 molecules per antibody conjugated via lysine residues. Once T-DM1 binds HER2 receptors, the T-DM1-receptor complex is internalized, and the antibody undergoes proteolytic degradation in lysosomes, which liberates cytotoxic DM1-containing catabolites within the target cell [2]. Consistent with this, in animal studies, the tolerated dose of DM1 was found to be at least twofold higher when administered as T-DM1 versus unconjugated

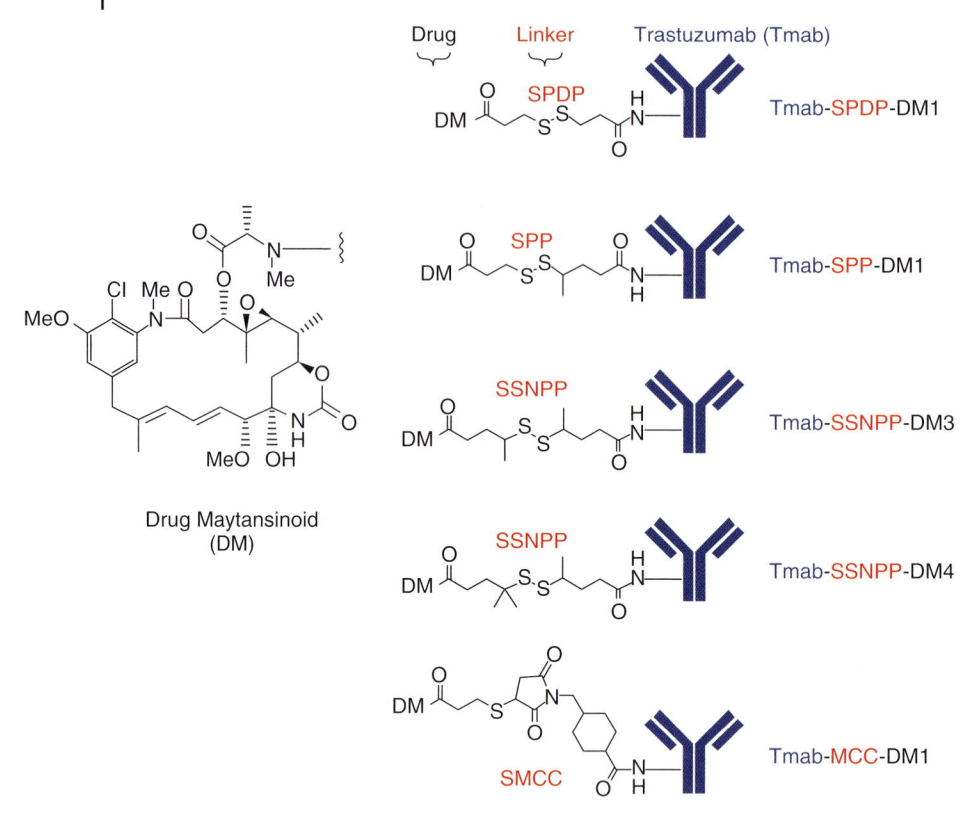

Figure 11.2 Structure of disulfide and thioether linkers evaluated in the design of T-DM1, arranged according to linker stability (least to most). (Adapted by permission from the American Association for Cancer Research: [22].)

DM1 [23]. Based on these observations, T-DM1 linked by MCC was selected for clinical development.

11.3
Strategies for Bioanalysis

The complex molecular structure of T-DM1 also posed several scientific challenges to bioanalysis. It could be expected that following administration, DM1 would be present in systemic circulation, predominantly in its conjugated form, due to the stability of the linker. After catabolism in target cells, it was expected that a combination of large and small molecule catabolites would be produced, including unconjugated trastuzumab antibody, unconjugated DM1, DM1 catabolites containing the linker, and so forth. Each of these analytes likely represents an aspect of the *in vivo* behavior of T-DM1; thus, it was important to select a thorough yet feasible set of analytes for simultaneous measurement to characterize the safety, efficacy, and risk/benefit of T-DM1. Any of these analytes could affect the pharmacokinetic exposure–efficacy relationship or the pharmacokinetic exposure–safety profile of T-DM1 and might also represent important parameters for defining population pharmacokinetics or drug interactions, all of which could impact dosing. In addition, exposure to T-DM1 and/or many of these analytes could conceivably evoke antitherapeutic antibody (ATA) responses. Assay selection would necessitate inclusion of assays most appropriate for monitoring both small and large molecules. Thus, choosing the most relevant analytes and assays depended on careful consideration of the linker and cytotoxic drug characteristics, as well as the antibody properties [24].

Three major assays were used for monitoring T-DM1 pharmacokinetics and exposure in nonclinical studies and clinical trials:

T-DM1 conjugate pharmacokinetic enzyme-linked immunosorbent assay (ELISA) measured the total concentration of trastuzumab antibody conjugated to one or more DM1 moieties in human plasma (but not unconjugated trastuzumab). The assay used an anti-DM1 antibody for capture and recombinant HER2 ECD for detection of any trastuzumab antibody molecules that were captured on the plate because of the presence of their bound DM1 moieties.

Total trastuzumab pharmacokinetic ELISA measured the total concentration of trastuzumab antibody, including both conjugated and unconjugated antibody.

DM1 pharmacokinetic liquid chromatography tandem mass spectrometry (LC-MS/MS) measured the total amount of DM1 in human serum (excluding that conjugated to trastuzumab antibody via the MCC linker). Because DM1 contains a sulfhydryl group, it has the potential to form disulfide bonds, leading to DM1 dimers and adducts between DM1 and glutathione, cysteine, albumin, or other molecules. To avoid under quantification of DM1, samples were first treated to cleave disulfide bonds; free sulfhydryl groups were then blocked to prevent formation of new disulfide bonds prior to LC-MS/MS quantification.

These assays allowed for the measurement of several biologically important parameters:

- Levels of T-DM1 capable of delivering DM1 to HER2-positive cells
- Rates of deconjugation, inferred from levels of DM1, total trastuzumab, and T-DM1
- Pharmacokinetics of unconjugated DM1 and any accumulation of DM1 with repeat dosing of T-DM1
- Pharmacokinetics of unconjugated trastuzumab, the other therapeutic component of T-DM1, known to have antitumor activity and toxicities of its own.

These clinical assays were complemented with mass spectrometric methods for quantifying levels of two circulating DM1-linker catabolites: MCC–DM1 and lysine–MCC–DM1. Taken together, these "small molecule" assays further enabled the measurement and assessment of potential accumulation of DM1 linker catabolites, and improved the understanding of the metabolism, clearance, and elimination pathways for T-DM1.

ATA immune responses were assessed in a tiered fashion. Screening assays identified all ATA responses to T-DM1, regardless of the moiety involved. The component of T-DM1 evoking the ATA was then identified using competitive binding assays between pairs of potential triggers, such as antibody and MCC–DM1 [24].

Overall, the development of the T-DM1 bioanalysis plan was complex and involved a number of assays optimized for both small and large molecules to assess the contributions of the different components of T-DM1 and its catabolites to its clinical and biological properties.

11.4
Strategies for Chemistry and Manufacturing Control

Multiple revisions were made to the T-DM1 manufacturing process during commercial development. T-DM1 was initially manufactured on a small scale suitable for supplying preclinical and Phase I needs and was formulated as a liquid. For midstage Phase II trials, manufacturing was scaled up and a lyophilized formulation was adopted to improve stability. Additionally, the concentration was increased from 5 to 20 mg ml^{-1} to optimize dosing volumes. In preparation for Phase III trials and commercial launch, manufacturing was further scaled up to meet expected demand. Rigorous *in vitro*, *in vivo* (preclinical and clinical), and bioanalytical studies were performed to ensure the biocomparability of the formulations used throughout development and for commercial launch, and to verify that changes in the manufacturing scale did not affect attributes critical to the quality of the product.

In vitro characterization at each stage included assays measuring the binding of T-DM1 to HER2 ECD, human Fcγ receptors, and complement C1q as well as evaluations for induction of ADCC and for antiproliferative activity against HER2-positive cell lines. Extended physicochemical characterization demonstrated that drug distribution profiles, sites of drug attachment, and levels of product-related variants were also consistent at each stage of development.

The pharmacokinetics of each formulation of T-DM1 (i.e., Phase I liquid and Phase II/III lyophilized) were assessed in rats (a nonbinding species) and cynomolgus monkeys (a binding species). Pharmacokinetic parameters in patients were also compared across Phases I–III trials to ensure that the T-DM1 product and formulation used in Phase III trials and for commercial launch were clinically comparable with those used in earlier studies.

Together, these studies confirmed the biocomparability of the T-DM1 used throughout preclinical and clinical development with respect to receptor binding, ADCC, antiproliferative activity, and pharmacokinetics. The range of comparability studies undertaken for this product reflects the complexity and heterogeneity of the T-DM1 molecule. It is important to note that many of these assays represent those that are used with biologic agents, reflecting the fact that the clinical properties of T-DM1 are mediated in part by its antibody component. However, additional methods related to the conjugated drug are also critical for a full evaluation of comparability.

11.5
Nonclinical Development

Preclinical studies in mouse tumor xenograft models and with cell lines *in vitro* helped to lay the groundwork for clinical trials, elucidating how the different components of T-DM1 relate to its biological properties. One series of experiments investigated the relative contributions of the agent's antibody and cytotoxic components to its antitumor activity. Comparisons of T-DM1 and trastuzumab in both trastuzumab-sensitive and insensitive models shed light on these questions. T-DM1 was found to exhibit greater potency than trastuzumab against cell lines and mouse tumor xenograft models that are known to be trastuzumab-sensitive [22]. T-DM1 was also found to have strong antitumor activity against HER2-positive cell lines and in tumor xenograft models that are insensitive to trastuzumab [16, 22] and to lapatinib [16]. These observations suggest that even in HER2-positive tumors that lose sensitivity to trastuzumab or lapatinib, possibly due to activation of alternate signaling pathways, the targeted delivery of a cytotoxic agent still provides substantial antitumor activity [22].

The basis of this activity in trastuzumab- and lapatinib-resistant tumors was further investigated through *in vivo* experiments in which trastuzumab-resistant HER2-positive mouse xenografts were treated with either T-DM1, trastuzumab alone, DM1 alone, or trastuzumab and DM1 administered separately. Of these treatments, T-DM1 exhibited by far the greatest potency in inhibiting tumor growth, suggesting its activity in trastuzumab-resistant tumors requires not only the cytotoxic effects of DM1, but also its selective delivery into HER2-positive cells (unpublished data).

Rodent studies performed during the initial work to select the optimal trastuzumab–maytansinoid conjugate suggested that the activity and toxicity of the ADC in these models depended on the stability of the linker [22]. Following selection of trastuzumab–MCC–DM1 (i.e., T-DM1) as the lead drug candidate, nonclinical studies were conducted to confirm the stability of the linker *in vivo* and to measure exposure to DM1 and other DM1-containing catabolites with potential for toxicity [22]. Studies of distribution and pharmacokinetics in rodents allowed quantification of nonspecific clearance and degradation mechanisms, such as lysosomal degradation and proteolytic cleavage [2]. The distribution and pharmacokinetics of T-DM1 and trastuzumab were investigated to assess the impact that conjugation of DM1 to the antibody might have on the overall behavior of T-DM1 [25]. Clearance of T-DM1 was expected to involve mechanisms typical of both small and large molecules. These included deconjugation of DM1 from the antibody, proteolytic degradation of the antibody, and metabolism of DM1 via cytochrome P450 (CYP) enzymes [22, 25–27].

The availability of HER2-binding and HER2-nonbinding species provided opportunities for studying the contribution of HER2-specific and HER2-nonspecific mechanisms to the pharmacokinetics and clearance of T-DM1. Rat and mouse are "nonbinding" species: trastuzumab does not bind their *neu* receptor (the rodent ortholog to human HER2). In contrast, cynomolgus monkeys and humans are "binding species," with similar affinity of trastuzumab to human HER2 and cynomolgus monkey HER2/neu (erbB2). The pharmacokinetics of T-DM1 in rats and mice (nonbinding species) were dose-proportional due to a lack of HER2 binding (data on file). Overall, conjugation of DM1 to trastuzumab did not change the pharmacokinetics or distribution of the antibody, as these were similar between T-DM1 and trastuzumab. The terminal half-life of T-DM1 in rodents was 3–5 days, with a slow clearance of 13–15 ml day^{-1} kg^{-1} [23]. In contrast, the pharmacokinetics and distribution of radiolabeled DM1 were substantially changed by conjugation to trastuzumab. DM1 exhibited a large volume of distribution and rapid clearance, whereas DM1 conjugated to trastuzumab exhibited much smaller distribution, which was limited to highly perfused tissues and plasma volume, with a higher blood-to-tissue ratio. After administration of T-DM1 to rats, plasma levels of DM1 and linker-containing DM1 catabolites peaked immediately and diminished thereafter [25]. Overall, plasma levels of DM1 were low (1.08–15.6 ng ml^{-1}), and no tissue accumulation of DM1 or DM1-linker catabolites was observed with repeated dosing of T-DM1 in rats [25]. Collectively, these observations demonstrate that the majority of DM1 in circulation remains conjugated to trastuzumab, with fewer than 5% being unconjugated.

Studies in rats show that DM1 and DM1-containing metabolites are eliminated primarily through bile and feces, with minimal urinary excretion [25, 26]. Experiments with human liver microsomes, selective CYP inhibitors, and recombinant CYP enzymes suggest that DM1 is metabolized primarily by cytochrome P450 3A4 (CYP3A4) with a smaller contribution from cytochrome P450 3A5 (CYP3A5) [26].

11.6
Clinical Pharmacology

Given the structural complexity of T-DM1, the pharmacology of T-DM1 was characterized during its clinical development. It was unclear whether a small or large molecule drug development paradigm would be appropriate. Therefore, a customized clinical pharmacology strategy incorporating relevant aspects of both small and large molecule drug development strategies were applied to characterize relationships between exposure and response (safety/efficacy). Multiple analyses were conducted from clinical trial data using both standard noncompartmental and population approaches [28–30]. The clinical trials were

11.6 Clinical Pharmacology

conducted exclusively in patients with cancer. The noncompartmental analyses demonstrated a predictable pharmacokinetic profile for the standard T-DM1 regimen at 3.6 mg kg^{-1} every 3 weeks (q3w) [28, 30]. The terminal half-life of T-DM1 was approximately 4 days, with a clearance of 7–13 ml day^{-1} kg^{-1} [28]. In patients with metastatic breast cancer, the rate of T-DM1 clearance decreased with dose, reflecting a dual mechanism of clearance, including both HER2-nonspecific mechanisms (seen in rodents) and HER2-specific mechanisms (not seen in rodents; data on file) The population pharmacokinetic analyses, which estimated typical parameter values and interindividual variability, provided support for the body weight-based dosing regimen for T-DM1 without the need to correct for other factors [29]. Mild to moderate renal impairment did not seem to affect the pharmacokinetics of T-DM1. In the population pharmacokinetic analysis, pharmacokinetic parameters were similar in patients with normal renal function (creatinine clearance, CLcr, 90 ml min^{-1}; $n = 361$), mild renal impairment (CLcr 60–89 ml min^{-1}; $n = 254$), and moderate renal impairment (CLcr 30–59 ml min^{-1}; $n = 53$). It was not possible to assess the pharmacokinetics in patients with severe renal impairment because only one such patient had been treated with T-DM1. At present, a dedicated study (BO25499; NCT01513083) is ongoing to assess the impact of mild and moderate hepatic impairment on T-DM1 exposure.

When DM1 and DM1-linker catabolites were examined in patients from a Phase II trial, the results were similar to those in rats; specifically, the concentrations in plasma were low – in many cases below the limit of detection – and both DM1 and MCC–DM1 peaked shortly after the administration of T-DM1. Although DM1 and MCC–DM1 were observed immediately after dosing, there was a lag time associated with the appearance of Lys–MCC–DM1 [25]. As in rats, repeated administration of T-DM1 to human patients was not associated with plasma accumulation of DM1 or DM1-linker catabolites [25]. In both rats and humans, therefore, systemic exposure to DM1 and DM1-linker catabolites appears to be low, confirming the stability of the thioether linker *in vivo* [25].

A dedicated QT study was conducted for T-DM1 in clinical development to rule out any clinically meaningful QT prolongation. Overall, the statistical analysis of electrocardiogram (ECG) data and concentration–QTc analysis indicate that single-agent T-DM1 given at a dose of 3.6 mg kg^{-1} q3w does not have a clinically meaningful effect on the QTc interval in patients with HER2-positive metastatic breast cancer [31].

T-DM1 was also examined for exposure–response relationships using data from Phase II and III clinical trials (including T-DM1 area under the curve [AUC], peak plasma concentration [C_{max}], minimum plasma concentration [C_{min}], trastuzumab AUC, and DM1 C_{max}) [28, 32]. Analysis of the Phase III EMILIA study showed no clear trends between T-DM1 exposure (C_{max} and AUC) and efficacy (progression-free survival [PFS]; overall survival [OS]; or response rate) [32]. A slight trend was observed for improved OS by stratified T-DM1 C_{min} quartiles, although with overlapping 95% confidence intervals. Analysis of five Phase II and III clinical trials showed no relationships between

T-DM1 exposure and the risk of developing thrombocytopenia or hepatotoxicity, two of the most common adverse events (AEs) related to T-DM1 [33].

As an ADC, T-DM1 carries the potential for drug interactions stemming from either its small molecule DM1 component or its large molecule antibody component. Although DM1 is metabolized by CYP3A4 and CYP3A5, *in vitro* studies found that it did not induce or inhibit these enzymes, even at drug concentrations up to 10 times those observed in patients during clinical trials [26]. Consistent with these data, pharmacokinetic data from the Phase III EMILIA trial found no evidence of drug interactions in patients who received concomitant medications known to be inhibitors or inducers of CYP3A4 or the drug transporter P-glycoprotein (data on file). In addition, no drug–drug interaction was observed between T-DM1 and the anti-HER2 therapeutic antibody pertuzumab when given together in the Phase II study TDM4373g [30].

Many patients entering T-DM1 clinical trials had measurable levels of trastuzumab persisting in their circulation from earlier trastuzumab-containing regimens, raising the concern that residual trastuzumab might antagonize the antitumor effects of T-DM1. However, co-administration of trastuzumab and T-DM1 (either as a bolus or as a continuous infusion) in a trastuzumab-insensitive mouse xenograft model revealed no reduction in the antitumor activity of T-DM1 (data on file). In addition in patients with metastatic breast cancer, there was no relationship between baseline trastuzumab or HER2 ECD levels and response to T-DM1 [28], suggesting that T-DM1 therapy may overcome effects of residual circulating trastuzumab.

11.7
Clinical Trials and Approval

As an ADC that combines anti-HER2 and cytotoxic mechanisms, single-agent T-DM1 was expected to have the potential to augment the standard of care for patients with HER2-positive metastatic breast cancer. The MTD of T-DM1 was initially established in a Phase I trial in which successive cohorts of patients received the agent at escalating doses of 0.3, 0.6, 1.2, 2.4, 3.6, or 4.8 mg kg^{-1} q3w. Dose-limiting grade 4 thrombocytopenia was observed at a dose of 4.8 mg kg^{-1}, and the MTD was determined to be 3.6 mg kg^{-1} q3w [34]. A weekly regimen of T-DM1 was also evaluated in the Phase I study. Dose-limiting grade 3 thrombocytopenia and grade 3 elevated aspartate aminotransferase (AST) were observed at 2.9 mg kg^{-1} once per week (qw); therefore, the MTD was determined to be 2.4 mg kg^{-1} qw [35]. In this Phase I dose escalation study conducted in a limited number of patients, T-DM1 showed comparable activity and was well tolerated at both dosing regimens. Given its convenience to patients and its better compliance, however, T-DM1 3.6 mg kg^{-1} q3w was chosen for further evaluation in subsequent Phase II and III trials.

Two Phase III trials have compared single-agent T-DM1 with combination regimens that included an anti-HER2 agent plus conventional chemotherapy.

Randomized patients with HER2-positive metastatic breast cancer in EMILIA (NCT00829166) were to receive either T-DM1 (3.6 mg kg^{-1}, q3w) ($n = 495$) or the cytotoxic agent capecitabine plus the HER2-targeted kinase inhibitor capecitabine plus lapatinib (CL) ($n = 496$). All patients had been previously treated with trastuzumab and a taxane. The primary endpoints were PFS, OS, and safety [36].

The median PFS was 9.6 months for patients receiving T-DM1 and 6.4 months for those receiving CL (hazard ratio [HR] 0.65; 95% CI, 0.55–0.77; $P < 0.001$). Median OS at the second interim analysis was 30.9 months for the T-DM1 group and 25.1 months for the CL group (HR 0.68; 95% CI, 0.55–0.85; $P < 0.001$). Grade ≥ 3 AEs occurred in 41% of patients in the T-DM1 group and 57% in the CL group. The most common grade ≥ 3 AEs in the T-DM1 group were thrombocytopenia (12.9%), elevated AST (4.3%), and elevated alanine aminotransferase (2.9%). The most frequent grade ≥ 3 AEs in the CL group were diarrhea (20.7%), palmar–plantar erythrodysesthesia (16.4%), vomiting (4.5%), neutropenia (4.3%), and hypokalemia (4.1%). Few patients experienced serious cardiac events: 1.7% in the T-DM1 group and 1.6% in the CL group experienced left ventricular ejection fraction (LVEF) <50% and with a ≥ 15 percentage point decrease from baseline. Grade 3 ventricular systolic dysfunction occurred in one patient receiving T-DM1 and in none of the patients receiving CL [36]. These data provided the basis for approval of T-DM1 by multiple health authorities, including the FDA in the United States and the European Medicines Agency.

The Phase III open-label TH3RESA trial (NCT01419197) compared T-DM1 with a broader variety of treatments than in EMILIA, including trastuzumab-based therapy in combination with chemotherapy. Patients with HER2-positive metastatic breast cancer were randomized to T-DM1 (3.6 mg kg^{-1}, q3w) ($n = 404$) or the treatment of physician's choice (TPC) ($n = 198$). All patients had previously received trastuzumab, lapatinib, and a taxane (median, four previous regimens). The primary endpoints were PFS and OS [37]. Among 184 patients who received treatment in the TPC arm, 83% received anti-HER2 agents (trastuzumab and/or lapatinib) generally in combination with conventional chemotherapy or hormonal therapy; the remaining 17% received single-agent chemotherapy. Median PFS was 6.2 months for the T-DM1 group and 3.3 months for the TPC group (HR 0.528; 95% CI, 0.422–0.661; $P < 0.0001$). Median OS at the first interim analysis was not reached for patients in the T-DM1 arm and was 14.9 months for those in the TPC arm (HR 0.552; 95% CI, 0.369–0.826; $P < 0.0034$). While the strong trend in OS favored the T-DM1 group, this difference did not cross the interim efficacy stopping boundary (HR < 0.370 or $P < 0.0000016$). The overall rate of grade ≥ 3 AEs was 32% in the T-DM1 group and 43% in the TPC group. The rates of AEs leading to discontinuation were 7% and 11% in the T-DM1 and TPC groups, respectively. The rates of AEs leading to dose reduction were 9% and 20%, respectively. The most common grade ≥ 3 AE observed in patients receiving T-DM1 was thrombocytopenia (5%). The most common grade ≥ 3 AEs occurring in patients receiving TPC were neutropenia (16%), febrile neutropenia (4%), and diarrhea (4%). Serious cardiac events were rare, with 1% of patients in the T-DM1 group and 1%

of patients in the TPC group exhibiting LVEF <50% and a ≥15 percentage point decrease from baseline [37].

The superior safety and efficacy of T-DM1 observed in these Phase III trials suggest that T-DM1 possesses an improved therapeutic index compared with conventional combinations of anti-HER2 agents plus chemotherapy. The improved therapeutic index of T-DM1 is also documented in a pooled pharmacokinetic analysis of patients enrolled in four clinical studies, which found no correlation between T-DM1 exposure, efficacy, and key AEs [28].

The improved therapeutic index described above is likely to arise from selective delivery of DM1 into HER2-positive target cells – a fundamental design feature of ADCs. The favorable therapeutic index of T-DM1 is consistent with nonclinical studies that show the high stability of its thioether linker, negligible levels of systemic deconjugation, and an absence of DM1 accumulation [22, 25]. It is also consistent with nonclinical results, showing that the tolerated dose of DM1 is twice as high when administered as T-DM1 compared with DM1 [23].

Hepatic enzyme elevations (one of the more frequent AEs observed with T-DM1) have been observed with other ADCs that contain cytotoxic maytansine derivatives [38]. Because DM1-containing catabolites are metabolized by CYP enzymes [26] and eliminated through the biliary–fecal route [25], it is possible that transient low levels of DM1 catabolites in circulation induce hepatic injury. It is also possible that DM1 is released into the hepatic microenvironment following internalization and catabolism of T-DM1 by Fc receptor–bearing Kupffer cells in the liver. Increased thrombocytopenia is also likely related to the DM1 component of T-DM1. *In vitro* studies found that DM1 had no direct effect on platelet function but did reduce platelet production by megakaryocytes [39].

It is important to note that both hepatic enzyme elevations and thrombocytopenia were generally reversible and noncumulative with repeated dosing of T-DM1 [36, 37, 40]. Other hematologic events that occur frequently with conventional chemotherapy, such as neutropenia and febrile neutropenia, were observed at lower rates in T-DM1-treated patients compared with those receiving combination regimens containing chemotherapy. The increased safety of T-DM1 is likely related to the targeted delivery of cytotoxic DM1 to HER2-positive cells.

In February 2013, the FDA approved T-DM1 under the brand name Kadcyla®. Several Phase II and III trials are exploring the role of T-DM1 in other breast cancer settings, evaluating it as (neo)adjuvant therapy and as first-line treatment for metastatic disease as well as in combination with other HER2-targeted agents such as pertuzumab (Perjeta®). In addition, clinical development in other HER2-positive cancers, such as metastatic gastric cancer, are also underway or in the planning stages.

11.8
Summary

The ADC T-DM1 represents a novel approach to treating patients with HER2-positive cancer. T-DM1 enables selective delivery of DM1, a potent cytotoxic agent, into HER2-positive target cells and, like trastuzumab, inhibits HER2 signaling, prevents HER2 shedding, and induces ADCC [16]. Development of the complex, heterogeneous T-DM1 molecule posed several challenges. *In vitro* and *in vivo* experiments delineated the contribution of small and large molecule components to the antitumor properties, pharmacokinetics, and unique safety profile of the agent. The agent's stable thioether linker minimized deconjugation and systemic exposure to DM1 – leading to an improved therapeutic index, with lower rates of grade ≥3 toxicities and greater efficacy compared with standard-of-care regimens that combine anti-HER2 agents with conventional chemotherapy.

List of Abbreviations

ADC	antibody–drug conjugate
ADCC	antibody-dependent cellular cytotoxicity
AE	adverse event
AST	aspartate aminotransferase
ATA	anti-therapeutic antibody
AUC	area under the curve
CI	confidence interval
CLcr	creatinine clearance
CL	capecitabine plus lapatinib
C_{max}	peak plasma concentration
C_{min}	minimum plasma concentration
CYP	cytochrome P450
CYP3A4	cytochrome P450 3A4
CYP3A5	cytochrome P450 3A5
ECD	extracellular domain
ECG	electrocardiogram
ELISA	enzyme-linked immunosorbent assay
ErbB2	v-erb-b2 avian erythroblastic leukemia viral oncogene homolog 2 (also referred to as Her-2/neu)
FDA	US Food and Drug Administration
HER2	human epidermal growth factor receptor 2
HER2/neu	human epidermal growth factor receptor 2 expressed by nonhuman primates (also referred to as ErbB2)
HR	hazard ratio
LC-MS/MS	liquid chromatography tandem mass spectrometry
LVEF	left ventricular ejection fraction
MCC	4-(maleimidomethyl)-cyclohexane-1-carboxylate
MTD	maximum tolerated dose
OS	overall survival
PFS	progression-free survival
PI3K	phosphoinositide 3-kinase

QT	QT interval
QTc	corrected QT interval
qw	once per week
q3w	every 3 weeks
T-DM1	trastuzumab emtansine
TPC	treatment of physician's choice

References

1. Chari, R.V. (2008) Targeted cancer therapy: conferring specificity to cytotoxic drugs. *Acc. Chem. Res.*, **41**, 98–107.
2. Erickson, H.K., Lewis Phillips, G.D., Leipold, D.D., Provenzano, C.A., Mai, E., Johnson, H.A., Gunter, B., Audette, C.A., Gupta, M., Pinkas, J., and Tibbitts, J. (2012) The effect of different linkers on target cell catabolism and pharmacokinetics/pharmacodynamics of trastuzumab maytansinoid conjugates. *Mol. Cancer Ther.*, **11**, 1133–1142.
3. Flygare, J.A., Pillow, T.H., and Aristoff, P. (2013) Antibody-drug conjugates for the treatment of cancer. *Chem. Biol. Drug Des.*, **81**, 113–121.
4. Sassoon, I. and Blanc, V. (2013) Antibody-drug conjugate (ADC) clinical pipeline: a review. *Methods Mol. Biol.*, **1045**, 1–27.
5. Sievers, E.L. and Senter, P.D. (2013) Antibody-drug conjugates in cancer therapy. *Annu. Rev. Med.*, **64**, 15–29.
6. Alley, S.C., Okeley, N.M., and Senter, P.D. (2010) Antibody-drug conjugates: targeted drug delivery for cancer. *Curr. Opin. Chem. Biol.*, **14**, 529–537.
7. Teicher, B.A. and Chari, R.V. (2011) Antibody conjugate therapeutics: challenges and potential. *Clin. Cancer Res.*, **17**, 6389–6397.
8. Mylotarg (gemtuzumab ozogamicin) market withdrawal http://www.fda.gov/Safety/MedWatch/SafetyInformation/SafetyAlertsforHumanMedicalProducts/ucm216458.htm (accessed 23 February **2014**).
9. Slamon, D.J., Clark, G.M., Wong, S.G., Levin, W.J., Ullrich, A., and McGuire, W.L. (1987) Human breast cancer: correlation of relapse and survival with amplification of the HER-2/neu oncogene. *Science*, **235**, 177–182.
10. Ross, J.S., Slodkowska, E.A., Symmans, W.F., Pusztai, L., Ravdin, P.M., and Hortobagyi, G.N. (2009) The HER-2 receptor and breast cancer: ten years of targeted anti-HER-2 therapy and personalized medicine. *Oncologist*, **14**, 320–368.
11. Pathmanathan, N., Provan, P.J., Mahajan, H., Hall, G., Byth, K., Bilous, A.M., and Balleine, R.L. (2012) Characteristics of HER2-positive breast cancer diagnosed following the introduction of universal HER2 testing. *Breast*, **21**, 724–729.
12. Paik, S., Hazan, R., Fisher, E.R., Sass, R.E., Fisher, B., Redmond, C., Schlessinger, J., Lippman, M.E., and King, C.R. (1990) Pathologic findings from the National Surgical Adjuvant Breast and Bowel Project: prognostic significance of erbB-2 protein overexpression in primary breast cancer. *J. Clin. Oncol.*, **8**, 103–112.
13. Gabos, Z., Sinha, R., Hanson, J., Chauhan, N., Hugh, J., Mackey, J.R., and Abdulkarim, B. (2006) Prognostic significance of human epidermal growth factor receptor positivity for the development of brain metastases after newly diagnosed breast cancer. *J. Clin. Oncol.*, **24**, 5658–5663.
14. Dawood, S., Broglio, K., Buzdar, A.U., Hortobagyi, G.N., and Giordano, S.H. (2010) Prognosis of women with metastatic breast cancer by HER2 status and trastuzumab treatment: an institutional-based review. *J. Clin. Oncol.*, **28**, 92–98.
15. Slamon, D.J., Leyland-Jones, B., Shak, S., Fuchs, H., Paton, V., Bajamonde, A., Fleming, T., Eiermann, W., Wolter, J., Pegram, M., Baselga, J., and Norton, L. (2001) Use of chemotherapy plus a monoclonal antibody against HER2 for

metastatic breast cancer that overexpresses HER2. *N. Engl. J. Med.*, **344**, 783–792.

16. Junttila, T.T., Li, G., Parsons, K., Lewis Phillips, G.L., and Sliwkowski, M.X. (2011) Trastuzumab-DM1 (T-DM1) retains all the mechanisms of action of trastuzumab and efficiently inhibits growth of lapatinib insensitive breast cancer. *Breast Cancer Res. Treat.*, **128**, 347–356.

17. Barok, M., Tanner, M., Köninki, K., and Isola, J. (2011) Trastuzumab-DM1 is highly effective in preclinical models of HER2-positive gastric cancer. *Cancer Lett.*, **306**, 171–179.

18. Chari, R.V., Martell, B.A., Gross, J.L., Cook, S.B., Shah, S.A., Blättler, W.A., McKenzie, S.J., and Goldmacher, V.S. (1992) Immunoconjugates containing novel maytansinoids: promising anticancer drugs. *Cancer Res.*, **52**, 127–131.

19. Lopus, M. (2011) Antibody-DM1 conjugates as cancer therapeutics. *Cancer Lett.*, **307**, 113–118.

20. Lopus, M., Oroudjev, E., Wilson, L., Wilhelm, S., Widdison, W., Chari, R., and Jordan, M.A. (2010) Maytansine and cellular metabolites of antibody-maytansinoid conjugates strongly suppress microtubule dynamics by binding to microtubules. *Mol. Cancer Ther.*, **9**, 2689–2699.

21. Kovtun, Y.V., Audette, C.A., Mayo, M.F., Jones, G.E., Doherty, H., Maloney, E.K., Erickson, H.K., Sun, X., Wilhelm, S., Ab, O., Lai, K.C., Widdison, W.C., Kellogg, B., Johnson, H., Pinkas, J., Lutz, R.J., Singh, R., Goldmacher, V.S., and Chari, R.V. (2010) Antibody-maytansinoid conjugates designed to bypass multidrug resistance. *Cancer Res.*, **70**, 2528–2537.

22. Lewis Phillips, G.D., Li, G., Dugger, D.L., Crocker, L.M., Parsons, K.L., Mai, E., Blättler, W.A., Lambert, J.M., Chari, R.V., Lutz, R.J., Wong, W.L., Jacobson, F.S., Koeppen, H., Schwall, R.H., Kenkare-Mitra, S.R., Spencer, S.D., and Sliwkowski, M.X. (2008) Targeting HER2-positive breast cancer with trastuzumab-DM1, an antibody-cytotoxic drug conjugate. *Cancer Res.*, **68**, 9280–9290.

23. Poon, K.A., Flagella, K., Beyer, J., Tibbitts, J., Kaur, S., Saad, O., Yi, J.H., Girish, S., Dybdal, N., and Reynolds, T. (2013) Preclinical safety profile of trastuzumab emtansine (T-DM1): mechanism of action of its cytotoxic component retained with improved tolerability. *Toxicol. Appl. Pharmacol.*, **273**, 298–313.

24. Girish, S, Li, C. (2015) *Clinical Pharmacology and Assay Consideration for Characterizing Pharmacokinetics and Understanding Efficacy and Safety of ADCs*. Future Science Ltd, in press.

25. Shen, B.Q., Bumbaca, D., Saad, O., Yue, Q., Pastuskovas, C.V., Khojasteh, C., Tibbitts, J., Kaur, S., Wang, B., Chu, Y.W., LoRusso, P.M., and Girish, S. (2012) Catabolic fate and pharmacokinetic characterization of trastuzumab emtansine (T-DM1): an emphasis on preclinical and clinical catabolism. *Curr. Drug Metab.*, **13**, 901–910.

26. Wong, S., Bumbaca, D., Yue, Q., Halladay, J., Kenny, J.R., Salphati, L., Saad, O., Tibbitts, J., Khojasteh, C., Girish, S., and Shen, B.Q. (2011) Nonclinical disposition, metabolism, and in vitro drug–drug interaction assessment of DM1, a component of trastuzumab emtansine (T-DM1). Presented at: AACR-EORTC-NCI International Conference: Molecular Targets and Cancer Therapeutics, San Francisco, CA, November 12–16, 2011, Abstract A136.

27. Lu, D., Joshi, A., Wang, B., Olsen, S., Yi, J.H., Krop, I.E., Burris, H.A., and Girish, S. (2013) An integrated multiple-analyte pharmacokinetic model to characterize trastuzumab emtansine (T-DM1) clearance pathways and to evaluate reduced pharmacokinetic sampling in patients with HER2-positive metastatic breast cancer. *Clin. Pharmacokinet.*, **52**, 657–672.

28. Girish, S., Gupta, M., Wang, B., Lu, D., Krop, I.E., Vogel, C.L., Burris, H.A. III, LoRusso, P.M., Yi, J.H., Saad, O., Tong, B., Chu, Y.W., Holden, S., and Joshi, A. (2012) Clinical pharmacology of trastuzumab emtansine (T-DM1): an antibody-drug conjugate in development

for the treatment of HER2-positive cancer. *Cancer Chemother. Pharmacol.*, **69**, 1229–1240.

29. Gupta, M., Lorusso, P.M., Wang, B., Yi, J.H., Burris, H.A. III, Beeram, M., Modi, S., Chu, Y.W., Agresta, S., Klencke, B., Joshi, A., and Girish, S. (2012) Clinical implications of pathophysiological and demographic covariates on the population pharmacokinetics of trastuzumab emtansine, a HER2-targeted antibody-drug conjugate, in patients with HER2-positive metastatic breast cancer. *J. Clin. Pharmacol.*, **52**, 691–703.

30. Lu, D., Burris, H.A. III, Wang, B., Dees, E.C., Cortes, J., Joshi, A., Gupta, M., Yi, J.H., Chu, Y.W., Shih, T., Fang, L., and Girish, S. (2012) Drug interaction potential of trastuzumab emtansine (T-DM1) combined with pertuzumab in patients with HER2-positive metastatic breast cancer. *Curr. Drug Metab.*, **13**, 911–922.

31. Gupta, M., Wang, B., Carrothers, T.J., LoRusso, P.M., Chu, Y.-W., Shih, T., Loecke, D., Joshi, A., Saad, O., Yi, J.-H., and Girish, S. (2013) Effects of trastuzumab emtansine (T-DM1) on QT interval and safety of pertuzumab plus T-DM1 in patients with previously treated human epidermal growth factor receptor 2–positive metastatic breast cancer. *Clin. Pharmacol. Drug Dev.*, **2**, 11–24.

32. Wang B., Jin J., Wada R., Fang L., Lu D., Guardino E., Swain S.M., Untch M., Girish S. Exposure–efficacy relationship of trastuzumab emtansine (T-DM1) in EMILIA, a phase III study of T-DM1 versus capecitabine (X) and lapatinib (L) in HER2-positive locally advanced or metastatic breast cancer (MBC). *J. Clin. Oncol.*, 2013, **31** (Suppl.), abstract 644.

33. Jin J., Wang B., Gao Y., Samant M., Li C., Song C., Swain S.M., Untch M., Girish S. Exposure–safety relationship of trastuzumab emtansine (T-DM1) in patients with HER2-positive locally advanced or metastatic breast cancer (MBC). *J. Clin. Oncol.*, 2013, **31** (Suppl.), abstract 646.

34. Krop, I.E., Beeram, M., Modi, S., Jones, S.F., Holden, S.N., Yu, W., Girish, S., Tibbitts, J., Yi, J.H., Sliwkowski, M.X., Jacobson, F., Lutzker, S.G., and Burris, H.A. (2010) Phase I study of trastuzumab-DM1, an HER2 antibody-drug conjugate, given every 3 weeks to patients with HER2-positive metastatic breast cancer. *J. Clin. Oncol.*, **28**, 2698–2704.

35. Beeram, M., Krop, I.E., Burris, H.A., Girish, S.R., Yu, W., Lu, M.W., Holden, S.N., and Modi, S. (2012) A phase 1 study of weekly dosing of trastuzumab emtansine (T-DM1) in patients with advanced human epidermal growth factor 2-positive breast cancer. *Cancer*, **118**, 5733–5740.

36. Verma, S., Miles, D., Gianni, L., Krop, I.E., Welslau, M., Baselga, J., Pegram, M., Oh, D.Y., Diéras, V., Guardino, E., Fang, L., Lu, M.W., Olsen, S., Blackwell, K., and EMILIA Study Group (2012) Trastuzumab emtansine for HER2-positive advanced breast cancer. *N. Engl. J. Med.*, **367**, 1783–1791 (Erratum in *N. Engl. J. Med.*, (2013), **368**, 2442).

37. Krop, I.E., Kim, S.-B., González-Martín, A., LoRusso, P.M., Ferrero, J.M., Smitt, M., Yu, R., Leung, A.C., and Wildiers, H. (2014) Trastuzumab emtansine versus treatment of physician's choice for pretreated HER2-positive advanced breast cancer (TH3RESA): A randomized, open-label, phase 3 trial. *Lancet Oncol.*, **15**, 689–699.

38. Tolcher, A.W., Ochoa, L., Hammond, L.A., Patnaik, A., Edwards, T., Takimoto, C., Smith, L., de Bono, J., Schwartz, G., Mays, T., Jonak, Z.L., Johnson, R., DeWitte, M., Martino, H., Audette, C., Maes, K., Chari, R.V., Lambert, J.M., and Rowinsky, E.K. (2003) Cantuzumab mertansine, a maytansinoid immunoconjugate directed to the CanAg antigen: a phase I, pharmacokinetic, and biologic correlative study. *J. Clin. Oncol.*, **21**, 211–222.

39. Mahapatra, K., Darbonne, W.C., Bumbaca, D., Shen, B.Q., Du, X., Tibbitts, J., Olsen, S., Sliwkowski, M.X., Lewis Phillips, G.D., Girish, S., Hartley, D., Dambach, D., Ramakrishnan, V., and Uppal, H. (2011) Trastuzumab emtansine (T-DM1)-induced thrombocytopenia results from impaired platelet production in a HER2-independent manner.

Presented at AACR-EORTC-NCI International Conference: Molecular Targets and Cancer Therapeutics, San Francisco, CA, November 12–16, 2011, Abstract A135.

40. Diéras V., Harbeck N., Budd G.T., Greenson J.K., Guardino A.E., Samant M., Chernyukhin N., Smitt M.C., and Krop I.E. (2014) Trastuzumab emtansine in human epidermal growth factor receptor 2-positive metastatic breast cancer: an integrated safety analysis. *J. Clin. Oncol.*, **32**, 2750–2757.

Dr. Sandhya Girish is a Senior Scientist and an Associate Director in the department of clinical pharmacology at Genentech. She is the global clinical pharmacology lead for antibody drug conjugates (ADCs). Prior to Genentech, she was a Senior Scientist with Novartis and provided strategic support in drug discovery and development of new molecular entities. As a study director and a co-principal investigator for pharmacokinetics and drug metabolism studies at SRI International in Menlo Park, she directed numerous *in vitro* and *in vivo* preclinical PK and metabolism studies. She has over 70 poster presentations, invited talks, and publications, and has received several awards for her contributions. She obtained her MS and PhD degrees from Northeastern University, Boston and Bachelors in Pharmacy from India.

Gail Lewis Phillips joined Genentech, Inc. in 1985 after completing graduate training in the Department of Pharmacology at the University of Texas Southwestern Medical Center. For more than 25 years, she has performed translational research in the area of therapeutic targeting of HER2 for breast and other types of HER2-positive cancers. Her work was instrumental in the development and approval of 2 HER2-targeted humanized antibodies, Trastuzumab and Pertuzumab, as well as the novel HER2-targeted antibody-drug conjugate Trastuzumab emtansine. Gail's work is represented in 50 publications as well as in numerous oral presentations at international conferences.

Fredric S. Jacobson is currently a Staff Scientist in the Protein Analytical Chemistry department at Genentech, Inc. His group is responsible for assay development and product characterization of protein therapeutics (focusing primarily on antibody–drug conjugates) in early clinical development through to market launch. He is also the Technical Development Team leader for Kadcyla®. From 1986 to 1997 he was a Scientist in the Cell Culture and Fermentation Process Development department at Genentech, where his group developed assays for recombinant protein product quality, culture medium nutrients and metabolites, and systems for on-line analysis and control of nutrient feeds. He received his PhD in Chemistry from MIT in 1981 working with Christopher Walsh, and was a post-doc in the lab of Bruce Ames at UC Berkeley from 1981 to 1986.

Jagath R. Junutula, a Senior Scientist, Molecular Oncology, Genentech, Inc., obtained his PhD in 1997 from the Indian Institute of Science, India, and did post-doctoral research in Germany as an Alexander von Humboldt fellow and at Stanford University. He has over 10 years of experience in oncology antibody discovery and development at Genentech. He developed a novel THIOMAB technology platform to generate specific antibody–drug conjugates (ADCs) to improve the therapeutic utility of these antibody therapeutics. He has authored 60 peer-reviewed publications/patents and has presented his work at numerous international conferences.

Ellie Guardino is the Clinical Leader for Global Development of T-DM1 at Roche. She is a Medical Oncologist with specialty in Breast Cancer and continues on the Faculty at Stanford University as Adjunct Clinical Faculty. She received her BS at UCLA, MD/PhD at Georgetown University, Internal Medicine Board Certification after Residency at Harvard University and is board certified in Medical Oncology following Fellowship training at Stanford University. Guardino's research is focused on cancer therapeutics, specifically developing the antibody drug conjugate, T-DM1 for breast and nonbreast solid tumors.

Index

a

acquired immunodeficiency syndrome (AIDS)   113
active pharmaceutical ingredient (API)   38
Acyl-CoA cholesterol acyltranferase (ACAT) inhibitors   11, 12
β_2-adrenergic receptor (β_2AR)   23
albumin/creatinine ratio (ACR)   145
aldosterone   7
Alimta   157 *see also* pemetrexed
α-amino-3-hydroxy-5-methyl-4-isoxazole-propionic acid (AMPA)   181
– competitive receptor antagonists   182
– HTS assays   183
– non-competitive receptor antagonists   182
aminopterin   160
angiotensin receptor blocker (ARB)   145, 146
antibody-drug conjugates (ADCs)
– chemotherapeutic component   213
– T-DM1, 213 *see also* trastuzumab emtansine (T-DM1)
antiepileptic drug (AED)   13, 14
– molecular targets   181
– perampanel   188
anti-HIV drugs   113
antimetabolite theory   159
antiretroviral drug zidovudine (AZT)   113
antiretroviral therapy (ART)   121
anti-therapeutic antibody (ATA)   217, 218
Apidra®   47
apremilast   6
– for psoriatic arthritis treatment   6
aromaticity   170
– SAR study   184
artemisinin   30
aspartate aminotransferase (AST)   222
AspB28-human insulin   46

avanafil
– adverse reactions   82
– clinical studies of   81
– on electroretinogram   79
– IC_{50} values of   75
– in vivo pharmacology   76
– isoquinoline derivatives, discovery of   65
– PDE inhibitory profiles   75
– pyrimidine nucleus structure modification   73
– scaffold-hopping approaches 67 *see also* scaffold-hopping approaches

b

basal insulin profile   39
benzimidazoles   16
benzimidazol-2-one derivatives, binding of   23
benzothiophene MK2 inhibitor   24, 25
benzylpyrazolone O-glucosides   94
biochemical efficiency (BE)   24
butterfly wing pigment   157

c

captopril   3
carboxylic acid
– in monoketo acid motif   117
C-aryl glucosides   98
– affinity of   102
– ring methylation of   103
C-glucosides
– disubstituted diarylmethane   104
– m-diarylmethane   97
– meta   98
– ortho   98
– SAR of   99
– synthetic route   98

Successful Drug Discovery, First Edition. Edited by János Fischer and David P. Rotella.
© 2015 Wiley-VCH Verlag GmbH & Co. KGaA. Published 2015 by Wiley-VCH Verlag GmbH & Co. KGaA.

chronic obstructive pulmonary disease (COPD)
– tiotropium bromide treatment   3
circulating metabolites   82
citalopram   10
cLogP   102, 105
cobicistat   121
competitive receptor antagonists of AMPA   182
continuous subcutaneous insulin infusion (CSII)   48

d
dapagliflozin
– clinical studies with   108
– and diet induced obese rats   107
– discovery of   4
– identification of   102
– profiling studies with   105
5,10-dideaza-5,6,7,8-tetrahydrofolic acid (DDATHF)
– antitumor agent   164
– convergent synthesis   166
– Lilly's results   163
– open-chain analog   169
– resolution of   167
– SAR results   170
– structure   167
– synthesis of   162, 167
diet induced obese (DIO) rats   106
– dapagliflozin and   107
dietary cholesterol absorption, inhibitor of 11
  see also ezetimibe
diketo acid   116
– bioisosteres of   117
– enolized/non-enolizable   118
– metal chelating functions of   122
diketo acid inhibitors   121
– HIV-1 integrase and   114
dipeptidyl peptidase-4 (DPP-4)   129
– ectodomain   139
– homodimer structure   139
– hydrogen bonds   140
– inhibition of   148
– inhibitory activity   131
– rationalization of   139
drospirenone   7, 8

e
elvitegravir   113
– discovery of   114
– monoketo acid integrase inhibitors and   116
– strand transfer by   120

emtricitabine   123
endogenous insulin secretion   38
enzyme-linked immunosorbent assay (ELISA)   217
epilepsy   183
erectile dysfunction (ED)   63 see also avanafil
– definition of   63
– PDE5inhibitor, physicochemical characteristics of   65
escitalopram   9, 10
ezetimibe   11, 13

f
fibroblast activating protein (FAP)   142
first in class medicines
– approved between 1999-2008   27
– examples of   27
first-in-class small-molecule new molecular entities   27, 29
folate-dependent enzymes   158
folic acid   160

g
gastrin receptor blockers   15
gemtuzumab ozogamicin   213, 214
glomerular filtration rate (GFR)   145
glucagon-like peptide (GLP)-1   129
– effects on   148
glucodynamic profiles   44
glucose
– and basal insulin profile, schematic representation of   39
– regulation by linagliptin   143
– renal processing of   89
glucose-dependent insulinotropic peptide (GIP)   129
glucosuric agents   90
glycated hemoglobin (HbA1C)   143

h
hemodynamics, avanafil effects on   79
hepatic enzyme elevations   224
hepatic impairment
– linagliptin   150
hepatitis C virus (HCV)   195
– clinical development   206
– HPLC assay   196
– NS3/4A protease   196
– peptide truncation study   197
– p-nitroanilide   196
– pre-clinical development   206
– protease catalytic domain   197
– PROVE-3 study   207
– structure-based inhibitor   200

Index | 233

- substrate-based inhibitor   197, 199
- THIQ   200, 201
high-fat diet (HFD)   144
high throughput screening (HTS)   130
- assay of AMPA   183
- structures of   130
HIV-1 integrase
- and diketo acid inhibitors   114
HPLC-based assay   196
Humalog® 45 *see also* insulin lispro
human epidermal growth factor receptor 2 (HER2)   214, 215
human *ether-à-go-go*-related gene (hERG) channel   135
human immunodeficiency virus (HIV)   113
human insulin   39, 40
- time-dependent, diffusion enabled dissociation of   44
hydroxybenzamide O-glucosides   90
hyperglycemia   88
- effects on   148

i
imatinib   3
insulin
- history of   38
- protracted glucodynamic profile   44
- self-association properties   45
- self-association tendencies of   41
insulin aspart   46
insulin degludec   53, 54
insulin detemir   51, 52
insulin glargine   50, 51
insulin glulisine   46
insulin lispro   46
- basal utility of   48, 49
- creation of   45
- dissociation of   47
- hexameric complex of   45
- NPH-like crystal formation   49
insulin lispro protamine suspension (ILPS)   49
integrase catalytic reactions   115
inter-patient variability, Lantus® formulation   51
intra-patient variability, Lantus® formulation   50
in vitro–in vivo extrapolation (*IVIVE*)
- perampanel   189
ionotropic glutamate receptors (iGluRs)   181
isoquinoline preparation   67

j
Janus kinase 3 (JAK3)   5

k
kinase domain receptor (KDR)   68

l
lamotrigine
- and BW 288U   14
- for epilepsy treatment   13
Lantus® formulation
- time course   50
- time extension strategy   50
leucopterin   158
Levemir® formulation 51 *see also* insulin detemir
linagliptin   129
- albuminuria-lowering effect of   149
- bioavailability of   147
- cardiovascular safety   150
- clinical pharmacodynamics   148
- clinical pharmacokinetics   146
- discovery of   130
- effects of   145
- glucose regulation by   143
- non-linear pharmacokinetics   147
- outcomes of   150
- patients with hepatic impairment   150
- patients with renal impairment   148
- physicochemical, pharmacologica, and kinetic characteristicsl   141
- structure of   130
long-acting insulin analog formulations   48
losartan   3
LY231514   172
LysB28ProB29-human insulin, 45 *see also* insulin lispro

m
maleimidomethyl-cyclohexane-1-carboxylate (MCC)   215
maximum tolerated doses (MTDs)   215, 222
m-diarylmethane C-glucosides   97
metabotropic glutamate receptors (mGluRs)   181
methotrexate (MTX)   159, 160
molecular mechanism of action (MMOA)   19
- artemisinin   30
- dose–response relationships   21
- features of   27
- IC50 and KI   22
- pharmacological action   21
- safety   22
- selection of   22
- target-based drug discovery approach   25

molecular mechanism of action (MMOA) (*contd.*)
– therapeutic index   21, 22
– value and impact of   19
monoketo acid   118
– metal chelating functions of   122
monoketo acid integrase inhibitors
– and elvitegravir   116
– structural optimization   119
multiple daily injections (MDI) therapy   37
multitargeted antifolate (MTA)   173

n

Neutral Protamine Hagedorn (NPH) formulation   48, 49
neutral protamine lispro (NPL)   49
Niemann-Pick C1-like1 (NPC1L1) protein   12
nitric oxide (NO)/cGMP signaling pathway   63
nitroglycerin (NTG)-induced hypotension, avanafil effects on   80, 81
N-methyl-D-aspartate (NMDA) receptor   30
non-competitive receptor antagonists of AMPA   183
NS3/4A protease   198
– activity   195
– crystal structure of   196
– structure-based inhibitor   200
– substrate-based inhibitor   197

o

o-Benzylphenol O-glucosides   95
O-glucosides
– benzylpyrazolone   94
– hydroxybenzamide   90
– o-benzylphenol   95
– SGLT SAR
– – distal substituted diarylmethanes   99
– – ortho hydroxybenzamides   92
– – ortho benzylphenols   96
– SGLT2 inhibitors   89
β-O-glucosides
– susceptibility of   93
Oil Red O staining   144
omeprazole   15, 16
oxadiazinone   184

p

pelvic nerve stimulation-induced tumescence   78
pemetrexed
– administration   173
– discovery of   175
– origin of   157
– structural studies   157
penile tumescence, potentiation of   76
peptide
– truncation study   197
perampanel
– antiepileptic drug   188
– *in vitro*–*in vivo* extrapolation   189
– pharmacokinetic evaluation   189
– pharmacological evaluation   187
– phase I   189
– phase II and III   190
PF-3644022   24, 25
phenotypic screening   26, 27
phenyltriazines   14
phlorizin   89
phosphodiesterase 4 (PDE4) inhibitors   6
phosphodiesterase 5 (PDE5) inhibitors
– blockade of   63
– physicochemical characteristics of   65
phosphodiesterases (PDEs)   64
phthalane derivative   9
picoprazole   16
piriqualone   182
2-pivaloyl-6-bromo-5-deazapterin   166
plasma glucose
– dose dependent reductions   106, 107
p-nitroanilide (pNA) assay   196
polymerase approaches   5
polynucleotidyl transferases   118
polyprotein   195
prandial insulin profile   39
prandial insulin secretion   48
proton-pump inhibitors, 15 *see also* omeprazole
pteridine   157
pyrazolone O-glucosides   94
pyrimethamine   13–15
pyrrole   170

q

4-quinolone-3-carboxylic acid   116
quinoxalinedione   182

r

rapid-acting insulin analogs   41
R-citalopram   10
renal function
– anticipated perturbation on   88
– SGLT2 transporters in   88
renal impairment
– linagliptin   148
renin-angiotensin-aldosterone system (RAAS)   149

Index

retinal function, avanafil effects on   79
ring-break framework modifications   68

s
scaffold-hopping approaches
– monocyclic type A series   68
– monocyclic type B series   68
– ring-break framework modifications   67
– 2-substituted-5-(3,4,5-trimethoxybenzoyl) pyrimidines   70
– 4-substituted-5-(3,4,5-trimethoxybenzoyl) pyrimidines   70
– 5-substituted-5-(3,4,5-trimethoxybenzoyl) pyrimidines   73
SCH 48461   12
S-citalopram   10
scurvy   20
SGLT1   88
SGLT2 potency
– modulation of   101
sildenafil   64
– on electroretinogram   79
– hemodynamic effects   79
– IC_{50} values of   75
sodium-glucose transporter type 2 (SGLT2) inhibitor
– dapagliflozin   4
– O-glucosides   89
– – benzylpyrazolone   94
– – hydroxybenzamide   90
– – o-benzylphenol   95
– recovery rate   88
– in renal function   88
sofosbuvir   5
Sovaldi®, 5 see also sofosbuvir
spironolactone   7, 8
spirorenone   7, 8
statins   11
Stendra®, 63 see also avanafil
Stribild   123
structure-activity relationship (SAR)   130, 138
– of C-glucosides   99
2-substituted-5-(3,4,5-trimethoxybenzoyl) pyrimidines   70

t
T-1032   65
tadalafil   65
talopram   9, 10

telaprevir
– clinical development   206
– compounds   204
– crystal structure   203
– pre-clinical development   206
– synthesis   205
tenofovir disoproxil fumarate   123
tetrahydrofolate coenzymes   160
tetrasubstituted pyrimidine derivatives   68
tezampanel   182
T↔R transition   41
timoprazole   15, 16
T6 insulin hexamer   42
tiotropium bromide, for COPD treatment   3
tofacitinib   5, 6
T-6932 pyrimidine derivative   69
transforming growth factor β (TGFβ)   146
trastuzumab emtansine (T-DM1)
– bioanalysis   216
– chemistry control   218
– clinical pharmacology   220
– manufacturing process   218
– for metastatic breast cancer   4
– molecular design   214
– nonclinical development   219
– parameters   217
– trials and approval   222
Tresiba® 53 see also insulin degludec
1,3,5-triaryl-1H-pyridin-2-one
– discovery of   184
– optimization of   185
trisubstituted pyrimidine derivatives   68
type 1 diabetes mellitus (T1DM), insulin treatment   37
type 2 diabetes mellitus (T2DM)   87
– insulin   87
– treatment   87

u
urinary excretion   108
urine albumin-to-creatinine ratio (UACR)   149

x
xanthin
– uracil substructure of   138
xanthopterin   158
xenograft models   219